ELSEVIER OCEAN ENGINEERING BOOK SERIES

VOLUME 2

WIND GENERATED OCEAN WAVES

Elsevier Science Internet Homepage

http://www.elsevier.nl (Europe)
http://www.elsevier.com (America)
http://www.elsevier.co.jp (Asia)

Consult the Elsevier homepage for full catalogue information on all books, journals and electronic products and services.

Elsevier Titles of Related Interest

WATSON
Practical Ship Design
(Ocean Engineering Book Series, Vol. I)
ISBN: 008-042999-8

MOAN & BERGE
ISSC '97, 13th International Ship and Offshore Structures Congress (3 Volume Set)
ISBN: 008-042829-0

VUGTS
BOSS '97, Behaviour of Offshore Structures (3 Volume Set)
ISBN: 008-042834-7

CHRYSSOSTOMIDIS
BOSS '94, Behaviour of Offshore Structures (3 Volume Set)
ISBN: 008-041916-X

GUEDES-SOARES
Advances in Safety and Reliability (3 Volume Set)
ISBN: 008-042835-5

FRANGOPOL, COROTIS & RACKWITZ
Reliability and Optimization of Structural Systems
ISBN: 008-042826-6

Related Journals

Free specimen copy gladly sent on request: Elsevier Science Ltd, The Boulevard, Langford Lane, Kidlington, Oxford, OX5 GB, UK

- Applied Ocean Research
- Coastal Engineering
- Computers and Structures
- Engineering Failure Analysis
- Engineering Structures
- Finite Elements in Analysis and Design
- Journal of Constructional Steel Research
- Marine Structures
- Ocean Engineering
- Structural Safety
- Thin-Walled Structures

To Contact the Publisher

Elsevier Science welcomes enquiries concerning publishing proposals: books, journal special issues, conference proceedings, etc. All formats and media can be considered. Should you have a publishing proposal you wish to discuss, please contact, without obligation, the publisher responsible for Elsevier's civil and structural engineering publishing programme:

James Milne PhD
Senior Publishing Editor
Engineering and Technology
Elsevier Science Ltd
The Boulevard, Langford Lane
Kidlington, Oxford
OX5 1GB, UK
Phone: +44 1865 843891 Fax: +44 1865 843920 E.mail: j.milne@elsevier.co.uk

General enquiries, including placing orders, should be directed to Elsevier's Regional Sales Offices – please access the Elsevier homepage for full contact details (homepage details at top of this page).

ELSEVIER OCEAN ENGINEERING BOOK SERIES

VOLUME 2

WIND GENERATED OCEAN WAVES

IAN R. YOUNG

University of Adelaide, Australia

OCEAN ENGINEERING SERIES EDITORS

R. Bhattacharyya & M.E. McCormick

1999

ELSEVIER

Amsterdam – Lausanne – New York – Oxford – Shannon – Singapore – Tokyo

ELSEVIER SCIENCE Ltd
The Boulevard, Langford Lane
Kidlington, Oxford OX5 1GB, UK

First edition 1999

Library of Congress Cataloging-in-Publication Data
Young, I. R.
Wind generated ocean waves / Ian R. Young. -- 1st ed.
p. cm. -- (Elsevier ocean engineering book series ; v. 2)
Includes bibliographical references (p. -) and index.
ISBN 0-08-043317-0 (hc)
1. Ocean waves. I. Title. II. Series.
GC211.2.Y68 1999
551.47'022--dc21 98-54991
CIP

British Library Cataloguing in Publication Data
A catalogue record from the British Library has been applied for.

ISBN: 0-08-043317-0 (hardcover)

∞ The paper used in this publication meets the requirements of ANSI/NISO Z39.48-1992 (Permanence of Paper).

Printed in The Netherlands

Acknowledgments

There are many people who have, either directly or indirectly, played a role in the compilation of this book. First and foremost I must acknowledge the very significant role played by my wife Heather and daughter Katrina. For many years they have tolerated (without once complaining) my rather obsessive interest in ocean waves. This interest has resulted in many long absences from home and weekends spent working. They have always given me total support and encouragement. Without this assistance, this book and much of the work described in it, could not have been achieved.

One of the distinct pleasures of working in the field of science/engineering is the interaction with ones colleagues. During my career, I have found such interactions to be both the most enjoyable and the most stimulating aspects of research. The collegiate spirit which exists within the Oceanography and Coastal Engineering professions is a major strength. Not only does it result in an exciting environment but one which has achieved much in a limited period of time. I have had the distinct pleasure of having been influenced by a number of people who have played very significant roles in the development of the field. The most notable amongst these has been Klaus Hasselmann. Through the application of his significant intellect Klaus has been able to make pioneering breakthroughs which have added order and understanding where there appeared to be only chaos. The individual areas in which he has contributed are too numerous to mention, an examination of the bibliography gives some insight into the magnitude of his contribution. To have worked closely with Klaus and Susanne Hasselmann has been a distinct honour. There are many other colleagues one could also mention. At the risk of omitting many, I will mention only three: Mark Donelan, Mike Banner and Linwood Vincent. All three have made significant contributions both to my own thoughts and to the wider profession.

I must also acknowledge the role played by Tom Hardy in reviewing the manuscript of this work. Tom's knowledgeable and critical eye has contributed significantly to the final product.

Ian Young

Contents

List of Major Symbols

a	wave amplitude
a_b	semi-excursion distance of water particle at bed
A	horizontal semi-axis of water particle ellipse
A_s	asymmetry of wave record
b	spacing between wave orthogonals
$b(f_1, f_2)$	bi-coherence
B	vertical semi-axis of water particle ellipse
c_h	Charnock parameter
C	wave phase speed
C_d	atmospheric drag coefficient
C_f	bed friction coefficient
C_g	group velocity
C_p	phase speed of components at the spectral peak frequency
d	water depth
D	diameter of bed material
$D(f, \theta)$	directional spreading function
E	wave energy per unit crest length
f	frequency $= 1/T$
f^*	non-dimensional frequency $= fu_*/g$
f_c	Coriolis parameter
f_p	frequency of spectral peak
f_w	bed friction factor
$F(f)$	one-dimensional or omni-directional wave frequency spectrum
$F(f, \theta)$	directional frequency spectrum
g	gravitational acceleration

G	quadruplet nonlinear interaction coupling coefficient
h_w	height of white-cap region of a wave
H	wave height
H_{max}	maximum wave height
H_s	significant wave height
H_{rms}	root-mean-square wave height
$\overline{H}$	mean wave height
k	wave number
K_p	pressure attenuation factor
K_r	refraction coefficient
K_s	shoaling coefficient
L	wave length
L_o	Obukhov scale height
L_w	length of white-cap region of a wave
$n(\mathbf{k})$	action density
p	pressure
$p(H)$	probability density function of wave height
p_w	pressure exerted on water surface by white-cap
p_0	central pressure of a hurricane
$\overline{P}$	energy flux of waves
$Q(k_x, k_y)$	wave number spectrum of waves
Q_b	fraction of waves breaking
$Q(s)$	normalization factor for directional spreading
r	bottom roughness
R	radius to maximum wind of a hurricane
R_b	bulk Richardson number
ΔR	fractional change in wind speed
s	directional spreading exponent
S	Jeffreys' sheltering coefficient
S	specific gravity of bed material
S_k	skewness of wave record
S_b	bottom interaction source term
S_{bf}	bottom friction source term
S_{brk}	depth limited breaking source term

S_{ds}	white-cap dissipation source term
S_{in}	atmospheric input source term
S_{nl}	quadruplet nonlinear interaction source term
S_{tot}	total source term
S_{tri}	triad nonlinear interaction source term
t	time
t	duration for which the wind blows
t^*	non-dimensional time $= tg/u_*$
T	wave period
T_a	temperature of air
T_r	length of recorded time series
T_v	mean virtual temperature of air at a height of 10 m
T_w	temperature of water
u	horizontal velocity (x direction)
u_b	velocity at the bed
u_*	friction velocity
$\overline{u}$	depth averaged velocity
U_{10}	wind speed measured at a height of 10 m
U_g	gradient wind speed
$U_{L/2}$	wind speed measured at a height of $L/2$
U_{min}	minimum wind speed which can generate waves
U_∞	free stream wind velocity outside boundary layer
v	horizontal velocity (y direction)
V_{fm}	velocity of forward movement of a hurricane
V_{max}	maximum wind speed within a hurricane
w	vertical velocity (z direction)
x	horizontal length co-ordinate
x	fetch length over which the wind blows
$X(f)$	Fourier transform
y	horizontal length co-ordinate
z	horizontal length co-ordinate
z_0	surface roughness length
α	JONSWAP Phillips parameter
α	bed slope

$\hat{\alpha}$	spectral steepness parameter
β	Donelan et al (1985) spectrum scale parameter
β_s	Donelan et al (1985) directional spreading factor
γ	energy growth rate
γ	JONSWAP peak enhancement factor
γ_d	Donelan et al (1985) peak enhancement factor
γ_{ds}	damping coefficient for white-capping
Γ	fractional increase in wave energy per radian
δ	boundary layer thickness
δ	non-dimensional water depth $= gd/U_{10}^2$
ε	non-dimensional energy $= \sigma^2 g^2/U_{10}^4$
ζ	vertical displacement of water particle
η	water surface elevation
ψ	phase angle of water wave
θ_m	mean wave direction
θ_s	Shields' entrainment parameter
$\overline{\theta}$	mean directional width of spectrum
κ	von Karman constant ≈ 0.4
κ	non-dimensional peak wave number $= k_p U_{10}^2/g$
λ_r	radar wave length
λ_w	Bragg scattering wave length
Λ	atmospheric turbulence wave number
ν	non-dimensional peak frequency $= f_p U_{10}/g$
ν_a	kinematic viscosity of air
ξ	non-dimensional deviation from mean growth rate
ξ	horizontal displacement of water particle
$\Pi(\mathbf{k}, \omega)$	pressure wave number-frequency spectrum
ρ_a	density of air
ρ_w	density of water
σ	JONSWAP peak width parameter
σ^2	variance or total energy of the wave record
ς	non-dimensional duration $= gt/U_{10}$
τ	shear stress
$\tilde{\tau}$	non-dimensional time scale for change in wind direction

ϕ	velocity potential
Φ_{mn}	cross-spectrum
χ	non-dimensional fetch $= gx/U_{10}^2$
ψ	integrated universal function
ω	radian measure of frequency
$\tilde{\omega}$	depth scaled non-dimensional frequency $= \omega^2 d/g$
$\overline{\omega}$	mean frequency
$\Omega(\omega)$	turbulent frequency pressure spectrum

Chapter 1

Introduction

The history of wind wave research is relatively short. Although there were basic developments last century (Airy, 1845; Stokes, 1847), a concerted effort really only began as a result of the military imperative of the Second World War. The work of Sverdrup, Munk and Bretschneider (Sverdrup and Munk, 1944a,b; Bretschneider, 1952a) provided the first observational data base, upon which, to base theories for the evolution of wind generated waves. This work was, however, largely empirical. A theoretical framework began to develop with the studies of wind-wave generation by Miles and Phillips (Miles, 1957; Phillips, 1957). A more complete understanding of the full evolution process, however, awaited the insight into nonlinear interactions provided by Hasselmann (Hasselmann, 1962).

By the mid 60s, the basic processes responsible for the evolution of the wind wave spectrum had been identified. An ability to accurately predict the evolution of the spectrum awaited an increase in the observational data base and advances in computational speed.

The goals of wind wave research are relatively well defined: to be able to predict the wind wave field and its effect on the environment. That environment could be natural (beaches, the atmosphere etc.) or imposed by human endeavour (ports, harbours, coastal settlements etc.). Although the goals are similar, the specific requirements of these various fields differ. For instance, coastal engineering activity may only require knowledge of the integral parameters of significant wave height and peak period, whereas studies of air-sea interaction will generally require a very detailed description of the full spectrum.

This book attempts to summarise the current state of this knowledge and to place this understanding into a common frame work. There are still many aspects of wind waves which are not fully understood. A notable example being wave breaking. Despite this, an impressive ability now exists to predict waves on both global and regional scales. Indeed, it is the author's belief that the most significant source of error in deep water wave models is the driving wind. Further advances in our understanding of the physics of wave evolution will result in only marginal improvements in the prediction capability. In finite depth regions, knowledge is still relatively poor and further work is clearly required.

The past 20 years has also brought enormous advances in our abilities to measure the ocean wave field. In particular, satellite based remote sensing instruments can now provide a global, real time view of the wave field. The analysis of this growing observational data base is beginning to yield the first global wave climatologies needed for activities such as shipping and ocean engineering. In addition, the assimilation of these data into wave prediction models will begin to address the deficiencies of the forcing wind fields.

This book attempts to take a balanced approach between the pragmatic engineering view of requiring a short term result and the scientific quest for detailed understanding. Thus, it attempts to provide a rigorous description of the physical processes involved as well as practical predictive tools.

A basic assumption which is made throughout the text is that, to first order, waves can be assumed linear. Nonlinearities can be considered as perturbations to this linear solution. It follows from this assumption that waves can be considered in a spectral sense. As a result, fundamentally nonlinear descriptions have been largely ignored. This approach stems from the pragmatic view that the spectral representation of waves is the only approach which has developed a comprehensive predictive capability, even in shallow water where nonlinearities increase.

Chapter 2

Wave Theory

2.1 Introduction

As a pre-curser to the investigation of the properties of waves under the active forcing of the wind, it is essential to consider the more idealized case of the wave field propagating in the absence of any forcing. Even this highly idealized situation presents formidable mathematical difficulties and thus requires many simplifying assumptions to form a tractable solution. As the restrictions of these simplifying assumptions are gradually reduced the complexity of the solution increases. Only the simplest of these solutions will be considered here. Interest will focus on linear or Airy wave theory (Airy, 1845). Despite the apparently restrictive simplifying assumptions associated with this theory, its range of application is extensive. Linear wave theory will form the basic theoretical rationale for the remainder of this book.

2.2 Small Amplitude or Linear Theory

2.2.1 Governing Equations

In order to form a tractable solution the following simplifying assumptions are made:

1. The water is of constant depth, d and wave length, L (or period, T).
2. The wave motion is two-dimensional, which leads to long crested waves with constant height along the crests.
3. The waves are of constant form, that is, they do not change with time.
4. The fluid (water) is incompressible.

5. Effects of viscosity, turbulence, and surface tension are neglected.

6. The wave height, H is small compared to the wave length, L and the water depth, d (i.e. $H/L << 1$ and $H/d << 1$).

The limitations which these assumptions place on the resulting theory will be investigated in Section (2.4).

The governing equations to be solved represent the equations for conservation of mass and momentum. Conservation of mass can be written in terms of the Laplace Equation

$$\frac{\partial^2 \phi}{\partial x^2} + \frac{\partial^2 \phi}{\partial z^2} = 0 \tag{2.1}$$

where x and z are the horizontal and vertical coordinates, respectively, of the two-dimensional solution domain. The velocity potential, ϕ is defined in terms of the horizontal and vertical components of velocity, u and w as

$$u(x, z, t) = -\frac{\partial \phi}{\partial x} \tag{2.2}$$

$$w(x, z, t) = -\frac{\partial \phi}{\partial z} \tag{2.3}$$

and t is time.

The conservation of momentum is represented by the Unsteady Bernoulli Equation

$$-\frac{\partial \phi}{\partial t} + \frac{p}{\rho_w} + gz = 0 \tag{2.4}$$

where p is the pressure, ρ_w the density of the fluid (water) and g is gravitational acceleration. It is desired to solve (2.1) and (2.4) within the domain represented by Figure 2.1 and subject to the boundary conditions defined below.

(i) Dynamic Boundary Condition at the free surface:

At the free surface, $z = \eta$, the pressure is atmospheric, $p = 0$. Therefore (2.4) becomes

$$-\frac{\partial \phi}{\partial t} + g\eta = 0 \text{ at } z = 0 \tag{2.5}$$

The boundary condition, (2.5) has been applied at $z = 0$ rather than $z = \eta$ since one of the simplifying assumptions was that the waves were of small amplitude (i.e. $\eta \rightarrow 0$).

(ii) Kinematic Boundary Condition at the free surface:

At the free surface there can be no transport of fluid through the free surface. Hence, the vertical velocity of the free surface must equal the vertical velocity of the fluid (i.e. $w = D/Dt[\eta(x,t)]$).

$$w = \underbrace{\frac{\partial \eta}{\partial t} + u\frac{\partial \eta}{\partial x}}_{\substack{\text{vertical velocity} \\ \text{of free surface}}} \quad \text{at } z = \eta \tag{2.6}$$

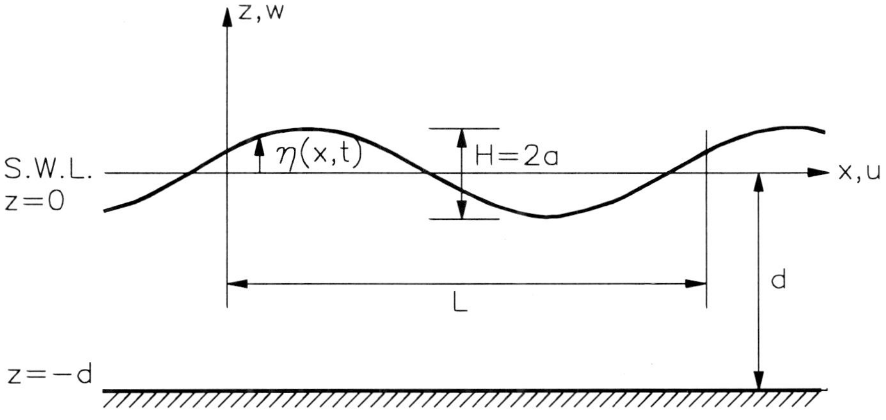

Figure 2.1: Definition sketch showing the solution domain for Linear Wave Theory

Substituting (2.3) for w, noting that due to the small amplitude assumption the slope of the water surface $\partial\eta/\partial x$ will be small and that the boundary condition can be assumed to apply at $z = 0$ reduces (2.6) to

$$-\frac{\partial \phi}{\partial z} = \frac{\partial \eta}{\partial t} \text{ at } z = 0 \tag{2.7}$$

(iii) Kinematic Boundary Condition at the bed:

In a similar fashion to the kinematic boundary condition at the free surface, there can be no flow through the solid bed

$$w = -\frac{\partial \phi}{\partial z} = 0 \text{ at z=-d} \tag{2.8}$$

Hence, the problem becomes the solution of (2.1) subject to the boundary conditions (2.5), (2.7) and (2.8). This system of equations, together with the solution domain is shown diagrammatically in Figure 2.2.

A solution to this set of equations can be found using the method of separation of variables (Airy, 1845; Stokes, 1847)

$$\phi(x, z, t) = \frac{ag}{\omega}\frac{\cosh\left[k\left(d + z\right)\right]}{\cosh\left[kd\right]}\cos\left(kx - \omega t\right) \tag{2.9}$$

where $a = H/2$ is the wave amplitude, $\omega = 2\pi/T = 2\pi f$ is the wave angular frequency and $k = 2\pi/L$ is the wave number. From (2.9) the basic properties of linear water waves can be developed. A number of these key properties are defined in Sections (2.2.2) to (2.2.9).

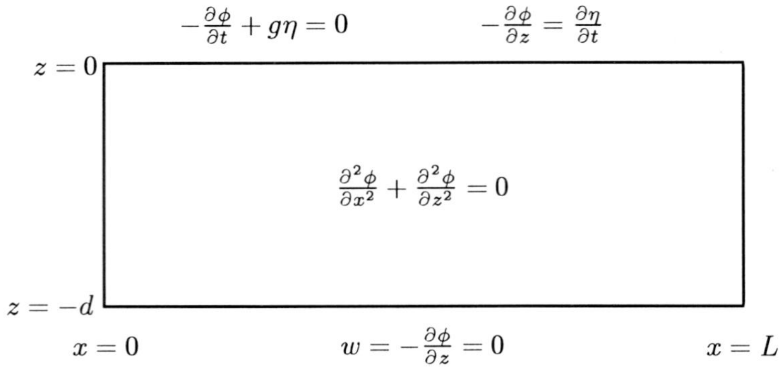

Figure 2.2: The governing equation, together with boundary conditions and the solution domain for Linear Wave Theory

2.2.2 The Wave Profile

Equation (2.5) states that at the free surface

$$\eta = \frac{1}{g}\left[\frac{\partial\phi}{\partial t}\right]_{z=0} \tag{2.10}$$

Substituting (2.9) for ϕ in (2.10) and differentiating with respect to t yields

$$\eta = a\sin(kx - \omega t) \tag{2.11}$$

Hence the water surface varies in a sinusoidal manner and is periodic in both space (wave length L) and time (period T).

2.2.3 Dispersion Relationship

Combining (2.5) and (2.7) yields

$$-\frac{\partial\phi}{\partial z} = \frac{1}{g}\frac{\partial^2\phi}{\partial t^2} \text{ at } z = 0 \tag{2.12}$$

Substituting (2.9) in (2.12) and evaluating at $z = 0$ yields

$$\omega^2 = gk\tanh(kd) \tag{2.13}$$

Equation (2.13) is called the "dispersion relationship" for linear waves. It states that there is a unique relationship between ω, k and d (or T, L and d). If two of the quantities are know, the third is uniquely defined.

2.2.4 Wave Phase Speed

By definition, the phase speed or speed of propagation, C of a wave is

$$C = \frac{L}{T} \text{ or } \frac{\omega}{k} \tag{2.14}$$

Substituting (2.13) in (2.14) for ω yields

$$C^2 = \frac{g}{k} \tanh(kd) \tag{2.15}$$

The phase speed, as defined by (2.15) is sometimes also called the "celerity". Equation (2.15) indicates that the phase speed varies with water depth. For given values of ω and k the wave will propagate faster in deep water than in shallow water. Also, C varies as a function of ω or k. Long period or long wave length waves will propagate faster than short period or short wave length waves. Hence, if a group of waves of varying wave length were to propagate away from an area where they were initially generated, they would propagate at different speeds. The longer wave length waves would propagate faster than the shorter wave length waves. The waves would gradually separate based on there respective wave lengths. The longer waves would lead followed by progressively shorter waves. Hence, the wave field would gradually disperse. This gives rise to the name, "dispersion relationship" attributed to (2.13).

2.2.5 Wave Length

Similarly, substituting (2.14) into (2.15) yields a transcendental relationship for the wave length, L

$$L = \frac{gT^2}{2\pi} \tanh\left(\frac{2\pi d}{L}\right) \tag{2.16}$$

The transcendental nature of (2.13), (2.15) and (2.16) means that iterative solution techniques are required. For this reason, a range of approximations have been proposed for the determination of C or k given ω and d. Hunt (1979) proposed a Padé approximation of the form

$$\frac{C^2}{gd} = \left[\tilde{\omega} + \left(1 + 0.666\tilde{\omega} + 0.445\tilde{\omega}^2 - 0.105\tilde{\omega}^3 + 0.272\tilde{\omega}^4\right)^{-1}\right]^{-1} \tag{2.17}$$

where $\tilde{\omega} = \omega^2 d/g$. Equation (2.17) is a remarkably accurate approximation to (2.16). Indeed, it is derived so as to be exact at both the short and long wave length limits. Young and Sobey (1980) have also suggested the use of a table look-up scheme for the routine evaluation of C within computer codes. Fenton and McKee (1990) have presented an extensive review of a number of different approximations which have been proposed. They conclude that there is little practical difference been the accuracy of the various approximations.

2.2.6 Water Particle Motion

The horizontal and vertical components of the fluid velocity (water particle velocity) are defined by (2.2) and (2.3). Substitution of (2.9) into these relationships yields

$$u = \frac{agk}{\omega} \frac{\cosh\left[k\left(d+z\right)\right]}{\cosh\left(kd\right)} \sin\left(kx - \omega t\right) \tag{2.18}$$

$$w = -\frac{agk}{\omega} \frac{\sinh\left[k\left(d+z\right)\right]}{\cosh\left(kd\right)} \cos\left(kx - \omega t\right) \tag{2.19}$$

The horizontal and vertical displacement of a fluid particle can be defined as

$$\xi = \int u dt \; ; \; \zeta = \int w dt \tag{2.20}$$

Substitution of (2.18) and (2.19) into (2.20) and integrating yields

$$\xi = \frac{agk}{\omega^2} \frac{\cosh\left[k\left(d+z\right)\right]}{\cosh\left(kd\right)} \cos\left(kx - \omega t\right) \tag{2.21}$$

$$\zeta = \frac{agk}{\omega^2} \frac{\sinh\left[k\left(d+z\right)\right]}{\cosh\left(kd\right)} \sin\left(kx - \omega t\right) \tag{2.22}$$

In (2.21) and (2.22) ξ and ζ are measured from the mean position.

Substituting the dispersion relationship (2.13) for ω^2 in (2.21) and (2.22), these relationships can be re-arranged to yield

$$\frac{\xi^2}{A^2} + \frac{\zeta^2}{B^2} = 1 \tag{2.23}$$

where

$$A = a \frac{\cosh\left[k\left(d+z\right)\right]}{\sinh\left(kd\right)} \tag{2.24}$$

$$B = a \frac{\sinh\left[k\left(d+z\right)\right]}{\sinh\left(kd\right)} \tag{2.25}$$

Equation (2.23) represents an ellipse with horizontal semi-axis A and vertical semi-axis B. Therefore, for linear wave theory, fluid particles move in closed elliptical orbits. In deep water $A = B$ and the orbit becomes a circle. This motion is shown diagrammatically in Figure 2.3.

Figure 2.4 shows the ratio of the horizontal to vertical components of velocity and displacement, u/w and ξ/ζ, respectively as a function of the parameter $k(d + z)$. As shown in Figure 2.3, the orbits progressively become more elongated with decreasing depth, d and submergence z. Indeed, at the bottom, $z = -d$, the kinematic boundary condition (2.8) requires that $w = 0$ and hence both u/w and $\xi/\zeta \rightarrow \infty$ for small $k(d + z)$.

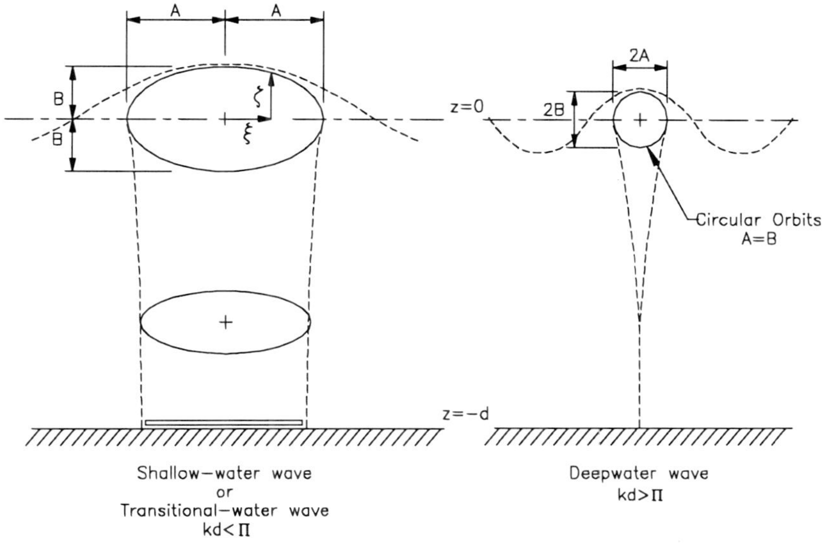

Figure 2.3: Diagrammatic representation of the motion of a fluid particle beneath a wave as predicted by linear wave theory [after CERC (1984)].

2.2.7 Pressure

Substitution of the velocity potential, ϕ from (2.9) into the Unsteady Bernoulli Equation (2.4) and differentiating yields

$$\frac{p}{\rho_w} = \frac{g\cosh\left[k\left(d+z\right)\right]}{\cosh\left(kd\right)}\underbrace{a\sin\left(kx-\omega t\right)}_{=\eta}-gz \tag{2.26}$$

Noting that from, (2.11), $\eta = a\sin(kx - \omega t)$, (2.26) becomes

$$\frac{p}{\rho_w g} = \eta\underbrace{\frac{\cosh\left[k\left(d+z\right)\right]}{\cosh\left(kd\right)}}_{=K_p}-z \tag{2.27}$$

or

$$\frac{p}{\rho_w g} = K_p\eta - z \tag{2.28}$$

where K_p is called the pressure response factor. The first term in (2.28) represents either a positive or negative (depending on the phase of the wave) deviation from hydrostatic pressure. In addition, the value of K_p decreases with increasing $|z|$ and hence the pressure fluctuations associated with the presence of the surface waves decrease with increasing distance below the water surface. For deep water conditions ($d \rightarrow \infty$), the value of K_p reduces from 1 at $z = 0$ to 0.04 at $z = -L/2$.

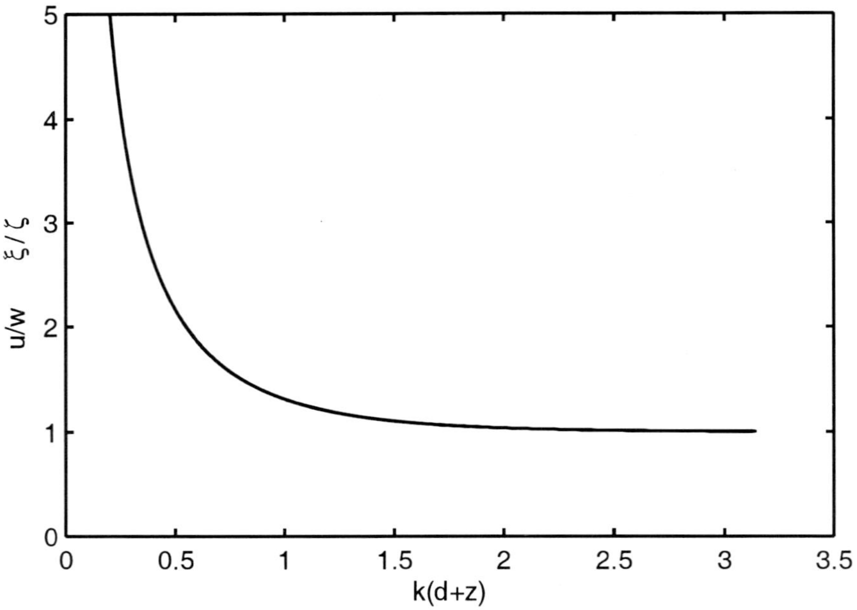

Figure 2.4: Ratio of the horizontal to vertical components of velocity, (u/w) and displacement (ξ/ζ) as a function of $k(d+z)$.

Hence, the influence of the surface waves is negligible at depths of submergence greater than approximately $L/2$.

2.2.8 Classification with respect to Depth

The hyperbolic functions which appear in the relationships of linear wave theory provide a convenient mechanism for the classification of waves with respect to water depth. The values of these hyperbolic functions as the argument kd approaches zero and infinity, respectively are shown in Table 2.1.

The asymptotic limits shown in Table 2.1 enable the linear wave relationships to be simplified for the limits of small and large values of kd. Arbitrary limits have been set for the values of kd (or $2\pi d/L$) at which the asymptotic relationships can be used with reasonable accuracy. These values are shown in Table 2.2. Use of the asymptotic relationships result in errors not exceeding 5% for the hyperbolic functions at the limits shown. The practice of classification of waves with respect to water depth is largely historical. Before the advent of modern computers, the evaluation of the hyperbolic functions and the transcendental relationships associated with linear wave theory was a tedious process. For the limits of deep and shallow water, these relationships could be significantly simplified. The classification system, and the resulting simplifications still find application today. Many mathe-

Function	Value of function as $kd \to 0$	$kd \to \infty$
$\sinh kd$	kd	$e^{kd}/2$
$\cosh kd$	1	$e^{kd}/2$
$\tanh kd$	kd	1

Table 2.1: Asymptotic limits of the linear wave theory hyperbolic functions.

Range of kd	Range of d/L	Type of waves
0 to $\pi/10$	0 to $1/20$	Shallow water waves (long waves)
$\pi/10$ to π	$1/20$ to $1/2$	Intermediate depth waves
π to ∞	$1/2$ to ∞	Deep water waves (short waves)

Table 2.2: Depth classification of linear waves in terms of the relative depth parameter kd.

matical manipulations involving linear wave relationships yield analytical solutions with the assumption of either deep or shallow water wave conditions. In addition, the classification system provides a physical interpretation of when bottom effects may influence wave properties. An example of this might be to what extent data collected in deep water might be applicable in finite depth situations. Table 2.2 also indicates that the depth, d is not the sole parameter which determines whether waves are in deep or shallow water. It is the relative depth parameter, kd, which provides this measure. The wave number, k (or wave length L) determines the depth to which the orbital velocities, accelerations and pressure are significant. Hence, it is the product kd which defines whether interaction with the bottom will influence the wave properties. A full summary of the linear wave relationships and their deep and shallow water simplifications is shown in Table 2.3.

RELATIVE DEPTH	SHALLOW WATER $kd < \pi/10$	TRANSITIONAL WATER $\pi/10 < kd < \pi$	DEEP WATER $kd > \pi$
Wave profile	Same as $\longrightarrow$	$\eta = a \sin(kx - \omega t)$	Same as $\longleftarrow$
Phase speed	$C = \sqrt{gd}$	$C = \frac{g}{\omega} \tanh(kd)$	$C = \frac{g}{\omega}$
Wave length	$L = T\sqrt{gd}$	$L = \frac{gT^2}{2\pi} \tanh\left(\frac{2\pi d}{L}\right)$	$L = \frac{gT^2}{2\pi}$
Angular frequency	$\omega^2 = gk^2 d$	$\omega^2 = gk \tanh(kd)$	$\omega^2 = gk$
Group velocity	$C_g = \sqrt{gd}$	$C_g = \frac{1}{2}\left[1 + \frac{2kd}{\sinh(2kd)}\right] C$	$C_g = \frac{g}{2\omega}$
Velocity components	$u = a\sqrt{\frac{g}{d}} \sin\psi$	$u = \frac{agk}{\omega} \frac{\cosh[k(d+z)]}{\cosh(kd)} \sin\psi$	$u = a\omega e^{kz} \sin\psi$
	$w = a\omega\left(1 + \frac{z}{d}\right) \cos\psi$	$w = -\frac{agk}{\omega} \frac{\sinh[k(d+z)]}{\cosh(kd)} \cos\psi$	$w = -a\omega e^{kz} \cos\psi$
Particle displacements	$\xi = \frac{a}{\omega}\sqrt{\frac{g}{d}} \cos\psi$	$\xi = \frac{agk}{\omega^2} \frac{\cosh[k(d+z)]}{\cosh(kd)} \cos\psi$	$\xi = ae^{kz} \cos\psi$
	$\zeta = a\left(1 + \frac{z}{d}\right) \sin\psi$	$\zeta = -\frac{agk}{\omega^2} \frac{\sinh[k(d+z)]}{\cosh(kd)} \sin\psi$	$\zeta = -ae^{kz} \sin\psi$
Subsurface pressure	$\frac{p}{\rho_w g} = \eta - z$	$\frac{p}{\rho_w g} = \eta \frac{\cosh[k(d+z)]}{\cosh(kd)} - z$	$\frac{p}{\rho_w g} = \eta e^{kz} - z$

Table 2.3: Depth classification of linear waves in terms of the relative depth parameter kd. Note that in the above, $\psi = (kx - \omega t)$.

2.2.9 Wave Trains and Group Velocity

Equation (2.11) defines the water surface elevation of a single linear wave train. Consider now a water surface consisting of two wave trains defined by the general form (2.11). Each train has the same amplitude, a but different frequencies ω_1 and ω_2, respectively. The water surface elevation becomes

$$\eta = a\sin(k_1 x - \omega_1 t) + a\sin(k_2 x - \omega_2 t) \tag{2.29}$$

$$= \overbrace{2a\cos\left[1/2\,(k_1 - k_2)\,x - 1/2\,(\omega_1 - \omega_2)\,t\right]}^{\text{variable amplitude of wave train}} \times \sin\left[1/2\,(k_1 + k_2)\,x - 1/2\,(\omega_1 + \omega_2)\,t\right] \tag{2.30}$$

The first term in (2.30) represents a time and space varying amplitude of the wave train. The resulting wave train is characterized by interference patterns and the appearance of groups of waves as shown in Figure 2.5.

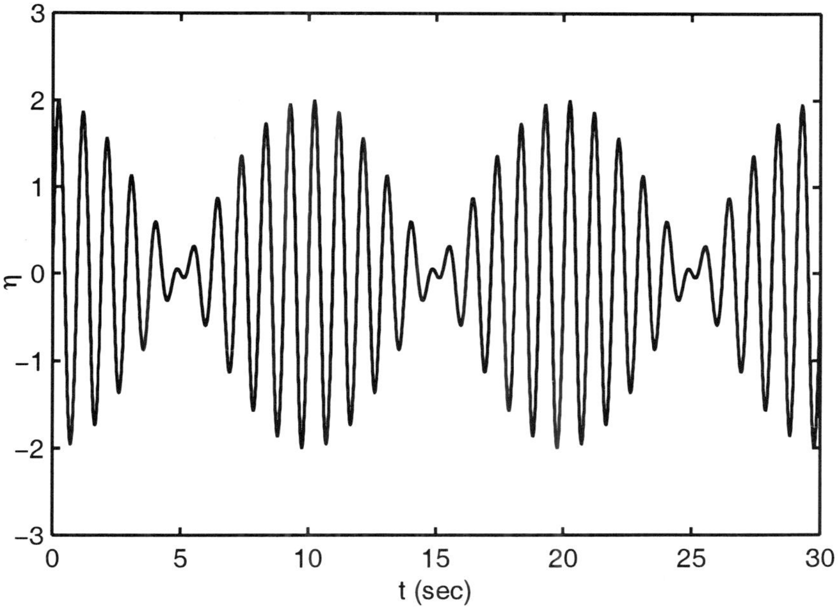

Figure 2.5: The formation of wave groups as a result of the superposition of two linear wave trains with different frequencies (Kinsman, 1965).

The water surface elevation, η is zero when the first term in (2.30) is zero. This occurs when

$$1/2\,(k_1 - k_2)\,x - 1/2\,(\omega_1 - \omega_2)\,t = (2m+1)\,\pi/2 \tag{2.31}$$

where m is an integer counter (i.e. $m = 0, 1, 2, \cdots$). Therefore, the envelope

(Figure 2.5) to the water surface elevation will be zero at $x = x_{node}$ given by

$$x_{node} = \left[\frac{\omega_1 - \omega_2}{k_1 - k_2}\right] t + \frac{(2m+1)\pi}{k_1 - k_2} \tag{2.32}$$

The speed of propagation of the nodal point and hence the wave "group" is

$$C_g = \frac{dx_{node}}{dt} = \frac{\omega_1 - \omega_2}{k_1 - k_2} \tag{2.33}$$

where C_g is called the group velocity. In the limit as ω_1 approaches ω_2, (2.33) becomes

$$C_g = \frac{d\omega}{dk} \tag{2.34}$$

Substituting the dispersion relationship (2.13) into (2.34) and differentiating yields

$$C_g = \underbrace{\frac{1}{2}\left[1 + \frac{2kd}{\sinh(2kd)}\right]}_{n} C \tag{2.35}$$

where

$n \longrightarrow 1/2$ in deep water
$n \longrightarrow 1$ in shallow water

Hence, the group velocity is half the phase speed in deep water and equal to the phase speed in shallow water. The variation in the ratio C_g/C is shown in Figure 2.6 as a function of the relative depth parameter kd.

In all but shallow water, the individual waves in the group are propagating faster than the group itself. The individual waves will appear at the rear of the group, propagate through the group and disappear at the front of the group. If a single group of waves propagating on an otherwise calm ocean is considered, individual waves will be observed to propagate through the group at their phase speed, C. The group will, however, remain intact and continue to propagate at the group velocity, C_g. Hence, the energy represented by the group will propagate at the speed of translation of the group rather than the individual waves that make up the group (Svendsen and Jonsson, 1976). Wave energy propagating from a generation site travels at the group velocity rather than the phase speed.

2.3 Wave Transformation

2.3.1 Basic Assumptions

Previously, only cases where the water depth was constant were considered. Indeed, the specification of the governing equation together with the appropriate boundary conditions [(2.1), (2.5), (2.7) and (2.8)] requires that $d =$ constant. To a reasonable

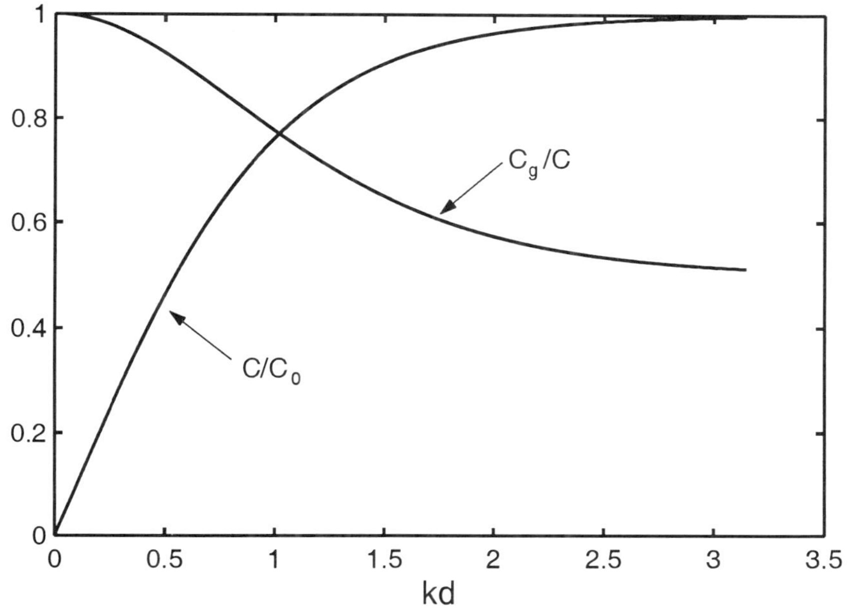

Figure 2.6: Variation in the ratio of group velocity to phase speed (C_g/C) and phase speed to deep water phase speed (C/C_0) as functions of the relative depth parameter kd.

level of accuracy, this condition may be relaxed provided that the depth varies only slowly on a scale given by the wave length. There is no absolute limit to how rapidly the bottom can vary whilst linear wave theory remains a reasonable approximation. The problem has been investigated by numerous authors including Rosseau (1952),Ogawa and Yashida (1959), Madsen and Mei (1969a,b). It appears that linear theory remains a reasonable approximation provided that the water depth changes by less than 20% to 30% within one wave length (Svendsen and Jonsson, 1976).

Consider the case of waves propagating towards the shore from deep to shallow water. Two points of different water depth can be specified along this propagation path and the number of waves passing each point can be counted over some finite period of time. As the individual waves must be conserved, the number of waves passing each point must be the same. The wave period, T, as defined in Section (2.2.2) is the time taken for successive waves to pass a fixed point. As the number of waves passing the two selected points is the same during the finite time period, the wave period at the two points must also be the same. Hence, as the water depth varies, the wave period, T remains constant. Since the wave period is constant, it follows from (2.15) and (2.16) that as waves propagate from deep to shallow water, the phase speed, C and the wave length, L will both decrease.

The basic assumptions to be invoked in considering wave transformation can be summarized as:

1. The bottom slope varies slowly.
2. Energy transmission between adjacent orthogonals (i.e. line perpendicular to the wave crest) is constant.
3. The waves do not break.
4. The wave period, T and hence frequency, f or ω are constant.

2.3.2 Wave Shoaling

Initially, consider the two-dimensional problem of waves at normal incidence to straight parallel depth contours as shown in Figure 2.7. The energy of a wave per

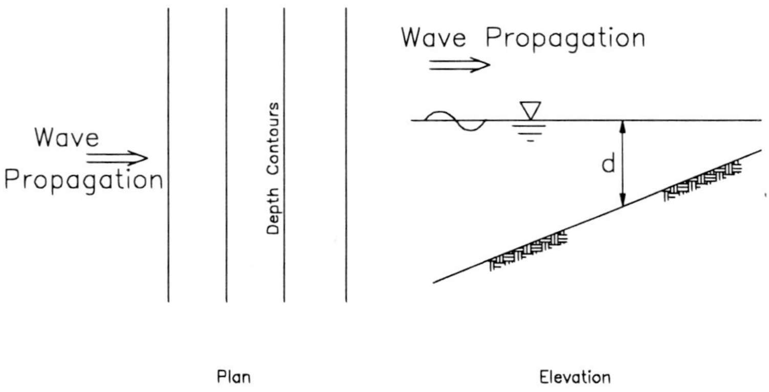

Figure 2.7: Definition sketch showing the two-dimensional shoaling of linear water waves.

unit crest length can be shown to be [Dean and Dalrymple (1991) and many others]

$$E = \frac{1}{8}\rho_w g H^2 L \tag{2.36}$$

The energy per unit area or specific energy is

$$\overline{E} = \frac{1}{8}\rho_w g H^2 \tag{2.37}$$

The energy flux, $\overline{P}$, is the rate at which the wave energy is advected. From Section (2.2.9), the energy is advected at the group velocity, C_g.

$$\overline{P} = \overline{E}C_g = \overline{E}nC \tag{2.38}$$

The energy flux, $\overline{P}$ is also called the wave power. Since the amount of energy entering a region must equal the amount leaving (conservation of energy) it follows that

$$\overline{E}_0 n_0 C_0 = \overline{E} n C \tag{2.39}$$

where the zero subscripts refer to deep water values. From (2.37) and (2.39) it follows that

$$\frac{H}{H_0} = K_s = \sqrt{\frac{n_0 C_0}{nC}} \tag{2.40}$$

The quantity K_s is called the *Shoaling Coefficient*. As shown in Figure 2.8, with decreasing water depth the shoaling coefficient decreases slightly below one before increasing rapidly. As stated in Section (2.3.1), the wave length also decreases with decreasing water depth. Hence, as the waves propagate into progressively shallower water, the wave slope, H/L will increase. As the steepness increases, the waves will reach a point where their form is unstable and they will begin to break.

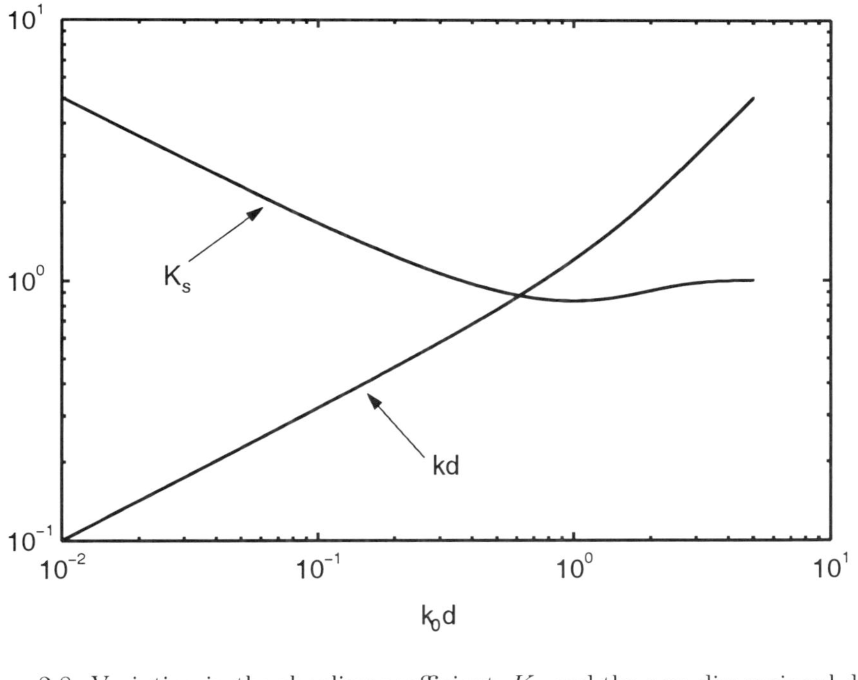

Figure 2.8: Variation in the shoaling coefficient, K_s and the non-dimensional depth parameter, kd as a function of $k_0 d$, where k_0 is the deep water wave number.

2.3.3 Wave Refraction

Equation (2.15) indicates that the phase speed, C varies as a function of the water depth, d. Therefore, for a wave which is approaching the depth contours of Figure 2.7 at an angle other than normal, the water depth will vary along the wave crest. Therefore, the phase speed will also vary along the wave crest. As a result, the crest will tend to bend towards alignment with the depth contours. Wave orthogonals which are normal to the wave crest will not be straight lines but will bend as the crest bends. As a result, adjacent orthogonals may converge or diverge depending on the local bathymetry as shown in Figure 2.9.

A simple extension of (2.38) to this three-dimensional case yields

$$\overline{P} = \overline{E}nCb = \text{constant} \tag{2.41}$$

where b is the spacing between adjacent orthogonals (see Figure 2.9). Equation (2.41) assumes that there is no lateral transport of energy along the wave crests. It follows that the wave height variation is

$$\begin{aligned} \frac{H}{H_0} &= \sqrt{\frac{n_0 C_0}{nC}}\sqrt{\frac{b_0}{b}} \\ &= K_s K_r \end{aligned} \tag{2.42}$$

where K_r is termed the refraction coefficient. Whereas K_s can be evaluated from linear wave relationships in deep water and at the point of interest, K_r requires knowledge of the paths adjacent orthogonals have taken in propagating from deep water to the point of interest.

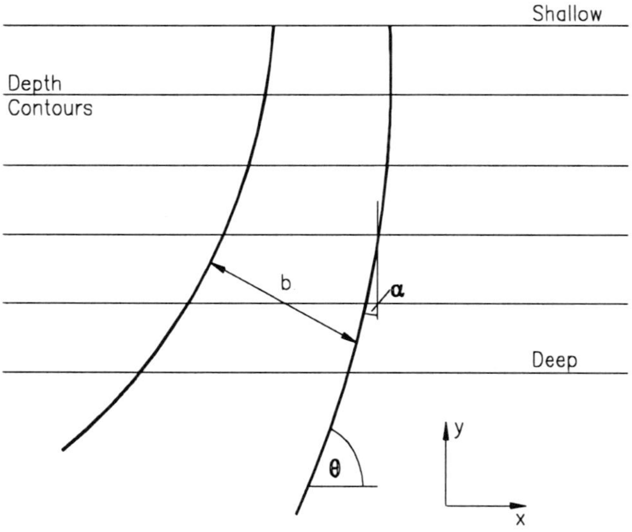

Figure 2.9: Definition sketch showing the refraction of waves over parallel depth contours. The wave orthogonals are shown by the bold lines.

Munk and Arthur (1952) have shown that the orthogonal or wave ray is defined by

$$\begin{aligned}\frac{\mathrm{d}s}{\mathrm{d}t} &= C \\ \frac{\mathrm{d}x}{\mathrm{d}t} &= C\cos\theta \\ \frac{\mathrm{d}y}{\mathrm{d}t} &= C\sin\theta\end{aligned} \tag{2.43}$$

where x and y are the standard coordinate directions, θ is the local direction of wave propagation and s is the coordinate direction along the wave ray (see Figure 2.9). In general, (2.43) must be solved numerically by stepping along the wave ray. For the case of parallel bed contours, however, an analytical solution in the form of Snel's law results

$$\frac{\sin\alpha}{C} = \text{constant} \tag{2.44}$$

where α is the angle between the wave ray and a normal to the bed contour. Historically, (2.44) has been applied to cases other than parallel bed contours and allowed the graphical construction of refraction diagrams. This process was extremely time consuming and prone to inaccuracy. It has now been replaced by numerical solutions to (2.43).

Snel's law is commonly encountered in the field of geometric optics and the analogy between the refraction of light and that of water waves is compelling. For light, it is the change in the refractive index of the medium which is responsible for refraction. For water waves, the change in water depth has the same effect.

2.4 Limitations of Linear Wave Theory

The wave theory developed in Section (2.2) assumed that the wave height was so small that the dynamic and kinematic boundary conditions at the free surface [(2.5) and (2.7)] could be applied at the still water level, $z = 0$ rather than at the wave disturbed surface $z = \eta$. In nature, the wave steepness, H/L seldom exceeds 0.05 to 0.08 (Svendsen and Jonsson, 1976) and hence the small amplitude assumption is often valid. There are, however, some applications where the simplifying assumptions of Linear Wave Theory become significant. In such cases it is necessary to resort to the use of a non-linear or *finite amplitude wave theory*. Such theories require the free surface boundary conditions to be applied at the free surface, which is, initially unknown as the wave form is part of the solution. The problem is made tractable by the use of *perturbation methods*. Such an analysis assumes that the non-linearities represent only small corrections to linear wave theory.

Physically the difference between linear and finite amplitude theories is that finite amplitude theories consider the influence of the wave itself on its properties. Therefore, in contrast to linear theory, the phase speed, wave length, water surface profile and other properties are functions of the actual wave height.

There are a number of different finite amplitude wave theories which have been proposed, but the two most commonly adopted theories are: *Stokes' Wave Theory* (Stokes, 1847, 1880; Miche, 1944) and *Cnoidal Wave Theory* (Korteweg and De Vries, 1895; Keulegan and Patterson, 1940; Wiegel, 1960). Stokes' theory is applicable in deep water whereas Cnoidal theory applies in finite depth situations.

Linear wave theory predicts that the crest and trough heights of the wave are equal. That is, the wave is evenly distributed about the still water level. In contrast, finite amplitude theories predict waves with peaked crests and flat troughs. The crests are further above the still water level than the troughs are below this level. Hence, in applications such as determining the deck elevation of an offshore structure, the use of finite amplitude wave theory would be important. Linear wave theory predicts that water particles move in closed orbits. Hence, there is no net transport of fluid. In contrast, finite amplitude theory predicts a small net fluid transport in the direction of wave propagation.

Despite the sophistication of finite amplitude wave theory, it still predicts waves which are of essentially a single period. There is a clear wave period and length. In reality, this seldom occurs in the ocean. Successive waves typically vary in height, period and propagation direction. Therefore, even high order finite amplitude theory is limited in such cases.

2.5 Spectral Representation of Waves

2.5.1 Frequency or Omni-directional Spectrum

Even the most casual observer of ocean waves would have realized that the simplifying assumptions of a constant wave period and waves of constant form place significant limitations on linear or finite amplitude wave theories. Waves typically appear confused with successive heights, periods and wave lengths varying significantly. Indeed, in some cases the actual direction of propagation is also difficult to define with confidence. The water surface appears confused rather than conforming to an ordered sinusoidal form as predicted by (2.11). An example of this is shown by the typical recorded wave record shown in Figure 2.10.

In many areas of physics it is common to represent records such as that of Figure 2.10 by the use of a spectral or Fourier model. Under this approximation, the water surface elevation is approximated by the linear superposition of sinusoidal forms as defined by (2.11).

$$\eta(t) = \sum_{i=1}^{N} a_i \sin(\omega_i t + \phi_i) \tag{2.45}$$

where a_i, ω_i and ϕ_i are the amplitude, frequency and phase of the ith component in the summation. An example of the manner in which a complex water surface

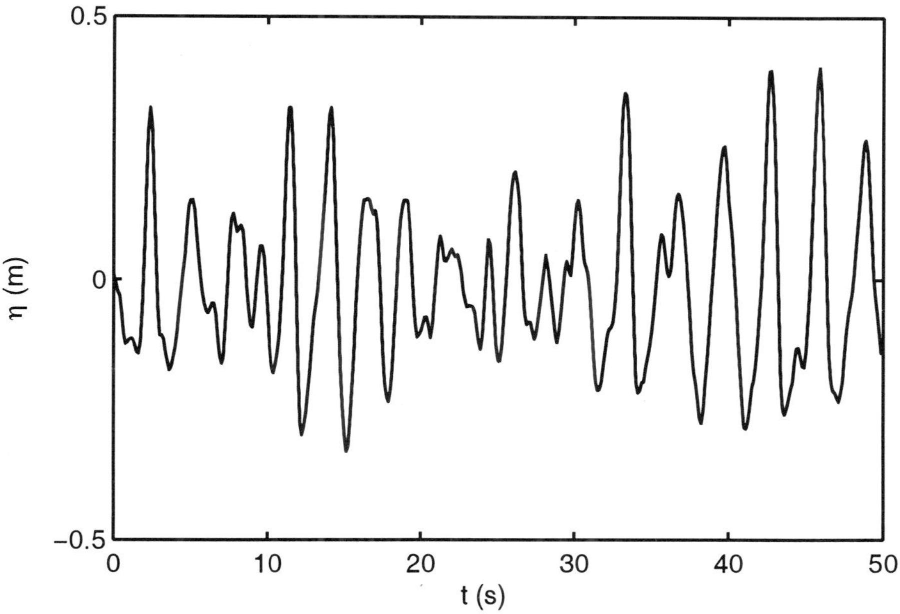

Figure 2.10: A typical wave record measured at Lake George, Australia in a water depth of approximately 2m. The wind speed was, U_{10}=16.3m/s. Note the variation in height and period of individual waves in the record.

record can be constructed from the summation of only 3 sinusoids is shown in Figure 2.11.

Each of the sinusoidal components in the summation (2.45) will satisfy all the properties of linear wave theory as developed in Sections (2.2.3) to (2.2.7).

From (2.37), the average energy of the wave profile can be represented as

$$\overline{E} = \frac{\rho_w g}{8} \sum_{i=1}^{N} H_i^2 \tag{2.46}$$

or

$$\frac{\overline{E}}{\rho_w g} = \frac{1}{2N} \sum_{i=1}^{N} a_i^2 = \sigma^2 \tag{2.47}$$

where σ^2 is the variance of the record.

Therefore, the amplitude components, a_i^2 are related to the energy of the record, the distribution of which as a function of frequency could be represented by plotting a_i^2 verses frequency as shown in Figure 2.12a. This amplitude spectrum is discrete, represented only at the frequencies ω_i of the summation (2.45). In the limit as $N \rightarrow \infty$ the amplitude spectrum can be transformed into the continuous spectrum,

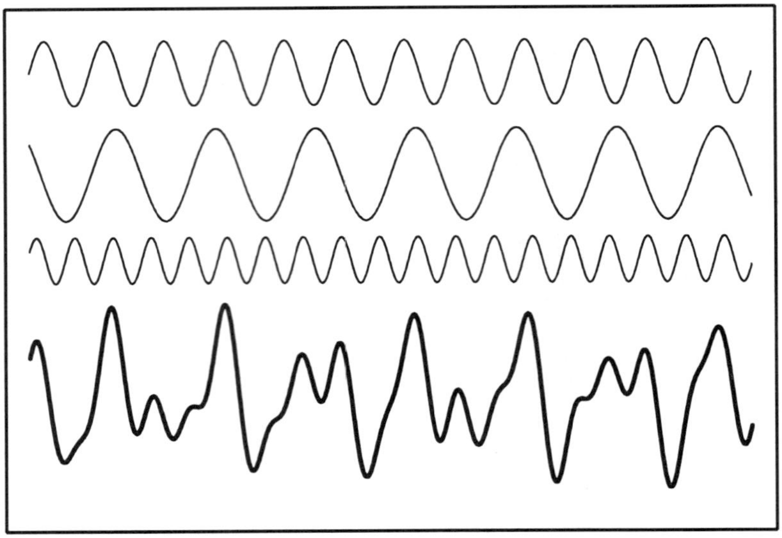

Figure 2.11: Construction of a complex water surface elevation record from the summation of three sinusoidal forms as represented by (2.45). The three sinusoidal records at the top (thin lines) sum to yield the record at the bottom (thick line).

$F(f)$, where

$$F(f)\Delta f = \frac{a_i^2}{2} \tag{2.48}$$

The spectrum, $F(f)$ is called the frequency or omni-directional [as no direction is associated with the spectrum - see Section (2.5.2)] or variance [as the area under the spectrum is the variance f the record] spectrum.

$$\sigma^2 = \int_0^\infty F(f)\mathrm{d}f \tag{2.49}$$

The use of the variance spectrum has the advantage that complex water surface elevation records can be considered, whilst retaining the compelling simplicity of linear wave theory. Examination of (2.45), however, reveals that the phase information represented by the ϕ_i terms have been discarded in forming the spectrum. As is evident in Figure 2.11, the choice of the values of ϕ_i will influence the resulting water surface elevation. The selection of the values of ϕ_i will, however, have no influence on the spectrum. Hence, the spectrum defines the distribution of energy with frequency but does not describe the actual water surface elevation unless additional assumptions about the associated "phase spectrum" are adopted.

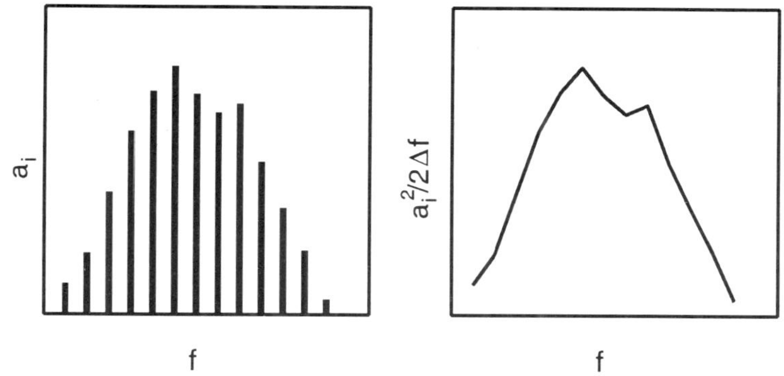

Figure 2.12: Illustrations of (a) the discrete amplitude spectrum (left) and (b) the continuous frequency or variance spectrum (right).

2.5.2 Directional Spectrum

A further extension of the Fourier model represented by (2.45) is to also include the possibility of wave components propagating in different directions

$$\eta(x, y, t) = \sum_{i=1}^{N} a_i \sin\left[k_i\left(x\cos\theta_i + y\sin\theta_i\right) - \omega_i t + \phi_i\right] \tag{2.50}$$

where θ_i is the angle between the x axis and the direction of propagation of the ith component in the summation. From this representation, a frequency-direction or directional spectrum, $F(f, \theta)$ can be defined in a similar manner to $F(f)$.

$$\sigma^2 = \int_0^{2\pi}\int_0^{\infty} F(f, \theta)\mathrm{d}f\mathrm{d}\theta \tag{2.51}$$

The directional frequency spectrum defines the distribution of energy with frequency and direction. From (2.50) it follows that this distribution could have been alternatively written in terms of wave number, thus forming a wave number spectrum $Q(k_x, k_y)$, where k_x and k_y are the wave number components in the x and y directions respectively. The wave number spectrum can be related to the directional frequency spectrum since

$$\sigma^2 = \int\int Q(k_x, k_y)\mathrm{d}k_x\mathrm{d}k_y = \int\int F(f, \theta)\mathrm{d}f\mathrm{d}\theta \tag{2.52}$$

Noting that $\mathrm{d}k_x\mathrm{d}k_y = |\boldsymbol{k}|\mathrm{d}k\mathrm{d}\theta$, this yields

$$F(f, \theta) = Q(k_x, k_y)|\boldsymbol{k}|\frac{\mathrm{d}k}{\mathrm{d}f} \tag{2.53}$$

where $\mathrm{d}k/\mathrm{d}f$ can be determined from the dispersion relationship (2.13).

Chapter 3

Stochastic Properties of Ocean Waves

3.1 Introduction

In Section (2.5) the statistical variability of the water surface elevation was modelled in terms of the Fourier series (2.45). In this chapter the statistical distribution of the Fourier amplitudes, a_i will be investigated.

3.2 Probability Distribution of Wave Heights

3.2.1 The Rayleigh Distribution

The basic theory to be adopted for the distribution of wave heights was originally developed by Rice (1954). It is assumed that the spectrum of the water surface elevation is narrow banded. That is, the energy is confined to a relatively narrow range of frequencies. For such a spectrum it can be shown that the probability density function for wave heights follows a Rayleigh distribution

$$p(H) = \frac{H}{4\sigma^2} e^{-\frac{H^2}{8\sigma^2}} \tag{3.1}$$

where $\sigma^2 = \int F(f)\mathrm{d}f$ is the variance of the record. Equation (3.1) must satisfy the requirement that $\int p(H)\mathrm{d}H = 1$.

Longuet-Higgins (1952) derived relationships between a number of characteristic wave heights based on (3.1). The mean and root-mean-square (rms) wave heights are, respectively

$$\overline{H} = \int_0^\infty Hp(H)\mathrm{d}H = \sqrt{2\pi\sigma^2} \tag{3.2}$$

$$H_{rms}^2 = \overline{H^2} = \int_0^\infty H^2 p(H)\mathrm{d}H = 8\sigma^2 \tag{3.3}$$

Adopting (3.3), the Rayleigh distribution (3.1) can be written in the alternate form

$$p(H) = \frac{2H}{H_{rms}^2} e^{-\frac{H^2}{H_{rms}^2}} \tag{3.4}$$

The cumulative form of (3.4) (i.e. probability that H is greater than a value $\hat{H}$) is

$$P(H > \hat{H}) = \int_{\hat{H}}^\infty p(H)\mathrm{d}H = e^{-\frac{\hat{H}^2}{H_{rms}^2}} \tag{3.5}$$

The average height of all waves greater than $\hat{H}$, written as $\overline{H}(\hat{H})$ is given by

$$\overline{H}(\hat{H}) = \frac{\int_{\hat{H}}^\infty H^2 e^{-\left(\frac{H}{H_{rms}}\right)^2} \mathrm{d}H}{\int_{\hat{H}}^\infty H e^{-\left(\frac{H}{H_{rms}}\right)^2} \mathrm{d}H} \tag{3.6}$$

3.2.2 Representative Measures of Wave Height

Rather than wishing to know the average height of all waves greater than a particular value, it is more common to consider the average height of the highest $1/N$ waves. This can be calculated from the Rayleigh distribution by determining the wave height which has a probability of exceedence given by $1/N$. A full analysis is given by Goda (1985). Typical results for various values of N appear in Table 3.1.

N	$H_{1/N}/\sigma$	$H_{1/N}/\overline{H}$	$H_{1/N}/H_{rms}$	Comments
100	6.67	2.66	2.36	
50	6.24	2.49	2.21	
20	5.62	2.24	1.99	
10	5.09	2.03	1.80	Highest 1/10th wave
5	4.50	1.80	1.59	
3	4.00	1.60	1.42	Significant wave
2	3.55	1.42	1.26	
1	2.51	1.00	0.87	Mean wave

Table 3.1: Representative wave heights calculated from the Rayleigh distribution [after Goda (1985)]

The significant wave height, H_s is defined in Table 3.1 as $H_{1/3}$, the average of the highest 1/3 of the waves. The term *significant wave height* is historical as this value appeared to correlate well with visual estimates of wave height from experienced observers. Table 3.1 indicates that $H_s = 4\sigma$. From (2.49), the variance, σ^2 can also be defined in terms of the spectrum as $\sigma^2 = \int F(f)\mathrm{d}f$. Therefore, the significant wave height can be calculated from the spectrum, its value is 4 times the square root of the area under the spectral curve.

The probability density function for wave height, as defined by the Rayleigh distribution (3.4) is shown in Figure 3.1. Also shown is the significant wave height, H_s. From Table 3.1, $H_s = 1.42H_{rms}$. The probability of a wave with height greater than H_s is represented by the shaded area in Figure 3.1 and is defined by the cumulative distribution (3.5). The probability, $P(H > H_s)$ takes a value of 0.135, that is, there is a 13.5% chance of a wave with height greater than the significant wave height. As stated previously, the significant wave height equates to the wave height which would be visually estimated by an experienced observer. On average, 13.5% of waves, or roughly 1 in 7, would be higher than this value. This agrees well, with the common sailors tale, that every 7th wave is large.

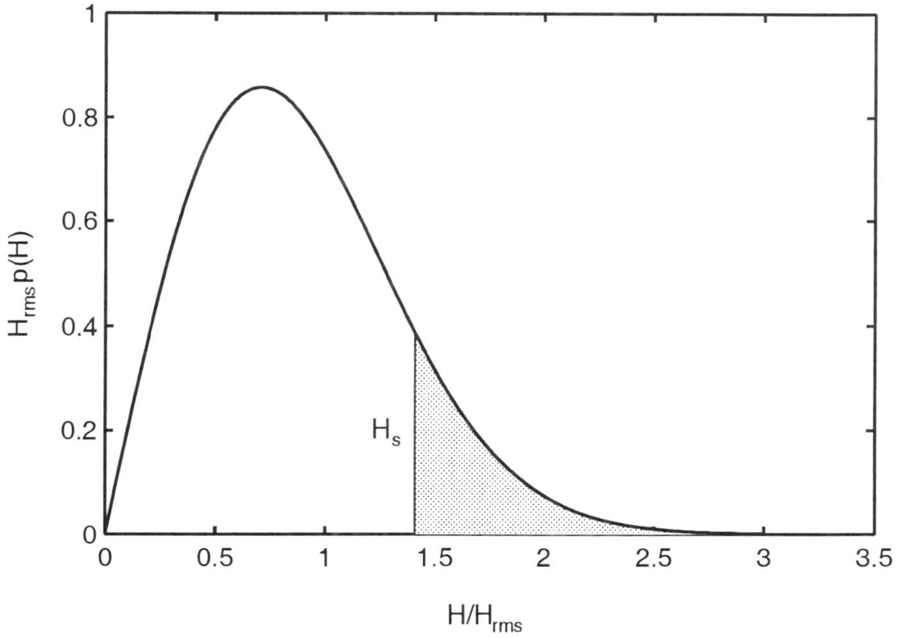

Figure 3.1: The probability density function, $p(H)$ for wave height as defined by the Rayleigh distribution (3.4). Also shown is the significant wave height, $H_s = 1.42H_{rms}$.

3.3 Global Distribution of Wave Properties

In Chapter 4 the detailed processes responsible for the generation and evolution of waves and methods for the prediction of waves will be investigated. Independent of these predictive capabilities, long term measurements of waves provides the opportunity to investigate regional variability of the wave climate. As an example, long term measurements at a point using an in situ instrument, such as a wave buoy, would provide information on the wave climate at that point. Such information is invaluable for engineering applications and long term measurement programs have been established at many sites around the world. By their nature, such measurements are point specific and cannot hope to give full global coverage. This type of coverage can, however, be obtained from Earth orbiting satellites. The details of such systems are discussed in Chapter 9.

Young (1994b), Young and Holland (1996a) and Young and Holland (1996b) have presented global statistics of significant wave height obtained from the Radar altimeter which was flown on the GEOSAT mission. This analysis extended more limited satellite climatologies presented by Mognard et al. (1983), Challenor et al. (1990) and Carter et al. (1991). The data set was, however, based on only 3 years data and hence limited in its application. Young and Holland (1998) extended the data set to approximately 9 years by combining data from 3 altimeter missions: GEOSAT, ERS1 and TOPEX. The data were processed to yield mean monthly values of H_s and exceedence probabilities for H_s (i.e. probability that H_s will be above a given value). Data were obtained for the full globe on a $2° \times 2°$ grid. Mean monthly values of H_s, obtained in this manner are shown in Figures 3.2 - 3.5 for the full year.

The 9 year data set also provides a means of determining the probability of obtaining wave heights of a particular value at any location. Figures 3.6 - 3.8 represent contour charts of the values of H_s which could be expected to be exceeded $10\% \cdots 90\%$ of the time, respectively.

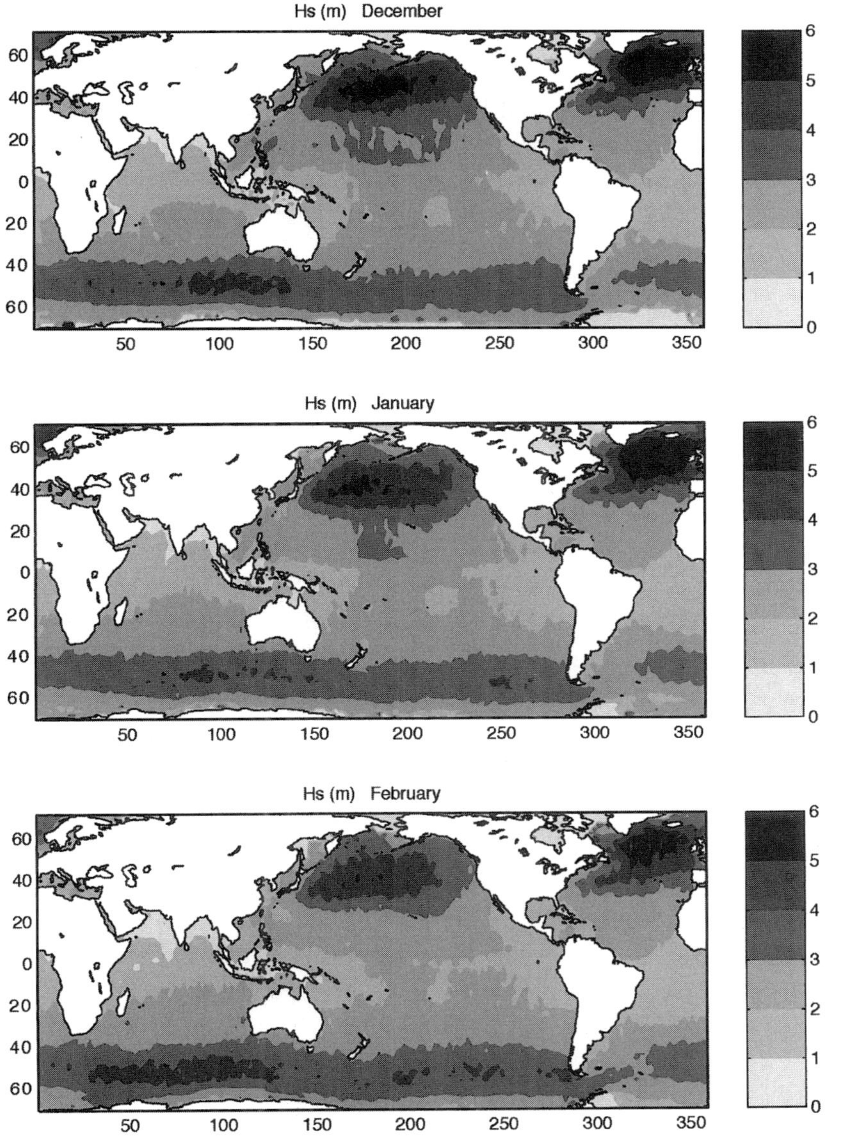

Figure 3.2: Contours of mean monthly values of H_s for the months of December, January and February (Northern Hemisphere Winter).

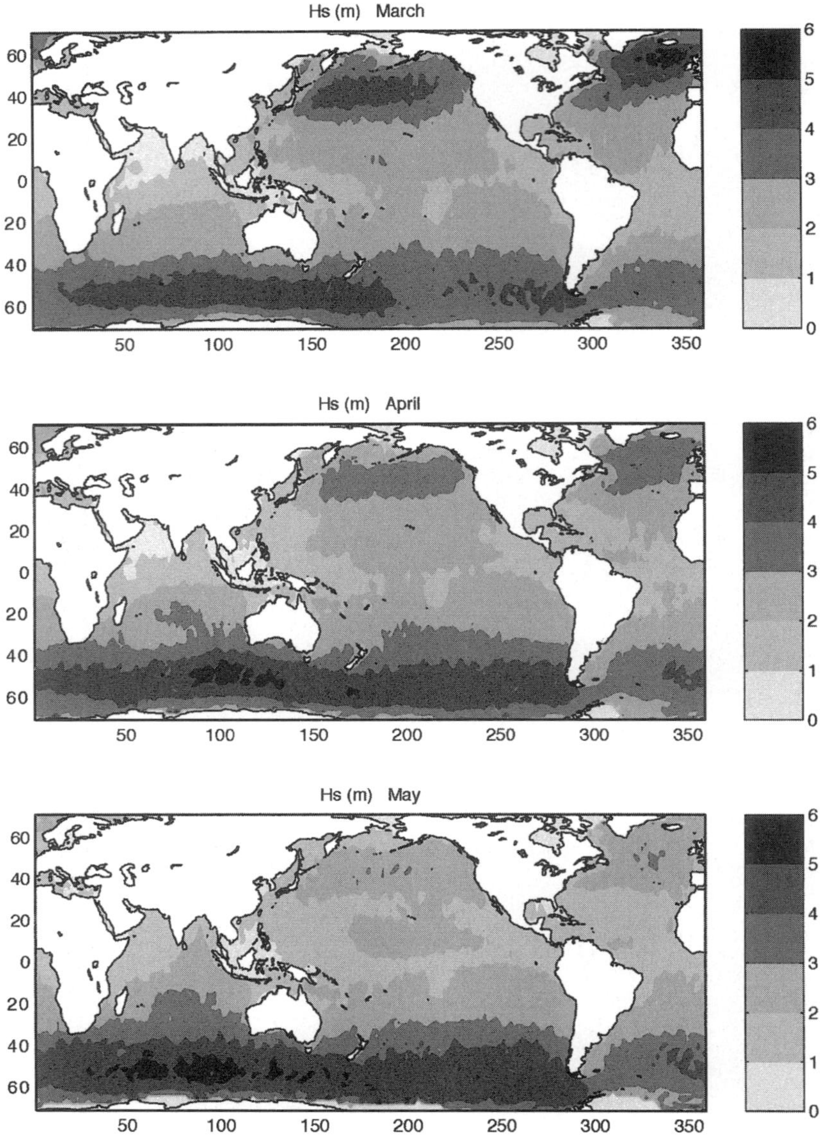

Figure 3.3: Contours of mean monthly values of H_s for the months of March, April and May (Northern Hemisphere Spring).

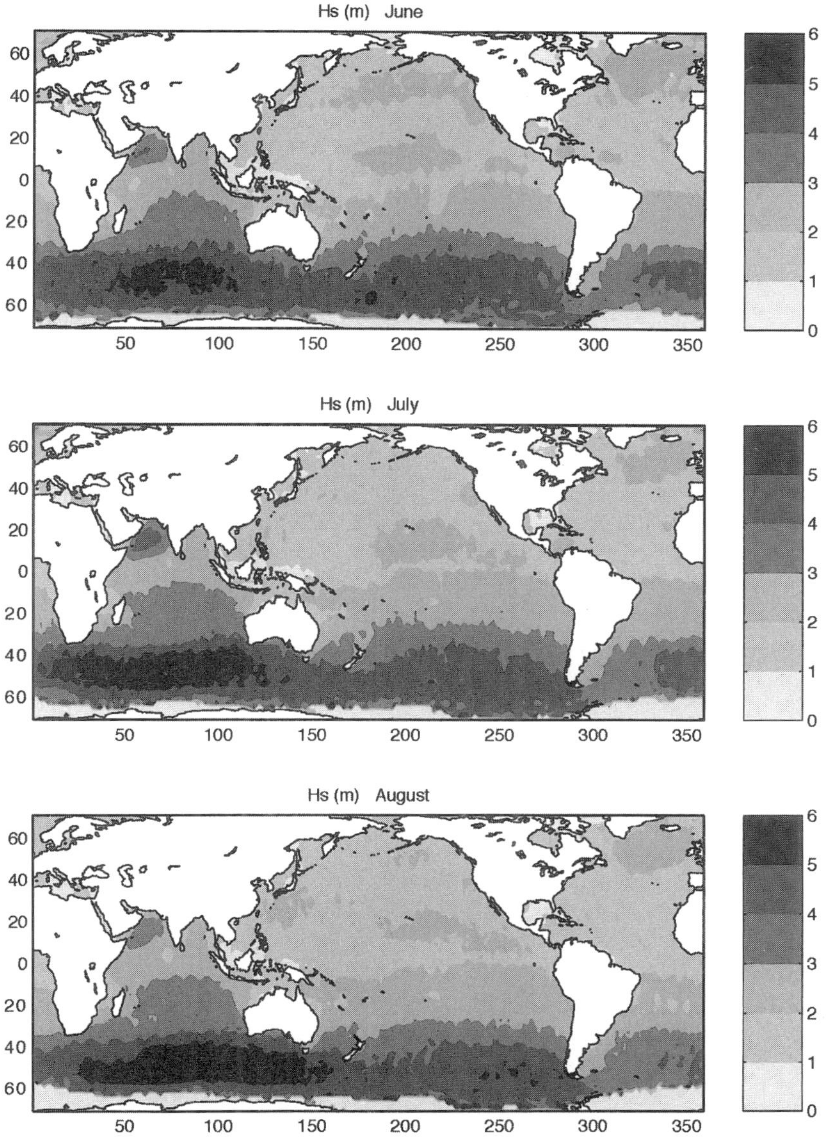

Figure 3.4: Contours of mean monthly values of H_s for the months of June, July and August (Northern Hemisphere Summer).

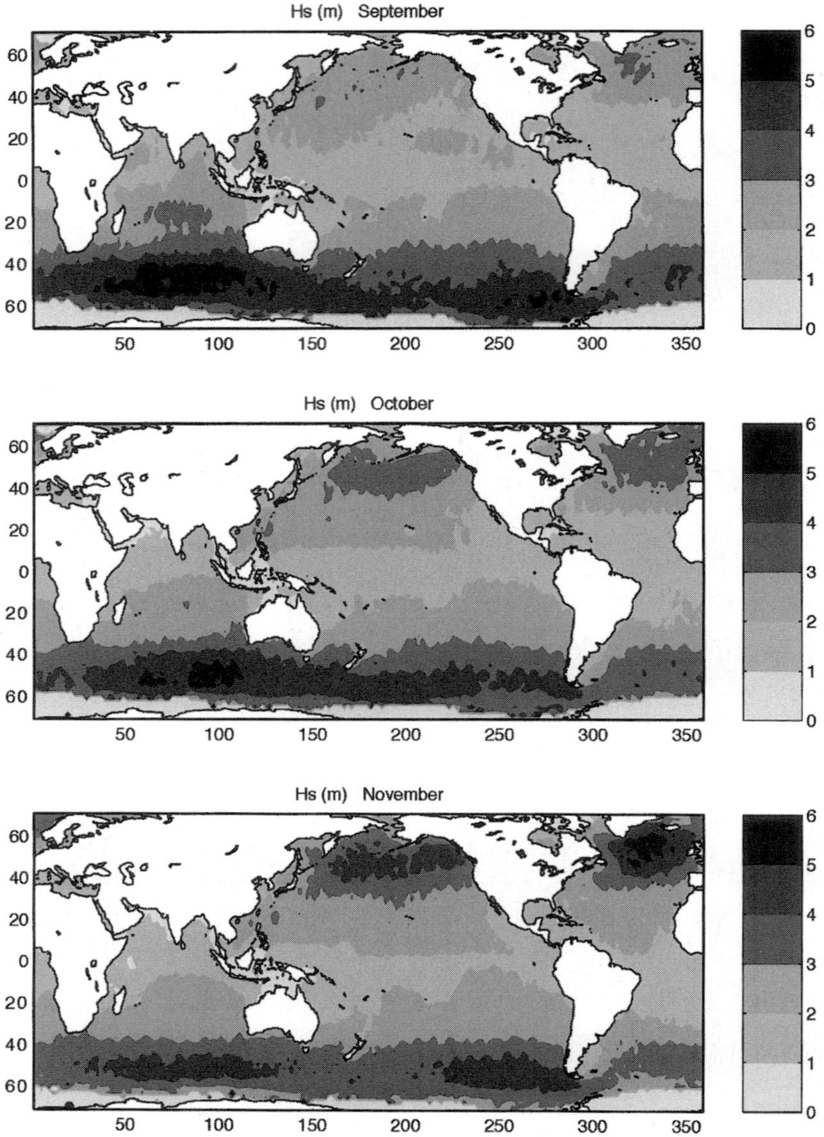

Figure 3.5: Contours of mean monthly values of H_s for the months of September, October and November (Northern Hemisphere Autumn).

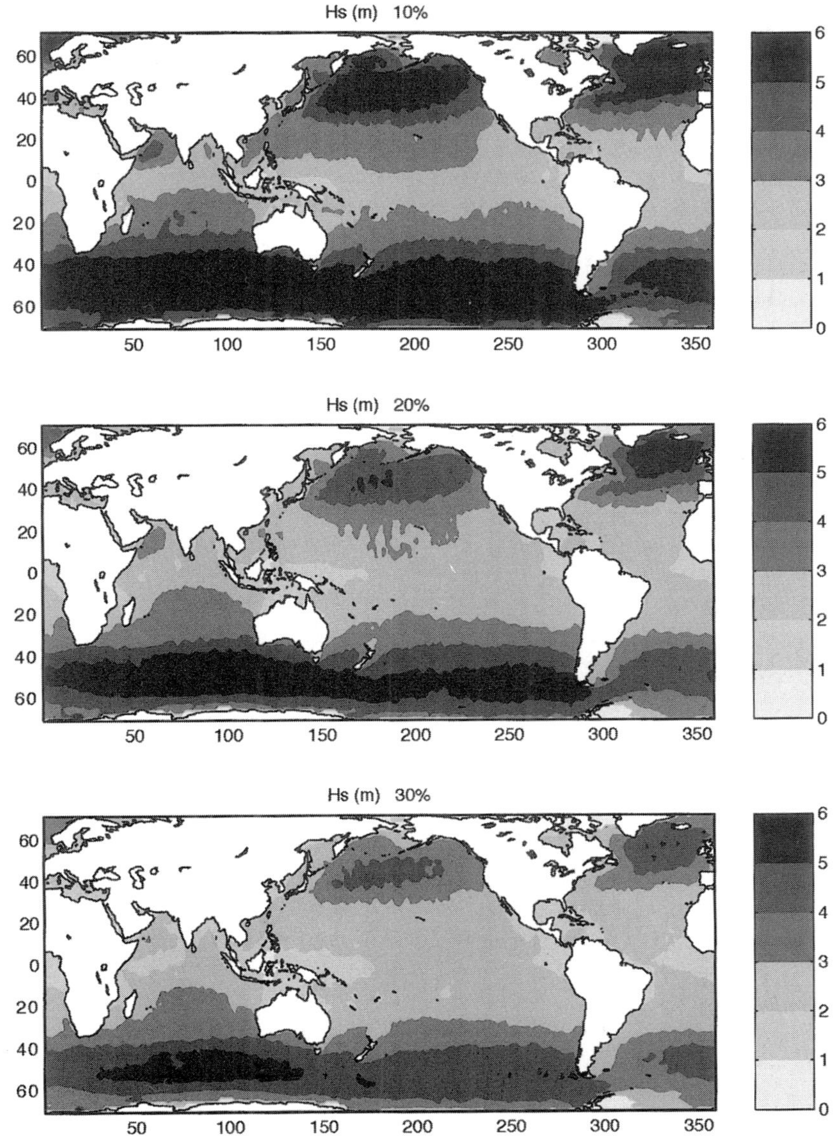

Figure 3.6: Contours of H_s which could be expected to be exceeded 10%, 20% and 30% of the time.

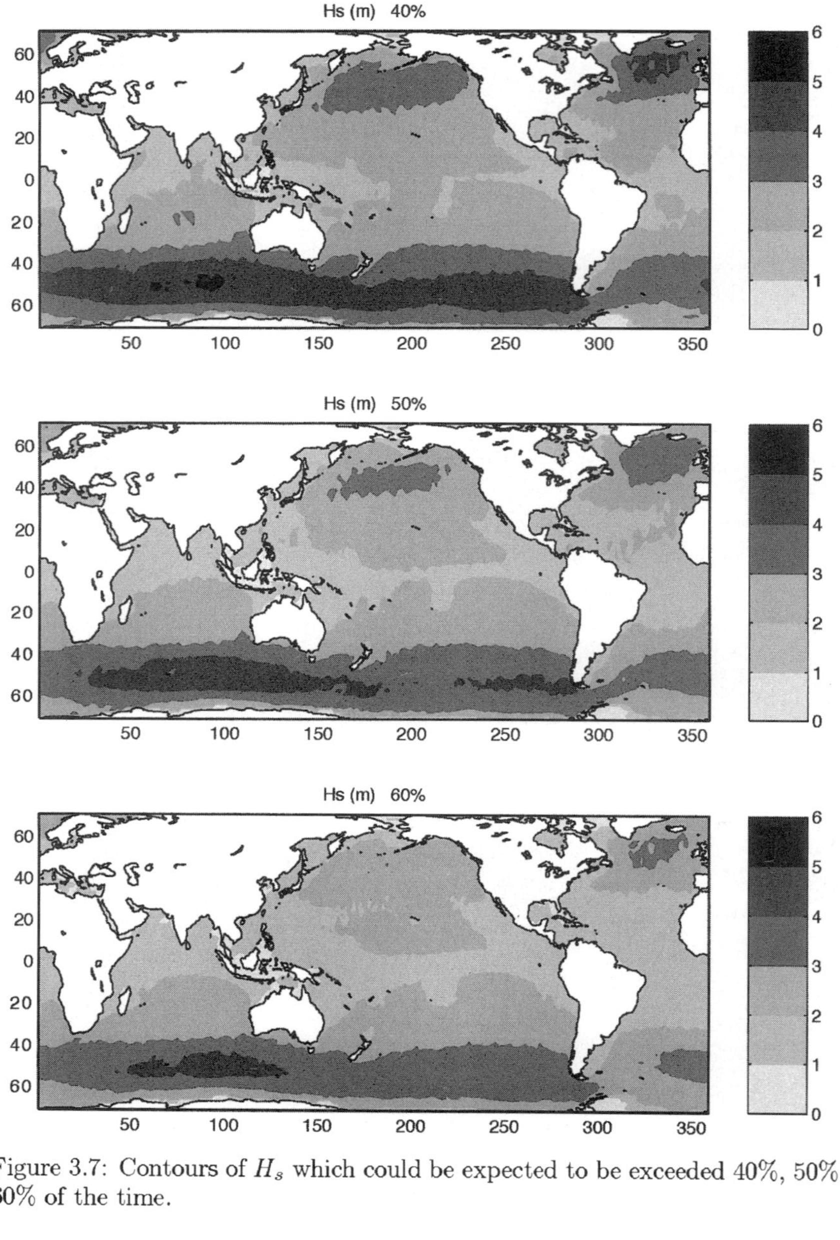

Figure 3.7: Contours of H_s which could be expected to be exceeded 40%, 50% and 60% of the time.

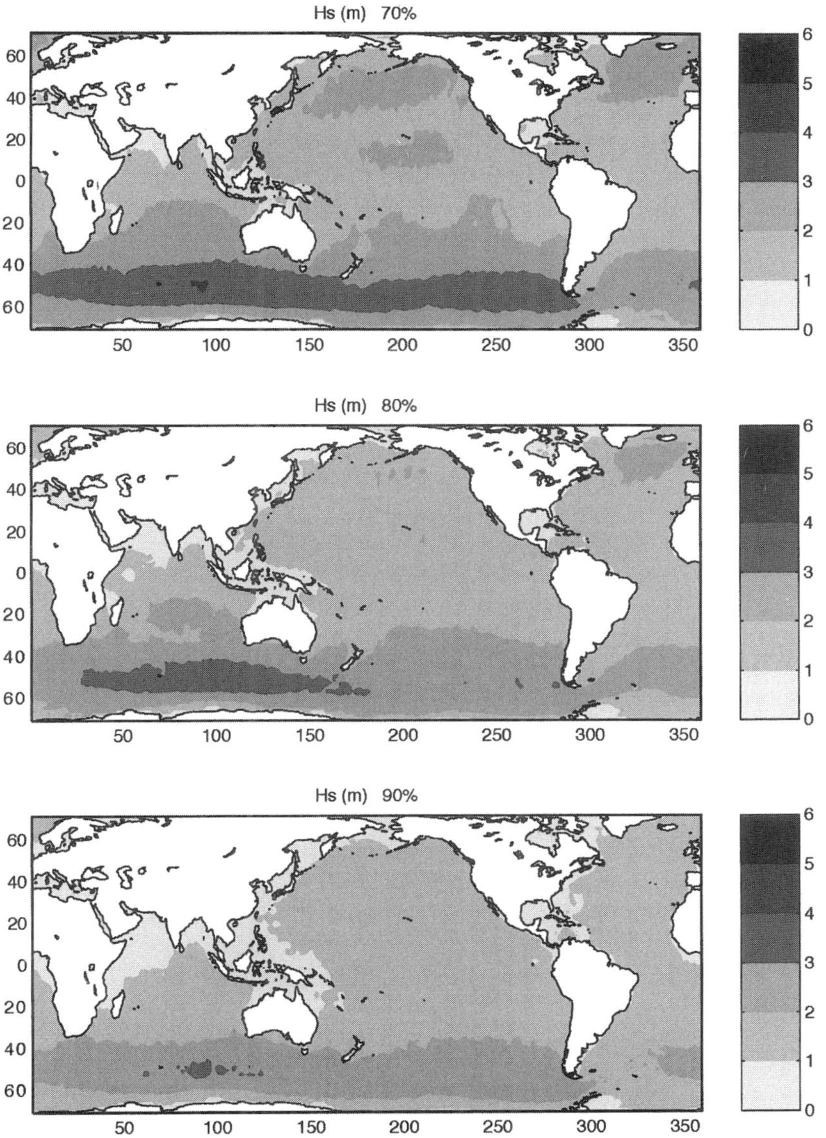

Figure 3.8: Contours of H_s which could be expected to be exceeded 70%, 80% and 90% of the time.

Examination of Figures 3.2 - 3.8 reveals many fascinating features of the global wave climate. The most striking feature of the figures is the north-south seasonal variability. Maximum waves occur at the higher latitudes in each hemisphere during their respective Winters. This is clear in Figure 3.9, which shows the yearly variation in H_s at points with the same absolute values of latitude from both hemispheres. The peak values for both hemispheres are approximately the same at 6m. The Southern Hemisphere, however, has much less seasonal variability with Summer wave heights falling to only 4m whereas in the Northern Hemisphere Summer wave heights are only 2.3m. This clearly reflects sailors observations that the Southern Ocean is rough all year round.

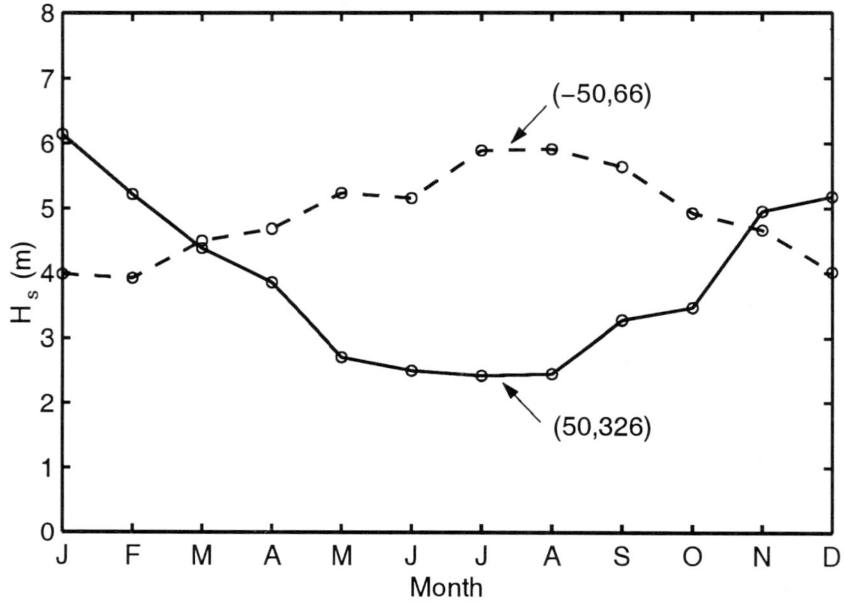

Figure 3.9: The annual variation in values of mean monthly H_s from points in the Southern Ocean ($-50°, 66°$, dashed line) and the North Atlantic ($50°, 326°$, solid line). These points represent the most extreme wave climates for each of the hemispheres.

In contrast to the extreme conditions at high latitudes, the equatorial regions are, not surprisingly, calm all year round. There is little seasonal variability in this region. As one moves either north or south from the equator both the extreme values and the seasonal variability increase.

Young and Holland (1996a) also presented global variations in wind speed obtained from the GEOSAT altimeter. Although the distribution of wind speed generally parallels that for wave height, there are notable exceptions. In particular, a clear band of higher wind speeds exists in both hemispheres at a latitude of approximately 20°. This band is associates with the trade winds. A similar band of higher wave heights does not exist for wave height in any of Figures 3.2 to 3.8. This occurs since waves at any particular point are a combination of locally generated wind seas and swell propagating from distant points. In the trade wind belts, swell propagating from high latitudes is sufficiently large to mask any local elevation in wave height caused by the stronger local trade winds.

The variation in wave height with latitude is clearly evident in Figure 3.10. This figure shows the value of H_s which could be expected to be exceeded 10% of the time along a line drawn north-south through the Pacific Ocean at longitude 209°E. The variation is approximately symmetric about the equator with the maximum waves occurring at latitude 50° in both hemispheres and the minimum at the equator.

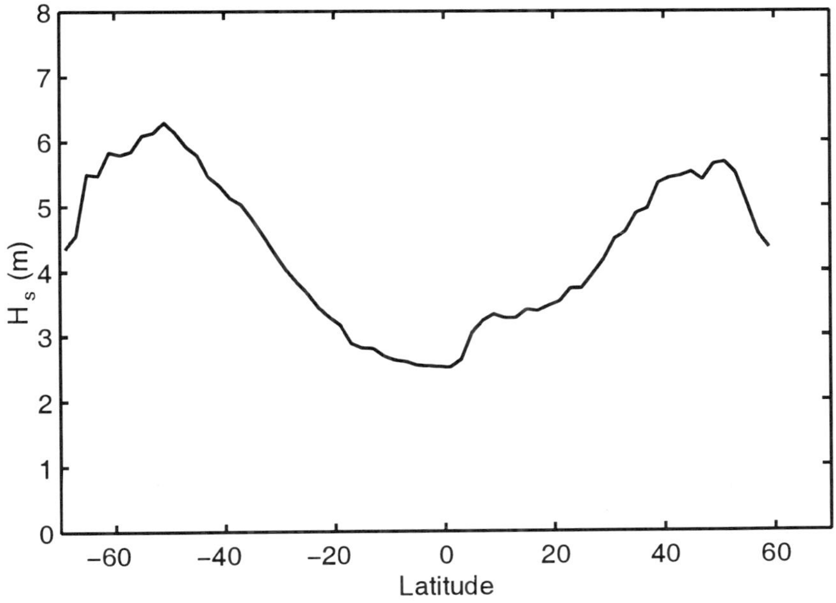

Figure 3.10: The variation in H_s with latitude which could be expected to be exceeded 10% of the time along a north-south line drawn through the Pacific Ocean at longitude 209°E.

There are many regional features of interest apparent in these global distributions of wave height. For the purposes of illustration, interest here is confined to examples of extreme events. As mentioned earlier, the Southern Ocean clearly has the most extreme, year round, wave conditions. Figure 3.11 shows the variation in mean monthly values of H_s along latitude −50°. Values for each of the four seasons are shown. Although the Southern Ocean is largely free of significant land masses, the wave height is not constant. As indicated above, the peak waves occur in Winter (July) in the area to the south of the Indian Ocean (between South

Africa and Australia). Winds in the Southern Ocean are generally from the west. Despite the fact that as one moves east from this point the fetch could be expected to increase, H_s actually decreases. The occurrence of the maxima and this gradual decrease in H_s appears to be associated with the distribution of wind speed rather than fetch. The results of Young and Holland (1996a) show that the wind speed follows a similar trend. This indicates that the progression of low pressure systems which track across the Southern Ocean tend, on average, to be most intense in the region between South Africa and Australia and gradually weaken as they propagate further east. In Spring and Autumn this intense wind and wave region weakens but moves further to the east.

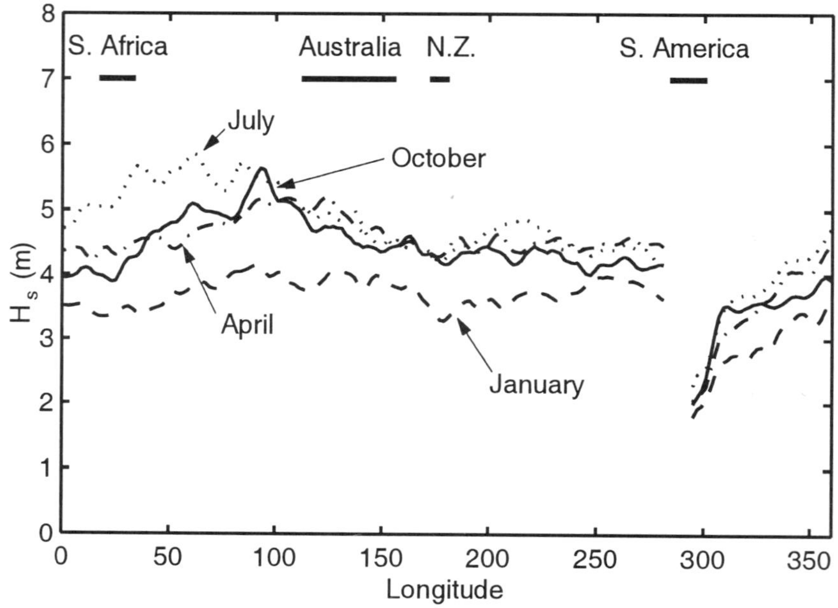

Figure 3.11: Variation in H_s as a function of longitude across the Southern Ocean at latitude $-50°$. Each of the four seasons is shown. Also shown at the top of the figure are the approximate locations and extent of the Southern Hemisphere land masses which lie to the north of this area.

It is clear from Figure 3.11 that H_s is significantly smaller to the east of South America than to the west. This is a consistent feature for many continents. Examination of the global distributions shows that the west coast wave climates of Africa, Australia and South America are all significantly higher that their respective east coast wave climates. This is largely due to the fact that conditions on the east coasts are fetch limited under the prevailing westerly winds (see Chapter 5). Young and Holland (1996a), however, show that the winds on the east coasts of these continents are also generally weaker than on their respective west coasts.

As stated earlier, the wave climate in equatorial and sub-equatorial regions is generally benign. One notable exception to this is the region of the Arabian Sea. In the period from June to September a very localized maximum develops close to the African Coast (see Figure 3.4). In this region mean monthly values of H_s reach 5m. Such values are typically only found at high latitudes. Figure 3.12 shows the variation in H_s with latitude along a line drawn at longitude 62°E, through the Arabian Sea. With the exception of the month of July, H_s consistently decreases with decreasing latitude. The localized maximum in the Arabian Sea is clearly apparent in July. This characteristic is caused by the strong south-west winds which develop in this region as a result of the Asian Summer Monsoon. These strong winds are confined to a jet relatively close to the coast and typically have mean monthly wind speeds of approximately 12m/s (Anderson and Prell, 1992).

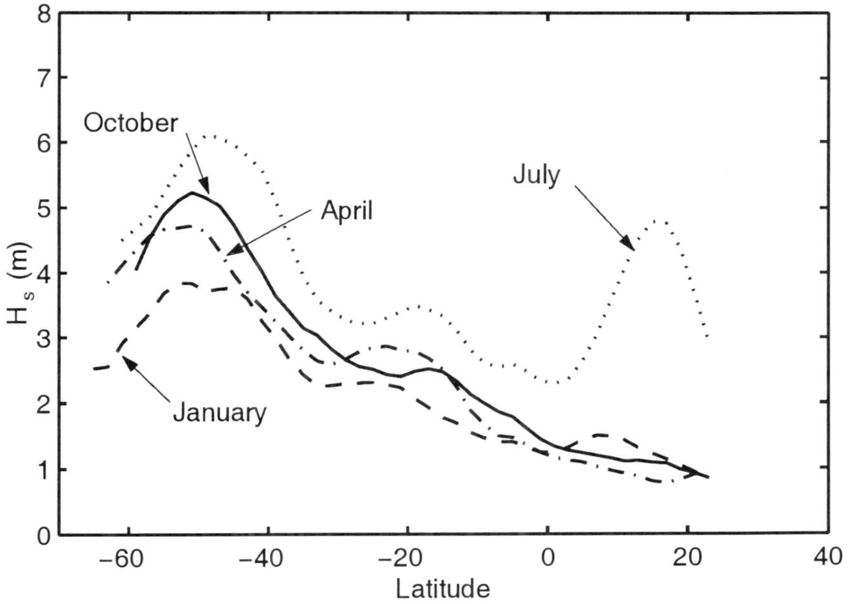

Figure 3.12: Variation in H_s with latitude along a north-south line drawn at longitude 62°E through the Arabian Sea. Values are shown for each season. Note the localized maximum which develops in the Arabian Sea during July.

3.4 Limitations of Global Statistics

Results such as those presented above are obtained purely from observation without any requirement to understand the physical processes responsible for the generation or evolution of ocean waves. Nevertheless, they provide a valuable data source. Extensive climatologies, such as those presented by Young and Holland (1998) are valuable resources for a host of applications. The use of this material should, however, be approached cautiously as the applicability of such data is limited by the manner in which it was obtained and processed. Its use should reflect these limitations.

The data set presented in Figures 3.2 to 3.8 is based on a total of 9 years of satellite observations. In addition, the satellite tracks are such that individual points are typically imaged, of order, once every 10 days. Ground tracks are typically separated by hundreds of kilometers. As a result, the data set will be typically too short to ensure that rare but extreme events are measured. In addition, intense but geographically confined events such as hurricanes may not be sampled by the satellites. Such events will have little influence on mean monthly values of H_s, such as presented in Figures 3.2 - 3.5. These same events will, however, be critical in determining exceedence probabilities which are typically used for engineering design. As an example, the data set may be unsuitable for determining the extreme probability distributions required for offshore structure design (i.e. return periods of 1 in 50 or 1 in 100 years). As a result, the exceedence probabilities presented in Figures 3.6 - 3.8 are shown at the most extreme level of only 10%.

Although the number of individual satellite observations used to construct these climatologies is enormous, in order to obtain stable statistical values, it has been averaged on a $2^\circ \times 2^\circ$ grid. Such a grid is relatively course and hence many localized features may not be resolved. This might include local variations in the wave climate close to shore or variations caused by currents.

As the period for which satellite data exists gradually increases, these limitations will decrease and the applications of such data will expand.

Chapter 4

Physical Mechanisms of Wave Evolution

4.1 Introduction

The aim of this chapter is two fold: firstly to define a mathematical model for the description of the evolution of the wind wave spectrum and secondly to describe the physical processes which influence this evolution. Two basic types of models have been adopted for the description of wave evolution. The first of these is termed *phase averaging* or spectral models (Gelci et al., 1956; Hasselmann, 1960; Pierson et al., 1966; Phillips, 1977; LeBlond and Mysak, 1978; Komen et al., 1994). As the name suggests, such models consider the evolution of the directional wave spectrum [see Section (2.5.2)]. The Gaussian random wave model is assumed and hence no attempt to made to predict the phases of the spectral components (2.50). The second approach is termed *phase resolving* models. These models include both Mild Slope (Berkhoff, 1972; Radder, 1979; Kirby and Dalrymple, 1986) and Boussinesq models (Peregrine, 1967; Madsen et al, 1991). These models attempt to represent the time varying water surface, resolving both the amplitudes and phases of the individual wave components.

The two classes of models should not be directly compared, they are complementary and their respective domains of application overlap only in certain areas. An excellent review has been provided by Battjes (1994). Phase resolving models have generally been used for the propagation of waves in near shore regions where there are relatively rapid changes in water depth and shoreline. Although attempts have been made to include processes such as generation and decay, the theoretical development in this area is not as developed as for phase averaging models. As the primary concern here is the representation of the influence of these forcing mechanisms, the phase resolving approach has been adopted for the mathematical formulation to follow. A detailed comparison of the two approaches appears in

Chapter 8.

4.2 Radiative Transfer Equation

The Radiative Transfer Equation has been commonly adopted to represent the temporal and spatial evolution of the directional spectrum as a result of physical processes, such as: atmospheric input from the wind, white-cap dissipation, bottom friction and weakly nonlinear interactions. Most derivations of this governing equation are not comprehensive. As the equation is central to the remainder of this book, an extensive derivation of the equation follows. This derivation closely follows Hardy (1993).

4.2.1 Kinematics

In Section (2.2.2) it was shown that for linear wave theory, the phase of an individual wave component propagating in two spatial dimensions is given by the relationship

$$\Psi = \mathbf{k} \cdot \mathbf{x} - \omega t \tag{4.1}$$

The wave number vector, $\mathbf{k}$ and the frequency, ω follow from (4.1) as

$$\mathbf{k} = \nabla \Psi \tag{4.2}$$

$$\omega = -\frac{\partial \Psi}{\partial t} \tag{4.3}$$

Equations (4.2) and (4.3) lead to the *conservation of crests equation*

$$\frac{\partial \mathbf{k}}{\partial t} + \nabla \omega = 0, \tag{4.4}$$

which indicates that spatial changes in the frequency must be associated with a local change in the wave number. As its name suggests, (4.4) demonstrates that for any two-dimensional area, the number of wave crests entering the area must equal the number exiting. Equation (4.2) also leads directly to the *curl of k* equation

$$\nabla \times \mathbf{k} = 0 \tag{4.5}$$

The frequency ω can be expressed as a function of the wave number vector, $\mathbf{k}$ and the water depth d through the dispersion relationship (2.13) as well as the position vector, $\mathbf{x}$ and time, t: $\omega = \omega(\mathbf{k}, d, \mathbf{x}, t)$. Substitution of ω into (4.4) yields

$$\frac{\partial \mathbf{k}}{\partial t} + \frac{\partial \omega}{\partial \mathbf{k}} \frac{\partial \mathbf{k}}{\partial \mathbf{x}} + \frac{\partial \omega}{\partial d} \frac{\partial d}{\partial \mathbf{x}} = 0 \tag{4.6}$$

Noting that, from (2.34)

$$\mathbf{C_g} = \frac{\partial \omega}{\partial \mathbf{k}} = \frac{\mathrm{d}\mathbf{x}}{dt}, \tag{4.7}$$

equation (4.6) becomes

$$\frac{\partial \mathbf{k}}{\partial t} + \mathbf{C_g} \cdot \nabla \mathbf{k} = -\frac{\partial \omega}{\partial d} \nabla d \tag{4.8}$$

Forming the scalar (dot) product of $\mathbf{C_g}$, (4.7) and (4.4) gives

$$\frac{\partial \omega}{\partial t} + \mathbf{C_g} \cdot \nabla \omega = 0 \tag{4.9}$$

In both (4.8) and (4.9) the total derivative operator appears, that is

$$\frac{\mathrm{d}}{\mathrm{d}t} = \frac{\partial}{\partial t} + \mathbf{C_g} \cdot \nabla$$

The total derivative describes the time rate of change of quantities advected along a wave ray at the group velocity, $\mathbf{C_g}$. Therefore, (4.9) states that as wave energy propagates along a wave ray, the frequency, ω will not change. Similarly, (4.8) states that the wave number, $\mathbf{k}$ will vary along the wave ray, only if there is a change in the water depth, d. Hence, in deep water, the wave number will also be invariant along the wave ray.

4.2.2 Wave Energy

Bretherton and Garrett (1969) show, that in the absence of forcing, the equation of wave action conservation takes the form

$$\frac{\partial}{\partial t}\left(\frac{E}{\omega_0}\right) + \nabla \cdot \left(\mathbf{C_g}_0 \frac{E}{\omega_0}\right) = 0 \tag{4.10}$$

The wave energy, E is defined by (2.36) and the zero subscript refers to quantities observed in a frame of reference moving with the wave. The quantity, E/ω_0 is called the *wave action*. Hence, (4.10) states that the wave action is conserved along wave rays. This result forms the basic premise upon which wave prediction models are based.

4.2.3 Wave Spectra

The concept of the wave spectrum was introduced in Section (2.5). Equation (4.10) is written in terms of a single or *monochromatic* wave, rather than a spectrum. The wave number spectrum, $Q(\mathbf{k})$ can be represented in terms of the total wave energy by (2.52)

$$Q\Delta\mathbf{k} = \frac{E}{\rho_w g} \tag{4.11}$$

where ρ_w is the density of water, g is gravitational acceleration and $\Delta\mathbf{k} = \Delta k_x \Delta k_y$ represents a small element of wave number space.

If the *wave action spectrum*, $n(\mathbf{k})$ is defined as

$$n(\mathbf{k}) = \frac{Q(\mathbf{k})}{\omega_0}, \tag{4.12}$$

then (4.10) can be rewritten as (LeBlond and Mysak, 1978; Hardy, 1993)

$$\frac{\partial}{\partial t}(n\Delta\mathbf{k}) + \nabla \cdot (\mathbf{C_g} n\Delta\mathbf{k}) = 0 \tag{4.13}$$

Following Phillips (1977), LeBlond and Mysak (1978) and Hardy (1993) it can be shown that, in the absence of forcing (4.13) becomes

$$\frac{\mathrm{d}n}{\mathrm{d}t} = \frac{\partial n}{\partial t} + \mathbf{C_g} \cdot \nabla n = 0 \tag{4.14}$$

Further, if currents are ignored and hence $\omega = \omega_0$, (4.14) becomes

$$\frac{\mathrm{d}Q}{\mathrm{d}t} = \frac{\partial Q}{\partial t} + \mathbf{C_g} \cdot \nabla Q = 0 \tag{4.15}$$

Hence, the wave number spectrum, $Q(\mathbf{k})$ is conserved along wave rays.

Considering the dependence of Q on $\mathbf{k}$, $\mathbf{x}$ and t, (4.15) can be expanded to yield (Hardy, 1993)

$$\frac{\mathrm{d}Q}{\mathrm{d}t} = \frac{\partial Q}{\partial t} + \frac{\partial Q}{\partial k_x}\frac{\mathrm{d}k_x}{\mathrm{d}t} + \frac{\partial Q}{\partial k_y}\frac{\mathrm{d}k_y}{\mathrm{d}t} + \frac{\partial Q}{\partial x}\frac{\mathrm{d}x}{\mathrm{d}t} + \frac{\partial Q}{\partial y}\frac{\mathrm{d}y}{\mathrm{d}t} = 0 \tag{4.16}$$

If forcing is now introduced, (4.16) becomes

$$\frac{\mathrm{d}Q}{\mathrm{d}t} = \frac{\partial Q}{\partial t} + \frac{\partial Q}{\partial k_x}\frac{\mathrm{d}k_x}{\mathrm{d}t} + \frac{\partial Q}{\partial k_y}\frac{\mathrm{d}k_y}{\mathrm{d}t} + \frac{\partial Q}{\partial x}\frac{\mathrm{d}x}{\mathrm{d}t} + \frac{\partial Q}{\partial y}\frac{\mathrm{d}y}{\mathrm{d}t} = S_{tot}, \tag{4.17}$$

where S_{tot} is a source/sink term which represents all processes which add or subtract energy to spectral components. Equation (4.17) is the basic equation used in many numerical wave prediction models (Sobey, 1986).

As many wave spectra are measured in terms of the frequency-direction spectrum $F(f,\theta)$ [see (2.52)], rather than the wave number spectrum $Q(\mathbf{k})$, (4.17) is often written in terms of $F(f,\theta)$. The wave number and frequency-direction spectra can be related through (2.53). Adopting the linear dispersion relationship, (2.53) becomes

$$Q(\mathbf{k}) = \frac{CC_g}{2\pi\omega}F(f,\theta) \tag{4.18}$$

Noting that from (4.9) the frequency remains constant along the wave ray, (4.17) and (4.18) yield

$$\frac{\partial}{\partial t}(CC_gF) + \frac{\partial}{\partial x}(CC_gF)\frac{\mathrm{d}x}{\mathrm{d}t} + \frac{\partial}{\partial y}(CC_gF)\frac{\mathrm{d}y}{\mathrm{d}t} + \frac{\partial}{\partial \theta}(CC_gF)\frac{\mathrm{d}\theta}{\mathrm{d}t} = CC_gS_{tot} \tag{4.19}$$

or

$$\frac{\mathrm{d}}{\mathrm{d}t}(CC_gF) = CC_gS_{tot} \tag{4.20}$$

which holds along wave rays defined by (2.43) (Munk and Arthur, 1952). Noting that wave energy is advected along the ray at the group velocity, C_g, (2.43) becomes

$$\begin{aligned}
\frac{\mathrm{d}x}{\mathrm{d}t} &= C_g \cos\theta \\
\frac{\mathrm{d}y}{\mathrm{d}t} &= C_g \sin\theta \\
\frac{\mathrm{d}\theta}{\mathrm{d}t} &= \frac{C_g}{C}\left[\sin\theta\frac{\partial C}{\partial x} - \cos\theta\frac{\partial C}{\partial y}\right]
\end{aligned} \tag{4.21}$$

Substitution of (4.21) into (4.20) gives (Sobey, 1986; Young, 1988a)

$$\begin{aligned}
&\frac{\partial}{\partial t}(CC_gF) + C_g\cos\theta\frac{\partial}{\partial x}(CC_gF) + C_g\sin\theta\frac{\partial}{\partial y}(CC_gF) + \\
&\frac{C_g}{C}\left[\sin\theta\frac{\partial C}{\partial x} - \cos\theta\frac{\partial C}{\partial y}\right]\frac{\partial}{\partial\theta}(CC_gF) = CC_gS_{tot}
\end{aligned} \tag{4.22}$$

4.2.4 Source Terms

The source term, S_{tot} represents all physical processes which transfer energy to, from or within the spectrum. Consistent with the linear superposition assumption inherent in the spectral representation, it can be represented as the summation of a number of independent processes (Hasselmann, 1968; Hasselmann et al, 1973)

$$S_{tot} = S_{in} + S_{nl} + S_{ds} + S_b + \cdots \tag{4.23}$$

where

$S_{in} \equiv$ atmospheric input from the wind
$S_{nl} \equiv$ nonlinear interactions between spectral components
$S_{ds} \equiv$ dissipation due to "white-capping"
$S_b \equiv$ dissipation due to interaction with the bottom

The four processes listed above are not exhaustive and there are undoubtedly many other factors which influence the evolution of waves. Knowledge in this area is far from complete. The processes listed above do, however, appear to account for the major fluxes of energy to the spectrum. The first three are all important in deep water and will be considered below. Interactions with the bottom are clearly only applicable in finite depth water and hence will be considered in Chapter 7.

4.3 Atmospheric Input, $\mathrm{S_{in}}$

The causal relationship between the wind and ocean waves is obvious to even the most casual of observers. In an excellent review of knowledge at the time Ursell

(1956) remarked: "*Wind blowing over the water generates waves in the water by physical processes which cannot be regarded as known*". Some 34 years later Donelan and Hui (1990) observed: "*...our ideas of the wind input into the waves is still rather primitive. The theoretical ideas of the fifties have not been capable of explaining the observed growth rates and essentially no new ideas have followed*". Although our understanding of the detailed processes which lead to a flux of energy from wind to waves is incomplete, the above statements may be a little pessimistic. As will be shown, our existing knowledge, incomplete as it is, yields models which have proven to be remarkably reliable.

It is reasonable to assume that waves are generated as a result of stress variations exerted on the water surface by the wind. These can be either normal (pressure) or shear stresses. Further, the stress variations could be assumed to be of two general types: turbulent eddies advected with the mean flow or those induced by the airflow over existing waves (Phillips, 1977). Little is know about fluctuations in the wind induced shear stress above waves, other than that it appears small in comparison to normal stress variations (Elliott, 1972a; Phillips, 1977). As a result, shear stress will not be considered here as a possible mechanism for the generation of waves.

4.3.1 Wave Induced Pressure Variations

The discussion here is focused on the fluctuation in the pressure exerted on the water due to the air flow above the "wavy" water surface. Mechanisms which may be responsible for the variation will be considered later. In particular, the important consideration is, whether such a variation in pressure can result in a flux of energy from the wind to the waves. Jeffreys (1924, 1925) has shown that such a variation in pressure can do work on the water surface, the magnitude of the energy flux being given by

$$\frac{\partial E}{\partial t} = \frac{1}{\rho_w g} \overline{p \frac{\partial \eta}{\partial t}} \tag{4.24}$$

where ρ_w is the density of water and p is the air pressure exerted on the water surface (units of Pa). The overbar in (4.24) represents an average with respect to time. Hence, the component of pressure which is correlated with the water surface slope (i.e. $\frac{\partial \eta}{\partial t}$) will result in an energy flux to the wave. This is often termed *the component of pressure in quadrature with the water surface.* For an energy flux to occur there must be a phase shift between the fluctuating pressure and the water wave. As shown in Figure 4.1, a positive energy flux (wave growth) will result if the phase shift is such that there is higher pressure on the windward side of the wave, compared to the leeward side.

If the magnitude and phase of the pressure fluctuations can be predicted, the energy transfer to the waves can be determined from (4.24). The development of a suitable theory which might explain the flow of air above the wave surface has

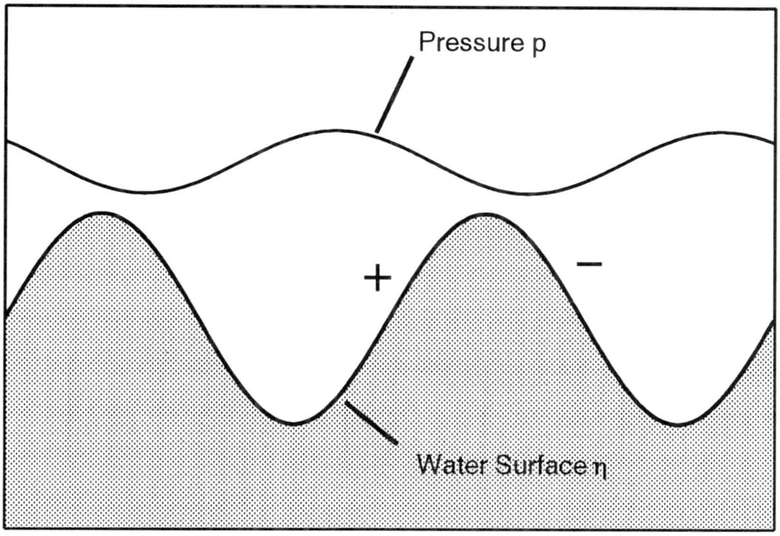

Figure 4.1: An example of the phase shift between the water surface and the atmospheric pressure which can give rise to an energy flux between the air and water. The case shown has a phase difference of 140° (pressure leading). In this case the pressure on the windward face of the waves is larger than the leeward face and the wave will receive energy from the wind. In the case shown above, the waves (and wind) propagate from right to left.

been a fertile area for research and a number of the significant developments will be considered below.

4.3.1.1 Potential Flow

In Chapter 2, flow beneath the air-water interface was considered assuming potential flow. This proved to be a surprisingly good approximation for many applications. As a starting point, one may make the same assumption on the air side of the interface. Potential theory for gravity waves on a density discontinuity between two fluids was first investigated by Stokes (1847). The theory for the case of a uniform mean velocity, U, in the upper fluid has been developed in many texts. The development below follows that of Lamb (1932).

The flow in both fluids is considered to be irrotational, incompressible and inviscid (a questionable assumption close to the interface). The velocity potential, ϕ is defined in terms of the horizontal and vertical components of velocity, u and w, respectively, as in (2.2) and (2.3). As in Chapter 2, the water surface is assumed sinusoidal. For mathematical brevity, it will be expressed using complex notation (note: $\exp i\theta = \cos\theta + i \sin\theta$)

$$\eta = a \exp\left[i\left(kx - \omega t\right)\right] \tag{4.25}$$

where a is the wave amplitude and only the real part of (4.25) is physically meaningful. The problem is, in essence, one of small oscillations about a state of steady motion, the velocity potential being defined as

$$\phi = -Ux + \phi' \tag{4.26}$$

where it is assumed that ϕ' is small and given by

$$\phi' = B \exp\left[-kz + i\left(kx - \omega t\right)\right] \tag{4.27}$$

and B is a constant. As in Chapter 2, the velocity potential, ϕ must satisfy the Laplace Equation (2.1) with the kinematic free surface boundary condition, being given to first order in η by

$$\frac{\partial \phi'}{\partial z} = \frac{\partial \eta}{\partial t} + U\frac{\partial \eta}{\partial x} \quad \text{at } z = 0 \tag{4.28}$$

Substitution of (4.25) and (4.27) into (4.28) yields

$$B = ia\left(C - U\right) \tag{4.29}$$

where $C = \omega/k$ is the wave phase speed. From (4.26), (4.27) and (4.29), the velocity potential becomes

$$\phi = -Ux + ia\left(C - U\right)\exp\left[-kz + i\left(kx - \omega t\right)\right] \tag{4.30}$$

The air pressure is obtained from the Bernoulli equation for unsteady irrotational flow

$$\begin{aligned} \frac{p}{\rho_a} &= \frac{\partial \phi'}{\partial t} - \frac{1}{2}\left[\left(U - \frac{\partial \phi'}{\partial x}\right)^2 + \left(\frac{\partial \phi'}{\partial z}\right)^2\right] - gz + \cdots \\ &= \frac{\partial \phi'}{\partial t} - U\frac{\partial \phi'}{\partial x} - gz + \cdots \end{aligned} \tag{4.31}$$

where ρ_a is the density of air. The terms omitted in (4.31) are either of second order or not relevant in the present context. The pressure, p, is obtained from (4.30) and (4.31) as

$$\frac{p}{\rho_a g} = -ae^{-kz}\left(1 - \frac{U}{C}\right)^2 \exp\left[i\left(kx - \omega t\right)\right] - z \tag{4.32}$$

Since the pressure must be a real quantity, (4.32) becomes

$$\frac{p}{\rho_a g} = -ae^{-kz}\left(1 - \frac{U}{C}\right)^2 \sin\left(kx - \omega t\right) - z \tag{4.33}$$

As the water surface elevation, $\eta = a\sin\left(kx - \omega t\right)$, (4.33) predicts that the pressure is in anti-phase with the water surface elevation. As shown by (4.24), this would imply no flux of energy from the air to the water. As this runs counter to our intuition that waves are generated by the wind, the initial assumptions required for the theory appear to have been invalid. Equation (4.33) also indicates that the wave induced pressure will decrease exponentially with height above the water surface. As will be shown later, this prediction of potential theory has been supported by measurements and forms a critical part of determining the energy flux from direct measurement.

4.3.1.2 Jeffreys' Sheltering Theory

Jeffreys (1925) produced the first plausible mechanism to explain the necessary phase shift of the atmospheric pressure required for an energy transfer from the wind to the waves. He assumed that the energy transfer was exclusively caused by form drag associated with flow separation. It was assumed that flow separation occurred on the leeward side of wave crests with re-attachment further down on the leeward slopes of the wave. Based on dimensional arguments Jeffreys (1925) showed that the pressure can be represented by

$$p = S\rho_a \left(U_\infty - C\right)^2 \frac{\partial \eta}{\partial x} \tag{4.34}$$

where the constant of proportionality, S is less than one and termed the *sheltering coefficient*. No boundary layer is considered in the theory and hence U_∞ represents the wind speed. It has been given the subscript ∞ to distinguish it from theories to be developed later where the variation of wind velocity through the boundary layer becomes critical. Considering (4.24) and (4.34) yields

$$\frac{\partial E}{\partial t} = \frac{1}{2\rho_w g} S\rho_a \left(U_\infty - C\right)^2 (ak)^2 C \tag{4.35}$$

The energy transfer predicted by (4.35) has a number of interesting features. Firstly, the energy transfer is exponential (i.e. $a^2 \propto E$) and secondly the mechanism is more efficient as the difference between the wind speed and phase speed increases. This is intuitively obvious if one considers an observer in a frame of reference moving with the wave. In this frame of reference, the observer and hence the wave will experience no wind when $U_\infty = C$, and hence no energy transfer would be expected. This limiting condition is satisfied by (4.35). Equation (4.35) also predicts that the energy transfer is a function of the wave slope as measured by ak. Since the mechanism is based on the assumption that flow separation occurs, this surface slope dependence seems logical.

The sheltering coefficient S was calculated by determining the rate of energy loss due to molecular viscosity and the minimum wind U_{min} that could maintain waves against this loss. This value was compared with observations of the minimum winds required to sustain waves. The calculated value of S varies as U_{min}^3. The choice of U_{min} is thus critical. Jeffreys chose U_{min} equal to 1.1 m/s, giving $S = 0.3$.

Ursell (1956) reviewed data available at the time on pressure variations over solid model waves in wind tunnels. The work of Stanton et al. (1932), Motzfeld (1937) and Thijsse (1951) did not support a value for S as large as that suggested by Jeffreys. Ursell (1956) concluded: "*The evidence of the three sets of measurements ... on the whole favours the conclusion that the pressure difference over a solid profile composed of a number of waves are an order of magnitude smaller than the differences postulated by Jeffreys*". As a result, the Sheltering Theory fell into disrepute.

It is now recognized that these experiments were almost totally irrelevant to the problem of wave generation (Phillips, 1977), although Ursell's conclusion is probably correct. More recent measurements by Banner and Melville (1976) indicated that air flow separation occurs above waves only with the onset of wave breaking. Hence, sheltering is unlikely to contribute significantly to the air-water energy flux.

4.3.1.3 Miles' Theory

The review of Ursell (1956) acted as a catalyst for new endeavours in the search for plausible mechanisms of wind wave generation. Miles (1957) assumed that the air was inviscid and incompressible and, in the absence of waves, can be described by a prescribed mean shear flow which varies only with height above the surface. It was assumed that there is a feedback from the waves such that wave-induced air pressure perturbations are two-dimensional and sufficiently small that they can be neglected in the nonlinear processes of the equations of motion. As turbulence in the air is neglected, the resulting theory is often termed *quasi-laminar*.

The surface elevation of each Fourier component is described by (4.25) and the perturbed aerodynamic pressure associated with the mean flow is taken as

$$p = (\alpha + i\beta)\, \rho_a U_r^2 k\eta \tag{4.36}$$

where U_r is a *reference* wind speed to be defined later [see (4.42)] and α and β are dimensionless coefficients. As interest is focused on the component of (4.36) which is in quadrature with the water surface, it is the coefficient β which must be determined.

Miles (1957) solved the hydrodynamic equations for the coupled system to determine the wave phase velocity C

$$C^2 = C_w^2 + \rho_a/\rho_w\, (\alpha + i\beta)\, U_r^2 \tag{4.37}$$

where C_w is the phase speed of free gravity waves [see (2.14), (2.15)]. Assuming that $C_w >> |\rho_a/\rho_w\, (\alpha + i\beta)\, U_r^2|$, C can be expanded in a perturbation series about C_w. Maintaining only the first two terms in the expansion, this yields

$$C \approx C_w \left[1 + \frac{1}{2} \rho_a/\rho_w\, (\alpha + i\beta) \left(\frac{U_r}{C_w} \right)^2 \right] \tag{4.38}$$

Substituting (4.38) into the equation for the water surface elevation (4.25) gives

$$\eta = a \exp\left[\frac{1}{2}\rho_a/\rho_w \beta k C_w \left(\frac{U_r}{C_w}\right)^2 t\right] \exp\left[i(kx - \omega t)\right] \tag{4.39}$$

Miles (1957) then defined the fractional increase in wave energy E per radian as

$$\zeta = \gamma/\omega = \left(\frac{1}{\omega E}\right)\frac{\partial E}{\partial t} \tag{4.40}$$

where γ is the growth rate. In a subsequent paper (Miles, 1960), this result was extended for arbitrary angles θ between the wind and waves. From (4.39), γ becomes

$$\gamma = (\rho_a/\rho_w)\,\beta\omega\left(\frac{U_r}{C_w}\cos\theta\right)^2 \tag{4.41}$$

In order to determine, γ, Miles (1957) solved the inviscid form of the Orr-Sommerfeld equation (Lin, 1955), showing that the energy transfer is a function of the curvature of the vertical wind shear at a *critical height* above the surface where $U = C_w$. Conte and Miles (1959) numerically integrated the Orr-Sommerfeld equation, assuming a logarithmic boundary layer profile

$$U(z) = U_r \log\left(\frac{z}{z_0}\right) = 2.5u_* \log\left(\frac{z}{z_0}\right) \tag{4.42}$$

where $u_* = (\tau/\rho_a)^{1/2}$ is the friction velocity, τ is the Reynolds' stress and z_0 is the roughness length. The final results of Miles (1959) presents β as a function of C_w/U_r for various values of the wind profile parameter $\Omega = gz_0/U_r^2$.

The growth rate predictions of Miles were of the same order of magnitude as observations but too low. The determination of the growth rate by this theory is, however, critically dependent on the shape of the near surface atmospheric boundary layer and the surface roughness as represented by z_0. Janssen (1991) has shown that the surface shear stress is determined by not only the wind speed but also the wave spectrum. The air and water form a coupled system. In particular, the high frequency components of the spectrum influence the roughness and hence increase the height of the critical layer. The results of the Miles Theory, as modified by Janssen (1991), are compared in Figure 4.2 with observational data compiled by Plant (1982). The agreement is impressive and provides strong support for the theory.

Since the surface roughness is influenced by the wave spectrum, the theory of Janssen (1991) predicts that β is only constant at high frequencies. The surface roughness at lower frequencies is still influenced by the high frequency components of the spectrum, but the value of β decreases. Hence, the values of γ shown in Figure 4.2 decrease below that which could be expected for a simple $(u_*/C)^2$ dependence. This behaviour is in excellent agreement with the data of Snyder et al (1981) as shown in Figure 4.2.

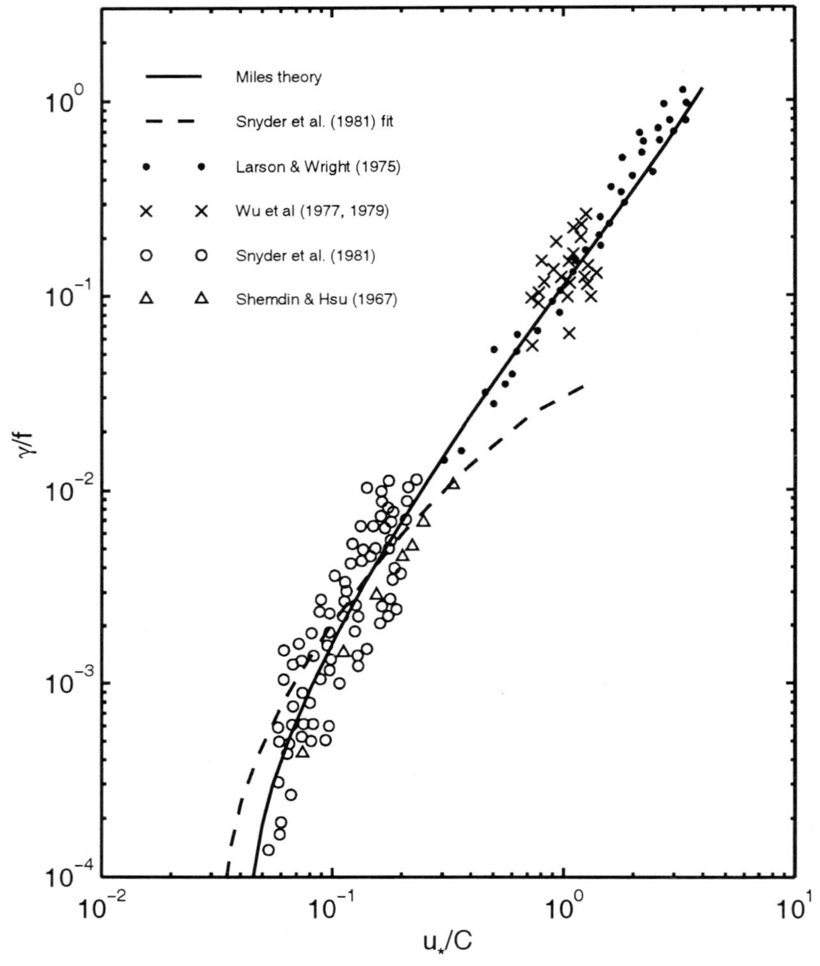

Figure 4.2: The dimensionless growth rate γ/f as a function of u_*/C as compiled by Plant (1982). The curve marked 'Miles theory' was evaluated by Janssen (1991). The data of Larson and Wright (1975) and Wu et al (1977, 1979) were compiled from measurements of wave evolution. Other data points are from direct measurements of the wave induced normal stress.

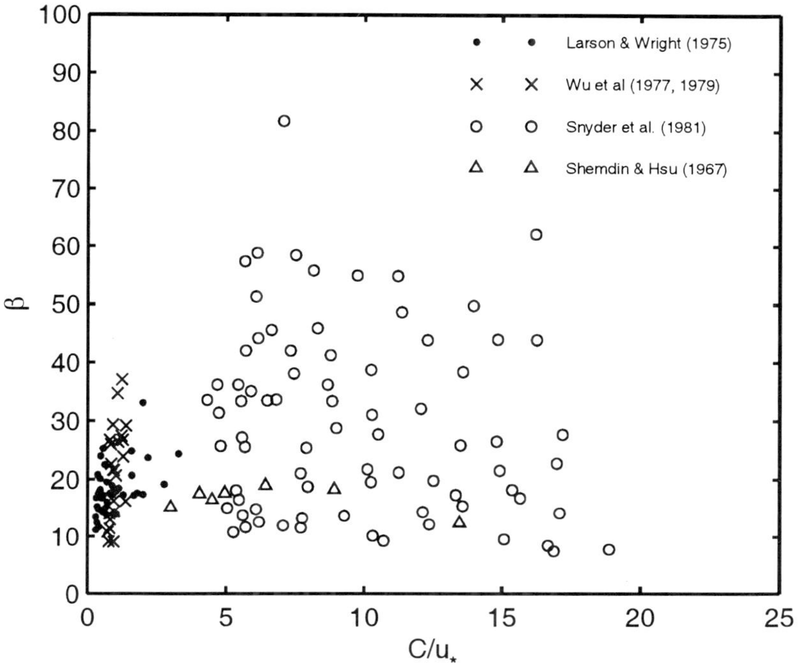

Figure 4.3: The same data plotted in Figure 4.2 but presented in terms of the parameter β [see (4.41)]. Equation (4.41) predicts that β = constant. Note the significant scatter when the data is presented on a linear scale rather than the logarithmic scales used in Figure 4.2.

Belcher et al (1994) have, however, suggested some caution should be exercised in interpreting a result such as Figure 4.2 due to the bias which may be introduced by the logarithmic scaling. Figure 4.3 shows the same data plotted on linear axes. The data are presented in terms of β, which should be constant if the $(u_*/C)^2$ scaling predicted by Miles Theory is correct. Although the data are consistent with the hypothesis that β is constant, the scatter is very large. Also, the data presented in Figure 4.2 for large u_*/C have been obtained from laboratory measurements of wave growth rates and not direct measurements of wave induced pressure. Other processes such as nonlinear interactions between spectral components may contribute to this observed evolution. Hence, some caution should be exercised in interpreting the data. Nevertheless, the theory of Miles, as modified by Janssen is consistent with the available observational data.

The neglect of turbulence within the Miles Theory is often raised as a shortcoming of the theory and further experimental corroboration of the theory is still required. Despite these concerns, the theory has proved successful in operational wave models such as WAM (WAMDI, 1988), the final source term for atmospheric input being of the form

$$S_{in}(f,\theta) = \frac{\rho_a}{\rho_w}\beta\omega\left(\frac{u_*}{C}\cos\theta\right)^2 F(f,\theta) \quad (4.43)$$

where $\beta = \beta(u_* \cos\theta, \Omega)$.

One method to address the role of turbulence is to numerically model the boundary layer flow. A critical element of such a modelling approach is to adopt a *turbulent closure model* for the Reynolds' stress. Numerous models have been proposed, including: Davis (1972), Gent and Taylor (1976), Townsend (1980), Al-Zanaidi and Hui (1984), Belcher and Hunt (1993) and Burgers and Makin (1993). Both the magnitude and phase of the wave-induced pressure appear sensitive to the closure scheme. Phillips (1977) concluded that: "*Closure schemes in turbulent shear flow are still rather ad hoc and different methods, which may be reasonably satisfactory in other flows, give very different results when applied to this problem. The situation is not one in which firmly established methods lead to results that one might seek, with some confidence, to verify experimentally. On the contrary, because of sensitivity of results to the assumptions made, the air flow over waves appears to provide an ideal context to test the theories of turbulent stress generation themselves*". More recently, Belcher et al (1994) have similarly concluded: "*We compare these theories and show how and why different turbulence models give qualitatively different results for key quantities, such as the atmospheric pressure*".

4.3.1.4 Surface Pressure Measurements

The theories outlined above for the generation of waves by wave induced normal stresses require experimental confirmation. The first moderately successful attempts to determine the exponential growth rate due to atmospheric forcing were conducted by Snyder and Cox (1966). They measured the actual growth rate of waves and assumed that all growth could be attributed to atmospheric input, neglecting other influences such as nonlinear interactions. They concluded that the atmospheric input term was of the form

$$S_{in}(f,\theta) = \frac{\rho_a}{\rho_w}\beta\omega\left(\frac{U_{10}}{C}\cos\theta - 1\right) F(f,\theta) \quad (4.44)$$

where β was determined to a constant of value 1. Barnett and Wilkerson (1967) obtained a similar result by using an air-born radar to measure wave growth.

A more direct and presumably reliable technique is to measure the wave induced stress at the water surface. The problems encountered in this type of measurement are, however, formidable. The recording instrumentation must be as close as possible to the water surface and ideally should be in a frame of reference oscillating with the water surface. If a stationary probe is used it can be no closer to the mean surface than the height of the highest crest. Whether stationary or oscillating, the air pressure probe can easily be contaminated with spray and even swamped. In addition, the aim is to measure a rather small phase difference from 180° in a pressure signal whose magnitude is quite small. Typically, the wave induced pressure is of order $\rho_a g a$ [see (4.33)], that is, of order 10^{-4} of the mean atmospheric pressure. As the measurements cannot be made at the air-water interface, they must be extrapolated to the interface. Hence, knowledge of the vertical structure of the air-pressure field is required.

The first measurements of this type were made by Longuet-Higgins et al (1963) using a large flat buoy, but were rather unsuccessful as no significant phase difference from 180° was found. Later laboratory experiments by Shemdin and Hsu (1967) and Shemdin (1969) over mechanically generated water waves and Kendall (1970) over flexible wavy walls were more successful, although there was significant scatter in the recorded data. Laboratory measurements should be treated with some caution as accurate modelling of not only the boundary layer structure but also turbulence levels may be important.

Dobson (1971), using a very small buoy in the shallow water of Burrard Inlet, Vancouver, made an extensive series of observations and found β values consistent with Snyder and Cox (1966). His analysis was, however, complicated by buoy motions and the effects of splashing. The measurements of Dobson (1971) were placed in doubt by Elliott (1972b) who employed a vertical array of fixed air-pressure sensors in order to extrapolate his observations to the mean surface. Elliott's analysis indicated that the air pressure field decayed approximately exponentially with height without change in phase, although slightly less rapidly than predicted by Potential Theory [see (4.33)]. He obtained growth rates which were significantly smaller than those of Dobson with $\beta \approx 0.2$. In a similar set of experiments conducted in the Bight of Aboco, Bahamas, Snyder (1974) found a still lower value of β equal to 0.1.

In order to reconcile the differences between the results of Dobson, Elliott and Snyder, a combined experiment was conducted (Snyder et al, 1981). This experiment has become known as the Bight of Aboco experiment and was conducted with meticulous care using both fixed and wave-following probes. The experiment showed that the results of Snyder (1974) were low by a factor of two due to the frequency response of his instrumentation and provided an extensive data set indicating thar β lies between 0.2 and 0.3. Such a range is consistent with both the results of Elliott (1972b) and the corrected values of Snyder (1974).

The Bight of Aboco experiment spanned the parameter range $1 < U_{10}/C < 3$. In addition, the data was recorded at relatively short fetch. In an effort to extend the parameter range, Hsiao and Shemdin (1983) conducted a further set of experiment at the Noordwijk Tower off the coast of The Netherlands in the North Sea. These experiments utilised a wave following probe and collected data in the range $1 < U_{10}/C < 7.4$. In contrast to previous field experiments, Hsiao and Shemdin (1983) found that the data had a quadratic dependence on U_{10}/C, and represented the atmospheric input source term as

$$S_{in}(f,\theta) = 0.12\frac{\rho_a}{\rho_w}\omega\left(\frac{U_{10}}{C}\cos\theta - 1\right)^2 F(f,\theta) \tag{4.45}$$

This result is significantly smaller than that of Snyder et al (1981) for the parameter range $1 < U_{10}/C < 3$. The input to the peak of the spectrum provided by (4.45) seems so small that observed growth rates could not be sustained. As a result, (4.45) has not been generally adopted by the wave modelling community.

A further attempt to investigate atmospheric input in open ocean conditions was made by Hasselmann and Bösenberg (1991). Measurements were made with a vertical array of stationary probes at the JONSWAP site (Hasselmann et al, 1973). The experiment produced almost identical results to those of Snyder et al (1981), with a linear dependence as in (4.44) with $\beta = 0.25 \pm 0.07$.

Figure 4.4 shows a comparison of the measured growth rates proposed by Snyder et al (1981) and Hsiao and Shemdin (1983) with the composite result of Plant (1982). The Plant result is included because it agrees closely with the Miles-Janssen theory (see Figure 4.2). As indicated above, the Hsiao and Shemdin (1983) data are significantly smaller than both the data of Snyder et al (1981) and theory in the energy containing range of the spectrum $1 < U_{10}/C < 3$. In this range it appears that the data are broadly in agreement with theory. At higher values of U_{10}/C further data are required to confirm the quadratic dependence predicted by theory.

For most engineering applications, detailed knowledge of the high frequency region of the spectrum is not required. Therefore, one might question whether details of the atmospheric input in this region is really required for these applications. Janssen (1991) has shown that the high frequency components of the spectrum have a critical role in determining the aerodynamic roughness of the surface and hence the atmospheric input to lower frequencies. Also, as will be shown in Section (4.4), the energy containing regions and the spectral tail are coupled through nonlinear interactions. The flux of energy from the tail of the spectrum to the peak has a critical influence on wave evolution. Hence, in order to predict even the crudest integral spectral properties, such as significant wave height, H_s and peak frequency, f_p, a detailed knowledge of the behaviour of the spectral tail is required.

Although the details of proposed atmospheric input terms differ, the general functional forms are similar. For illustrative purposes, the widely used result of Snyder et al (1981) (4.44), is adopted here. Figure 4.5 shows the one dimensional form, $S_{in}(f)$ evaluated from (4.44) for $U_{10} = 5, 10, 15$ m/s. The spectrum $F(f)$ has been assumed to be of the JONSWAP form [see Section (5.4.1)]. Panel (a) of the figure shows S_{in} as a function of U_{10}/C whereas panel (b) shows it as a function of f/f_p. The magnitude of the energy transfer is essentially governed by two factors, the values of U_{10}/C and the magnitude of the spectrum, F. High frequency or *young* waves have high values of U_{10}/C and are strongly forced by the wind. Their spectral energy is however relatively low. The net result is a source term which has a shape not dissimilar to the spectrum. The maximum value of S_{in} occurs in the vicinity of the spectral peak frequency for young and moderately forced waves.

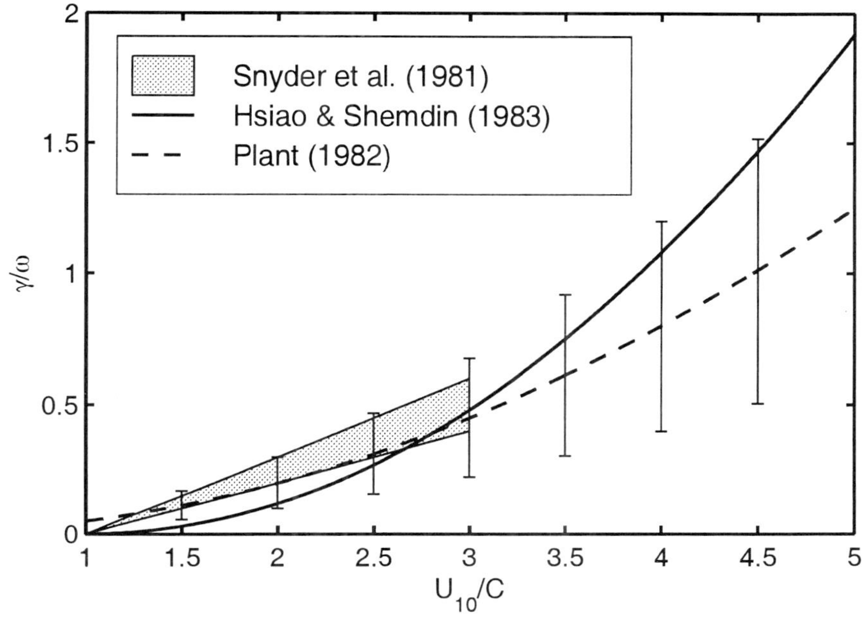

Figure 4.4: A comparison of the growth rate, γ where $S_{in} = \rho_a/\rho_w \omega\gamma F$ as a function of U_{10}/C. Results obtained from the field data sets of Snyder et al (1981) and Hsiao and Shemdin (1983) are compared with the composite result of Plant (1982). The Plant result agrees closely with theory as proposed by Miles (1959) and Janssen (1991). The error bars represent the span of values considered by Plant (1982). The Plant result was originally presented in terms of u_*/C, it has been converted to U_{10}/C for comparison assuming $U_{10} \approx 28u_*$.

For weakly forced or *old* waves for which U_{10}/C is small the source term is much flatter with a peak at a frequency higher than the spectral peak. This is clearly illustrated for the cases of $U_{10} = 5$ m/s in Figure 4.5. Although the maximum energy input is generally near the spectral peak, the input as a percentage of the spectral energy increases with frequency [i.e. scales as $\omega(U_{10}/C - 1)$ from (4.44)]. Hence, these high frequency waves will respond very rapidly to changes in the wind speed and direction. Indeed, remote sensing techniques take advantage of this fact to infer wind speed from the properties of the high frequency spectral tail.

The results shown in Figure 4.4 indicate that observations have a similar magnitude to the Miles-Janssen theory. These measurements do not, however, shed any light on the physical processes responsible for the wave induced pressure and hence energy input. As such, they do not conclusively confirm the physical mechanism responsible for atmospheric input to the waves. One situation which could further shed light on the actual mechanism would be to investigate cases where the waves are either in opposition to the wind or overrun the wind (i.e. $U_{10}/C < 1$). In such cases, no *critical layer* will exist and the theory of Miles predicts neither damping

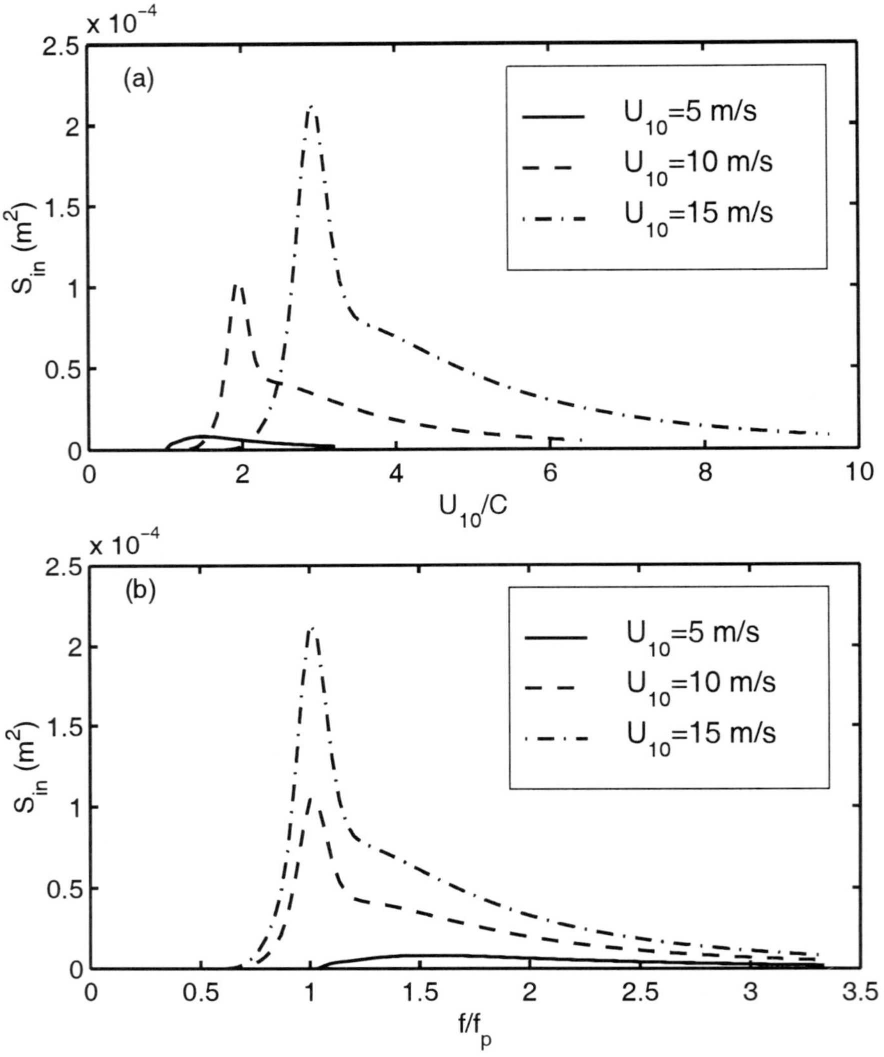

Figure 4.5: The integrated one dimensional atmospheric input term, $S_{in}(f) = \int S_{in}(f,\theta)\mathrm{d}\theta$ evaluated using the relationship of Snyder et al (1981) (4.44). The term has been evaluated for wind speeds of $U_{10} = 5, 10, 15$ m/s and a JONSWAP spectrum, $F(f)$, with parameters: $\alpha = 0.01$, $f_p = 0.3$ Hz, $\sigma = 0.08$ and $\gamma = 3.3$. Panel (a) shows S_{in} as a function of U_{10}/C where as panel (b) is expressed in terms of f/f_p.

nor growth of the waves.

Unfortunately, in this case, measurements seem to be inconclusive. Snyder et al (1981) and Hasselmann and Bösenberg (1991) both report phase differences of 180°, consistent with Miles theory. The same conclusion was reached in an extensive laboratory study by Young and Sobey (1985). More recently, however, Donelan (1997) has measured phase shifts which indicate decay in opposing winds, although the rate is smaller than the corresponding growth rates for following winds. The author has actually stood at the exit of the large flume at the Canada Center for Inland Waters used by Donelan et al (1985) and observed wind generated by mechanically generated waves. The conclusion is clear, the waves are doing work on the air and hence an energy transfer from the waves to the wind exists.

It appears that in such cases wave decay does occur, although it is weak. Indeed, this must be the case, as swell is commonly observed to propagate huge distances with little, if any, decay (Snodgrass et al, 1966). Such a decay is inconsistent with the theory of Miles, however, other influences, such as turbulence, may account for the weak decay. Indeed, the turbulent closure models mentioned earlier, generally predict some decay in these cases.

4.3.2 Generation by Turbulent Pressure Fluctuations

Independently of the work of Miles, but at the same time, Phillips (1957) developed a theory for the generation of waves due to resonance between the surface waves and the turbulent pressure field. The resulting source term is of the form

$$S_{in}(f,\theta) = \frac{2\pi^2\omega}{\rho_w^2 C^3 C_g}\Pi(\boldsymbol{k},\omega) \tag{4.46}$$

where $\Pi(\boldsymbol{k},\omega)$ is the three dimensional wave number, frequency spectrum of atmospheric pressure. In physical terms, Phillips' theory predicts that waves will grow by a resonance mechanism when the speed and length of the atmospheric pressure fluctuations match those of the water waves. Thus, the mechanism is frequency selective, transferring energy to those components of the wave spectrum which satisfy the resonance condition.

Assuming that turbulent pressure fluctuations are advected at a speed U_c, the resonance condition becomes, $\omega = kU_c \cos\alpha$, where α is the propagation direction of the waves measured relative to the wind direction[1]. Therefore, for surface gravity waves, the resonance angle becomes (Phillips, 1977)

$$\alpha = \cos^{-1}\frac{C}{U_c} \tag{4.47}$$

The convection velocity, U_c is usually taken as the wind velocity. Hence, (4.47) indicates that for $U_c = C$, $\alpha = 0$ and the energy transfer is a maximum in the mean

[1]This assumption is often termed *frozen turbulence*

wind direction. At higher frequencies, α increases and the directional distribution becomes bi-modal, with the resonance angles approaching $\pm 90°$.

In contrast to the exponential growth predicted by Miles (1957), (4.46) indicates a linear growth with time. Hence, it is generally assumed that the Phillips (1957) mechanism is only important in the initial stages of growth. Evaluation of the actual magnitude of (4.46) requires knowledge of the pressure spectrum, $\Pi(\boldsymbol{k}, \omega)$ for which there exists little experimental data. The determination of $\Pi(\boldsymbol{k}, \omega)$ requires both spatial and temporal knowledge of the atmospheric pressure field. Such data have been collected by Priestley (1965), Elliott (1972a) and Snyder et al (1981). The data sets are generally consistent, although none are truly comprehensive. Priestley (1965) parameterized the turbulent pressure spectrum as

$$\Pi(\boldsymbol{k}, \omega) = \frac{\Phi(\omega)}{\pi^2} \left[\frac{\nu_2}{\nu_2^2 + k^2 \sin^2 \theta} \right] \left[\frac{\nu_1}{\nu_1^2 + (k \cos \theta - \Lambda)^2} \right] \tag{4.48}$$

where $\Phi(\omega) = \int \Pi(\boldsymbol{k}, \omega) d\boldsymbol{k}$ is the turbulent pressure frequency spectrum, $\Lambda = \omega / U_c$ is the atmospheric turbulence wave number and θ is the angle between the wind direction and the wave propagation direction. Priestley (1965) represented $\Phi(\omega)$ as

$$\Phi(\omega) = 1.23 \Omega \omega^{-2} \tag{4.49}$$

The ω^{-2} dependence in (4.49) is consistent with the data of both Elliott (1972a) and Snyder et al (1981). The quantity Ω represents a scaling parameter, the functional dependence of which is poorly know. The available data suggests a dependence on U_{10} which has been represented by Snyder and Cox (1966) as

$$\Omega = 5.72 \times 10^{-3} U_{10}^4 \tag{4.50}$$

where U_{10} has units of m/s and Ω has units of $N^2 s^{-1} m^{-4}$. In contrast, Barnett and Wilkerson (1967) have proposed a U_{10}^6 dependence. The terms ν_1 and ν_2 are empirical coefficients given by

$$\nu_1 = 0.33 \Lambda^{1.28} \tag{4.51}$$

$$\nu_2 = 0.52 \Lambda^{0.95} \tag{4.52}$$

where both ν_1 and ν_2 have units of m^{-1}.

The parameterization described by (4.46) to (4.52) is shown in Figure 4.6. This figure shows a contour plot of $S_{in}(f, \theta)$ for $U_c = U_{10} = 10$ m/s. It is clear from the figure that the main energy input occurs at $U_{10}/C = 1$ and in the wind direction. At higher frequencies, the energy input decreases and the angles at which the resonance conditions are met increase. For $U_{10}/C > 3$, the distribution becomes tri-modal with resonance peaks at $0°$ and approximately $\pm 90°$.

An indication of the magnitude of the linear Phillips' term (4.46), and hence its potential importance for wind wave generation, can be achieved by comparison with the exponential Miles term (4.44). Such a comparison is shown in Figure 4.7

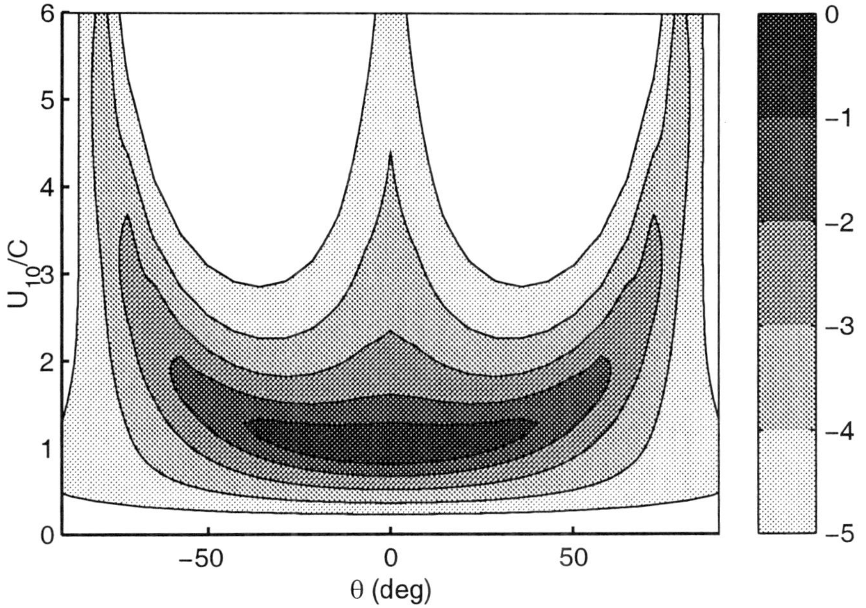

Figure 4.6: Contour plot of the atmospheric input term, S_{in}, due to turbulent pressure fluctuations proposed by Phillips (1957) and parameterized by Priestley (1965). The case shown is for $U_c = U_{10} = 10$ m/s. Values of S_{in} have been normalized to have a maximum of 1 and contours have been drawn with a logarithmic spacing. The contour levels shown on the colour bar represent the exponent of the level (i.e. 10^{-5} ... 10^{-3} ... 10^{0}). Before normalization, the maximum value of S_{in} was 2.96×10^{-5} m^2.

for the case of $U_{10} = 10$ m/s. The spectrum, $F(f)$ has been assumed to be of the JONSWAP form (5.51). It is clear from this figure that the exponential Miles' term is an order of magnitude larger than the linear Phillips' term. This difference would be even larger if a more energetic spectrum had been chosen, as this would increase the magnitude of the exponential term but have no impact on the linear term. The linear term does, however, peak at a much lower frequency than the exponential term and hence may play a role in the generation of low frequency spectral components. Figure 4.7 shows only the integrated form $S_{in}(f)$, and the directional distributions of the terms are markedly different. The linear term may also be important in the generation of high frequency spectral components propagating at significant angles to the wind.

It can be concluded that, in general, the Phillips' linear term is important only in triggering initial growth from a calm water surface. Once waves are generated, the wave induced pressure variations, which result in exponent growth, quickly dominate. Other processes, such as nonlinear interactions [see Section (4.4)] also become important as the spectrum develops.

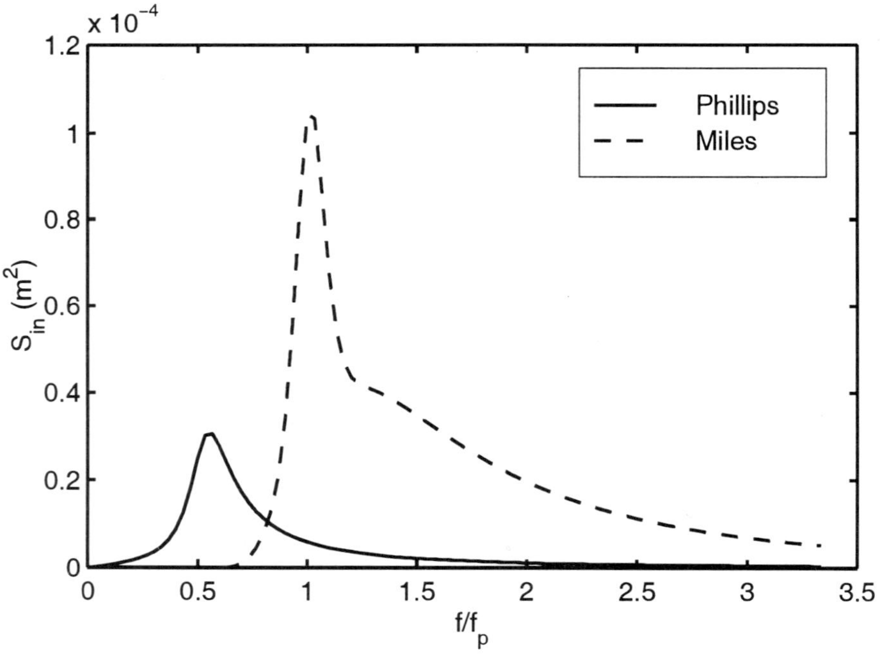

Figure 4.7: A comparison between the linear Phillips' term (4.46) and the exponential Miles' term as parameterized by Snyder et al (1981) (4.44). The integrated one dimensional form, $S_{in}(f) = \int S_{in}(f, \theta)\mathrm{d}\theta$, is shown. The case illustrated is for $U_{10} = 10$ m/s and a JONSWAP spectrum, $F(f)$, with parameters: $\alpha = 0.01$, $f_p = 0.3$ Hz, $\sigma = 0.08$ and $\gamma = 3.3$.

One final point of caution should be mentioned concerning the parameterization represented by (4.46) to (4.52). This result is based on a very small number of measurements of the atmospheric pressure spectrum over only a limited range of conditions and wind speeds. Indeed, Snyder et al (1981) indicate that the spatial correlation of pressure fluctuation may be quite different to that assumed by Priestley (1965). Whether this has a significant influence on the magnitude of S_{in} is unclear. There is, however, no observational evidence that the Phillips' mechanism plays a major role in the transfer of energy from the wind to the water. Hence, there have been few incentives for more detailed studies of the mechanism.

4.4 Nonlinear Quadruplet Interactions, S_{nl}

To first order, ocean waves can be regarded as the superposition of free and independent spectral components, as described in Section (2.5). At higher order, however, there exist interactions between spectral components which result in a transfer of energy between the components (Phillips, 1960). The lowest order at which such interactions may occur involves the interaction of three waves, *triad interactions*. In such cases, two waves interact nonlinearly and transfer energy to a third component. Such interactions are important in finite depth conditions [see Section (7.2.7)] but are insignificant in deep water (Hasselmann, 1962). A general perturbation theory for the nonlinear resonant interaction of waves in a random sea was developed by Hasselmann (1962, 1963a,b). He found that a set of four waves, called a *quadruplet*, could exchange energy when the following resonance conditions are satisfied:

$$\mathbf{k}_1 + \mathbf{k}_2 = \mathbf{k}_3 + \mathbf{k}_4 \tag{4.53}$$

$$\omega_1 + \omega_2 = \omega_3 + \omega_4 \tag{4.54}$$

where ω_j is the radian frequency and $\mathbf{k}_j$ the wave number vector $(j = 1, \cdots, 4)$. The frequency and the wave number are related through the dispersion relationship (2.13). The four interacting wave components described by (4.53) and (4.54) form the quadruplet.

Hasselmann (1963a) described the nonlinear interactions between wave quadruplets in terms of their action density, n, where $n(\mathbf{k}) = Q(\mathbf{k})/\omega$ and $Q(\mathbf{k})$ is the wave number spectrum, [see (2.52)]. Hasselmann assumed that the waves were only weakly nonlinear and hence applied a perturbation technique to determine the rate of change in action density at $\mathbf{k}_1$ due to all quadruplet interactions involving $\mathbf{k}_1$. Working independently of Hasselmann, Zakharov (1968) applied a Hamiltonian approach to yield the same result. The rate of change $\partial n_1/\partial t$, which is directly related to $S_{nl}(f, \theta)$ is

$$\frac{\partial n_1}{\partial t} = \int\int\int G(\mathbf{k}_1, \mathbf{k}_2, \mathbf{k}_3, \mathbf{k}_4) \times \delta(\mathbf{k}_1 + \mathbf{k}_2 - \mathbf{k}_3 - \mathbf{k}_4) \times \delta(\omega_1 + \omega_2 - \omega_3 - \omega_4)$$
$$\times [n_1 n_3 (n_4 - n_2) + n_2 n_4 (n_3 - n_1)] \, d\mathbf{k}_1 d\mathbf{k}_2 d\mathbf{k}_3 \tag{4.55}$$

where $n_j = n(\mathbf{k}_j)$ is the action density at wave number $\mathbf{k}_j$ and G is a complicated coupling coefficient (Herterich and Hasselmann, 1980; Van Vledder, 1990). The delta functions in (4.55) ensure that contributions to the integral only occur for quadruplets that satisfy the resonance conditions (4.53) and (4.54). The integral expression (4.55) is also known as the Boltzmann integral for wind waves, in analogy to similar expressions used in theoretical physics to describe the rate of change of particle density distributions in a system of interacting particles.

The resonance conditions define not only the frequencies of spectral components that can interact nonlinearly but also their propagation directions, since (4.53) is a vector expression. This relationship can be represented in the form of a vector diagram as shown in Figure 4.8. Hence, all components of the spectrum are potentially coupled and energy can be exchanged, not only between components of different frequency, but also amongst components propagating in different directions.

The nonlinear energy transfer represented by (4.55) conserves both the total energy and momentum of the wave field, merely redistributing it within the spectrum. As a consequence of the symmetry of the resonance conditions with respect to the pairs of wave numbers $(\mathbf{k}_1, \mathbf{k}_2)$ and $(\mathbf{k}_3, \mathbf{k}_4)$, the quadruplet interactions also

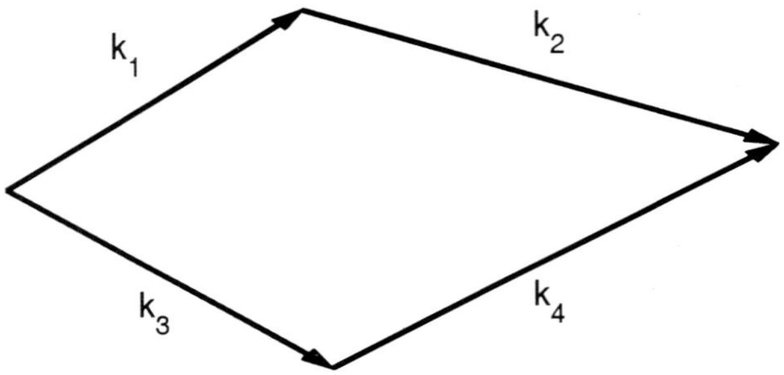

Figure 4.8: An example of four wave number vectors which satisfy the resonance condition (4.53). To contribute to the integral expression (4.55), the frequencies associated with each of these wave number must also satisfy (4.54).

conserve the wave action. The absolute value of the rate of change of the action density is equal for all wave numbers within the quadruplet

$$\frac{dn_1}{dt} = \frac{dn_2}{dt} = -\frac{dn_3}{dt} = -\frac{dn_4}{dt} \tag{4.56}$$

This result is know as the property of *detailed balance* (Hasselmann, 1966). In a more general form, detailed balance states that the absolute value of the change in action density, $|\Delta n_j|$, is the same for all components in a resonant set of wave numbers.

4.4.1 Solution Techniques for $\mathbf{S_{nl}}$

Numerous techniques have been proposed for the solution of (4.55). The major goals of all these techniques have been to: improve computational speed and to provide a better physical understanding of the source term. The significant computational time required for the solution arises for two reasons: the fact that (4.55) represents a six-dimensional integral and the complexity of the coupling coefficient, G.

Hasselmann (1963b), Sell and Hasselmann (1972), Webb (1978) and Masuda (1980) first removed the delta functions by projecting the six-dimensional integral onto the three-dimensional resonance space defined by (4.53) and (4.54). Although the details of the various projections differ, the goal was the same. The integral was then solved within this transformed integration space.

A conceptional difficulty arises in attempting to visualize the multi-dimensional interaction space. It would add significantly to the understanding of the problem, if the wave number quadruplets which contribute significantly to the integral could be visualized in some simple form. Based on a suggestion by M.S. Longuet-Higgins, Hasselmann (1963b) proposed the, so called, *figure of eight* graphical representation of the interaction space. For deep water conditions, where the dispersion relationship (2.13) reduces to $\omega^2 = gk$, (4.53) and (4.54) become

$$\mathbf{k}_1 + \mathbf{k}_2 = \mathbf{k}_3 + \mathbf{k}_4 = \mathbf{k} \tag{4.57}$$

$$\sqrt{k_1} + \sqrt{k_2} = \sqrt{k_3} + \sqrt{k_4} = \gamma\sqrt{k} \tag{4.58}$$

Normalizing, such that $k = 1$ and representing the angle between $\mathbf{k}_1$ and $\mathbf{k}$ as $\hat{\alpha}$ yields

$$\cos\hat{\alpha} = \frac{1 + k_1^2 - \left(\gamma - \sqrt{k_1}\right)^4}{2k_1} \tag{4.59}$$

For a fixed value of γ, the locus of all values of $\mathbf{k}_1$ which satisfy the resonance conditions is given by (4.59). The corresponding $\mathbf{k}_2$ is given by (4.57). Typical interaction curves for a given value of γ are illustrated in Figure 4.9.

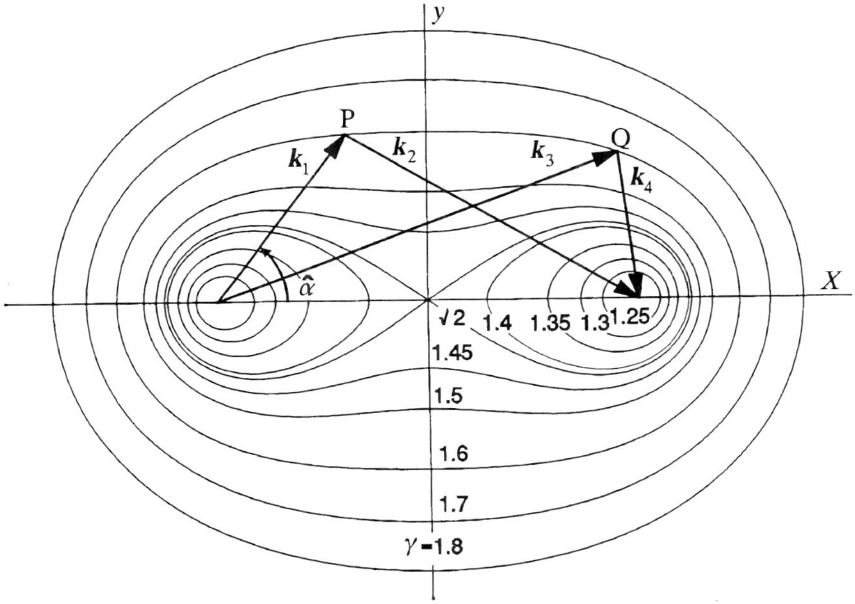

Figure 4.9: The so called, *figure of eight* diagram showing the interaction space for a given value $\mathbf{k} = \mathbf{k}_1 + \mathbf{k}_2 = \mathbf{k}_3 + \mathbf{k}_4$. Each curve is for a specific value γ [after Phillips (1960); Hasselmann (1963b)].

A similar and perhaps more straightforward representation of the interaction space was presented by Webb (1978). For $\mathbf{k}_1$ and $\mathbf{k}_3$ fixed, the locus of possible values of $\mathbf{k}_2$ and $\mathbf{k}_4$ which will close the vector polygon represented by (4.53) and also satisfy (4.54), trace out two 'egg-shaped' figures. Figure 4.10 shows the loci of $\mathbf{k}_2$ and $\mathbf{k}_4$ for one combination of $\mathbf{k}_1$ and $\mathbf{k}_3$. For this combination of $\mathbf{k}_1$ and $\mathbf{k}_3$, only wave number components which lie on these loci will satisfy the resonance conditions and contribute to the integral.

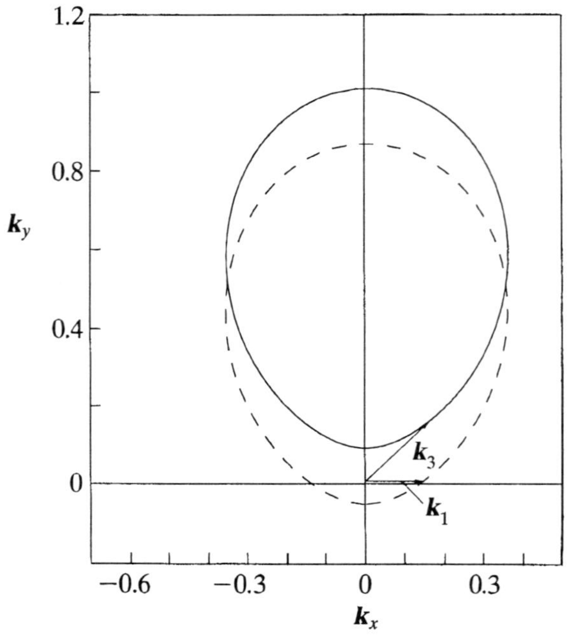

Figure 4.10: Loci of the vectors $\mathbf{k}_2$ (solid line) and $\mathbf{k}_4$ (dashed line) for a fixed combination of $\mathbf{k}_1$ and $\mathbf{k}_3$ ($k_1 = 0.14$ rad/m and $k_3 = 0.20$ rad/m). The $\mathbf{k}_1$ vector is in the x direction and $\mathbf{k}_3$ is 45° clockwise from $\mathbf{k}_1$ [after Young and Van Vledder (1993)].

The direct integration techniques mentioned above provided valuable insight into the basic properties of the nonlinear source term, S_{nl}, but were computationally too expensive for use in wave prediction models. As a result, approximations for narrow-peak spectra were proposed by Longuet-Higgins (1976), Fox (1976), Herterich and Hasselmann (1980) and Dungey and Hui (1979). Although these approximations capture many of the qualitative features of the interactions, such as the 'plus-minus' lobe structure, they do not reproduce other features, notably shape stabilization [see Section (4.4.2)]. Hence, they have not been widely adopted.

Hasselmann and Hasselmann (1981, 1985b) proposed a symmetric integration technique which significantly increased computational speed. This approach takes advantage of the property of detailed balance (4.56), to reduce the effective computation time by a factor of four. In addition, Hasselmann and Hasselmann (1981) show that this symmetric form also enables effective use to be made of all possible symmetries within the integral and coupling coefficient to further reduce computational effort. Additional savings are made by filtering the interaction space and hence neglecting quadruplets which have an insignificant contribution to the integral. Hasselmann and Hasselmann (1981) showed that only 5-10% of the interacting wave number quadruplets contribute 95% of the nonlinear transfer. The result is a decrease in computational time by a factor of up to 100 (Hasselmann and Hasselmann, 1981).

This increase in computational speed enabled the development of the first spectral model with a full solution to S_{nl}, EXACT-NL (Hasselmann and Hasselmann, 1985a). This model solved the energy balance equation (4.22) for the one-dimensional (one spatial dimension) cases of either fetch or duration limited growth [see Chapter 5]. A full two-dimensional solution of (4.22) was still not feasible, despite the computational efficiency of the solution for S_{nl}.

EXACT-NL has proven a useful vehicle for the investigation of a number of observed properties of wind generated waves. Komen et al (1984) have investigated the fully developed wind-sea spectrum [see Section (5.4)], Young et al (1987) and Van Vledder and Holthuijsen (1993), the response of the spectrum to turning winds [see Section (6.4)] and Weber (1988) the energy balance in finite depth water [see Section (7.2)].

The solution technique proposed by Webb (1978) reduced the multi-dimensional integral represented by (4.55) to a series of line integrals around the 'egg-shaped' loci resulting from the projection used to remove the delta functions. Although this operation simplifies the mathematical formulation, the number of possible loci along which line integrals need be performed is enormous. Tracy and Resio (1982) and Resio and Perrie (1991) showed, however, that for deep water conditions, these loci are geometrically similar. Hence, the number of line integrals could be considerably reduced, achieving computational times similar to that of (Hasselmann and Hasselmann, 1985a). A one-dimensional model, similar to EXACT-NL, but based on the solution technique of Resio and Perrie (1991) has been used by Banner and Young (1994) to investigate fetch limited spectral evolution [see Chapter 5)].

To incorporate the effects of nonlinear interactions into two-dimensional wave prediction models, various techniques have been proposed to parameterize S_{nl} in relatively simple and computationally efficient forms (Barnett, 1968; Ewing, 1971; Hasselmann et al, 1985a; Young, 1988a). Although these parameterizations capture many of the gross features of the nonlinear transfer, they lose most of the essential physics of the Boltzmann integral (4.55).

This problem was overcome, to some extent, by Hasselmann et al (1985a) in their *discrete interaction approximation* (DIA) to the Boltzmann integral. Whereas a full solution to (4.55) uses a very large set of wave number quadruplets with many different configurations, the DIA uses a rather small number of quadruplets which all have the same configuration. The chosen configuration has $\mathbf{k}_1 = \mathbf{k}_2$ with $\mathbf{k}_3$ and $\mathbf{k}_4$ lying at an angle to the first two wave numbers. The magnitudes of the quadruplet are given by

$$\begin{aligned} f_1 &= f_2 &= f \\ f_3 &= f(1+\lambda) &= f^+ \\ f_4 &= f(1-\lambda) &= f^- \end{aligned} \tag{4.60}$$

The four wave numbers in this quadruplet must also satisfy the resonance condition (4.53), (4.60) ensuring that (4.54) is satisfied. Hasselmann et al (1985a) set $\lambda = 0.25$ based on numerical experiments. The rates of change of the energy densities (δS_{nl}, δS_{nl}^+, δS_{nl}^-) within one such wave number quadruplet are given by

$$\begin{Bmatrix} \delta S_{nl} \\ \delta S_{nl}^+ \\ \delta S_{nl}^- \end{Bmatrix} = \begin{Bmatrix} 2 \\ -1 \\ -1 \end{Bmatrix} Cg^{-4} f^{11} \left[F^2 \left(\frac{F^+}{(1+\lambda)^4} + \frac{F^-}{(1-\lambda)^4} \right) - 2\frac{FF^+F^-}{(1-\lambda^2)^4} \right] \tag{4.61}$$

where g is the gravitational acceleration, C is a constant equal to 3×10^7 and F, F^+ and F^- are the energy densities (expressed in polar, frequency-direction space, $F(f, \theta)$) at the values of the interacting wave numbers.

To compute the nonlinear transfer for a given energy spectrum, all interactions between four wave numbers satisfying conditions (4.53) and (4.60) are considered, and for which the central wave number $\mathbf{k}$ ($= \mathbf{k}_1 = \mathbf{k}_2$) loops over all wave numbers of the spectrum which is represented on a discrete (f, θ) grid. This reduces the problem to a two-dimensional integral rather than a six-dimensional integral required for the full solution.

The interaction combination represented by (4.60) is only one of many which could have been selected. Indeed, more than one geometry of wave numbers could have been chosen, although this would involve a computational penalty. To date, no major systematic attempts have been made to optimize the choice.

Despite the fact that the number of interacting wave numbers considered by the DIA is only a small fraction of those used in the full solution, the approximation retains the essential physical properties of the Boltzmann integral. Hence, the DIA reproduces the basic properties of the nonlinear interactions. These properties of the full Boltzmann integral are discussed in Section (4.4.2). Naturally, the neglect of so many possible interactions has consequences. One example of this is that the DIA typically produces spectra with broader directional spreading than do models based on a full solution to S_{nl} as shown in Figure 4.11 (Young et al, 1987; Van Vledder, 1990). The DIA has, however, resulted in the development of the first, so called, *third generation* global wave model, WAM (WAMDI, 1988).

Recently, Lin and Huang (1998) have presented a full solution technique for S_{nl} based on the Hamiltonian approach of Zakharov (1968) which they claim to be significantly faster than previous solution techniques. Whether such an approach would enable the development of a two-dimensional model with a full solution to S_{nl} is, however, yet to be demonstrated.

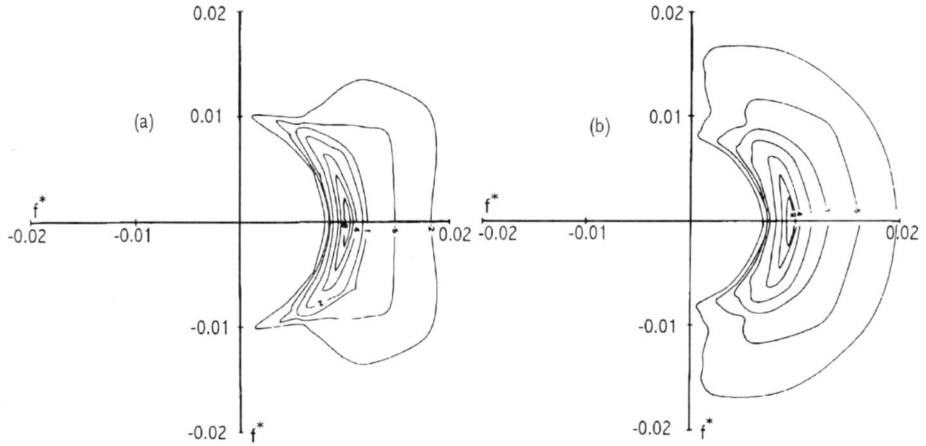

Figure 4.11: A comparison between the directional spectra produced from models with (a) (left) a full solution to the nonlinear source term, EXACT-NL and (b) (right) the DIA approximation to the nonlinear source term. The case shown is for duration limited growth with a non-dimensional time, $t^* = tg/u_* \approx 3.2 \times 10^4$. The spectra have been normalized such that the peak values are 10. The frequency axes are shown in non-dimensional form where $f^* = fu_*/g$. Note the broader directional spreading which occurs as a result of the DIA approximation to S_{nl} [after Young et al (1987)].

4.4.2 Properties of Quadruplet Interactions

4.4.2.1 Spectral Evolution

Because of the complexity of the Boltzmann integral (4.55) an understanding of the source term, S_{nl} and the role it plays in the evolution of wind-waves is best obtained through numerical experiment. This was achieved to some extent during JONSWAP (Hasselmann et al, 1973), where S_{nl} was evaluated for idealized fetch limited spectra at various fetches ranging from very 'young' to almost fully developed states. One problem with this approach is that the nonlinear transfer is very sensitive to the spectral shape. Hence, imposing a spectral shape which is, at best, an approximation to observed spectra may yield misleading results. Small differences in the spectral shape can result in quite large differences in the computed S_{nl}. This was shown by Komen et al (1984) where they attempted to determine the dissipation term required to balance the other source terms for an assumed 'fully developed' Pierson-Moskowitz (Pierson and Moskowitz, 1964) spectrum. As the Pierson-Moskowitz spectrum is only an approximation to the fully developed form and as some directional spread must be assumed, no plausible balancing dissipation could be found. Instead, they inverted the problem by using the one-dimensional EXACT-NL model (Hasselmann and Hasselmann, 1985a). They allowed the two-dimensional spectrum to develop under the action of the source terms and determined the dissipation term required to produce an asymptotic state close to the Pierson-Moskowitz form [see Section (4.5)].

Young and Van Vledder (1993) adopted this same approach to investigate the role played by S_{nl} in spectral evolution. For the case of deep water fetch limed growth, (4.22) becomes

$$C_g \cos\theta \frac{\mathrm{d}F}{\mathrm{d}x} = S_{tot} \tag{4.62}$$

where x is the fetch length [see Figure 5.1] and $S_{tot} = S_{in} + S_{nl} + S_{ds}$ is the total source term for deep water conditions. The source terms used in the model were: S_{in} as proposed by Snyder et al (1981), (4.44); S_{nl} as proposed by Tracy and Resio (1982) and S_{ds} as proposed by Komen et al (1984), (4.67).

Figure 4.12 shows the one-dimensional spectrum and the various source terms at two different fetches. A relatively 'young' sea at short fetch is shown in Figure 4.12a whereas a mature spectrum approaching fully developed conditions appears in Figure 4.12b. The balance between the three source terms determines the evolution of the spectrum and this will be considered in detail in Section (4.6). Here, only the behaviour of S_{nl} will be considered.

The nonlinear source term, S_{nl} has a characteristic *plus-minus signature.* A narrow positive lobe exists in the vicinity of the spectral peak, with a broader negative lobe at higher frequencies. Early calculations of S_{nl} (Hasselmann et al, 1973) were for prescribed spectra, generally of the JONSWAP form, and these calculations showed a narrower negative lobe with another positive region at higher frequencies. When the spectrum is allowed to develop under the source term balance, the spectrum evolves slightly differently and the broader negative lobe results. Hence, S_{nl} causes a transfer of energy from higher frequencies to frequencies near the spectral peak. At short fetches, the spectrum is very peaked [Figure 4.12a] and the positive lobe of S_{nl} is located at frequencies slightly less than the spectral peak frequency. This transfer helps to maintain the observed migration of the spectral peak to lower frequencies with increasing fetch. As the spectrum approaches full development [Figure 4.12b], it broadens. Responding to this change in spectral shape, the magnitude of S_{nl} decreases (i.e. 'older' waves are less nonlinear) and the position of the positive lobe moves to slightly higher frequencies (close to the spectral peak frequency). Hence, the growth of the spectrum slows and the migration of the spectral peak to lower frequencies halts.

4.4.2.2 Shape Stabilization

As shown in Figure 4.12, wind-wave spectra have a characteristic shape with a single peak and a high frequency face which is a form $\approx f^{-4}$ [see Section (5.4.2)]. Young and Van Vledder (1993) have shown that S_{nl} plays a central role in defining the spectral shape, a process they termed *shape stabilization.* This process is

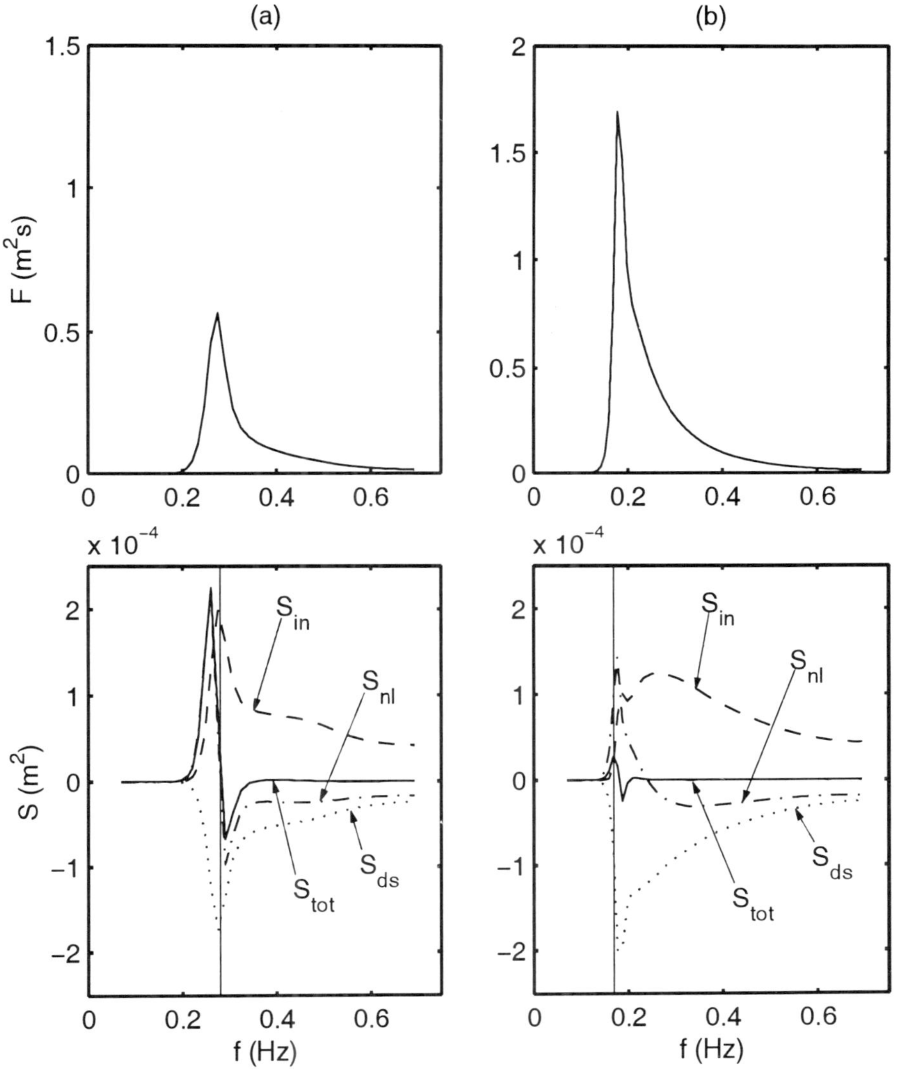

Figure 4.12: The one-dimensional spectrum (above) and the source terms (below) for fetch limited conditions and a wind speed $U_{10} = 10$ m/s. Two cases are shown: (a) at short fetch and (b) approaching fully developed conditions. In each case, the position of the spectral peak is shown by the thin vertical line [after Young and Van Vledder (1993)].

best illustrated by applying a perturbation to the spectrum and then observing the source term response. The one-dimensional model represented by (4.62) was allowed to evolve under the action of the wind. A perturbation was then induced by reducing the magnitude of all spectral components of the directional spectrum at a given frequency by a value of 50%. The model run was then continued to investigate the response to the perturbation. Figure 4.13 shows the response of the spectrum and the source terms to the perturbation as a function of time.

The major response to the perturbation is caused by S_{nl}. A large positive input is directed to the frequency where the spectral energy was reduced. This acts to fill the 'hole' created in the spectrum. As can be seen by the time series Figures 4.13a-d, the result is that the perturbation is gradually smoothed out. As total energy must be conserved by the nonlinear term, the positive energy transferred to the perturbed region must be balanced by a decrease in the transfer in other regions of the spectrum. Figure 4.14 provides a comparison of S_{nl} immediately before and after the introduction of the perturbation. The energy transfer to the region where the perturbation occurred comes from components of the spectrum immediately neighbouring the perturbed region. Hence, the stabilization can be viewed as a flow of energy from the neighboring wave numbers to the perturbed region. The introduction of the perturbation has almost no impact on the energy transfer near the peak of the spectrum. Although resonant quadruplet interactions potentially couple almost all components within the spectrum, the magnitude of the coupling, as measured by the coupling coefficient, G, decreases rapidly as the wave number vectors become separated in either magnitude or direction. As a result, spectral components with similar frequencies (or wave numbers) which propagate in similar directions interact nonlinearly most strongly. As shown in Figure 4.14, there is little interaction between the spectral peak and the tail region where the perturbation was introduced, as they are separated significantly in frequency space. The coupling of wave number components through quadruplet interactions is also important when considering processes such as: the interaction of swell and wind sea [see Section (6.2)] and the response of the spectrum to changes in wind speed or direction [see Sections (6.3) (6.4)].

4.4.2.3 Directional Spreading

Wind generated waves are not uni-directional. Typical wind wave spectra have a directional spread about the wind direction. This directional spreading is often represented in terms of parametric models such as the $\cos^{2s}\theta/2$ model (Mitsuyasu et al, 1975), where θ is the angle of wave propagation measured relative to the wind direction. Such a model predicts that the directional spectrum is a maximum in the wind direction and gradually decreases as the angle to the wind direction increases [see Section (5.5)]. Indeed, this model predicts that a small amount of wave energy propagates at angles greater than 90° to the wind. The atmospheric input term, S_{in} (4.44) predicts that atmospheric input exists at angles to the wind, as there is a component of the wind in these direction. Atmospheric input cannot, however, account for energy propagating at angles greater than 90°.

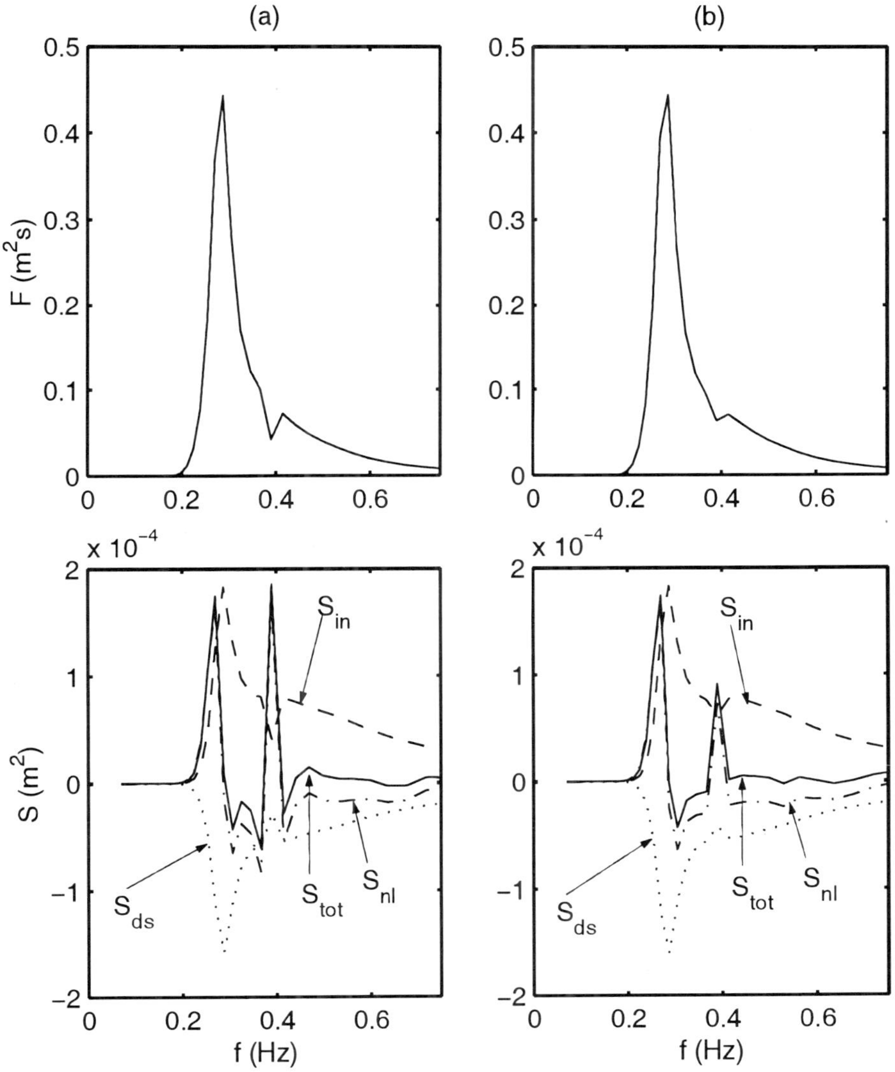
(a)
(b)
F (m²s)
0.5
0.4
0.3
0.2
0.1
0
0.2
0.4
0.6
x 10⁻⁴
S (m²)
2
1
0
−1
−2
S_in
S_ds
S_tot
S_nl
f (Hz)

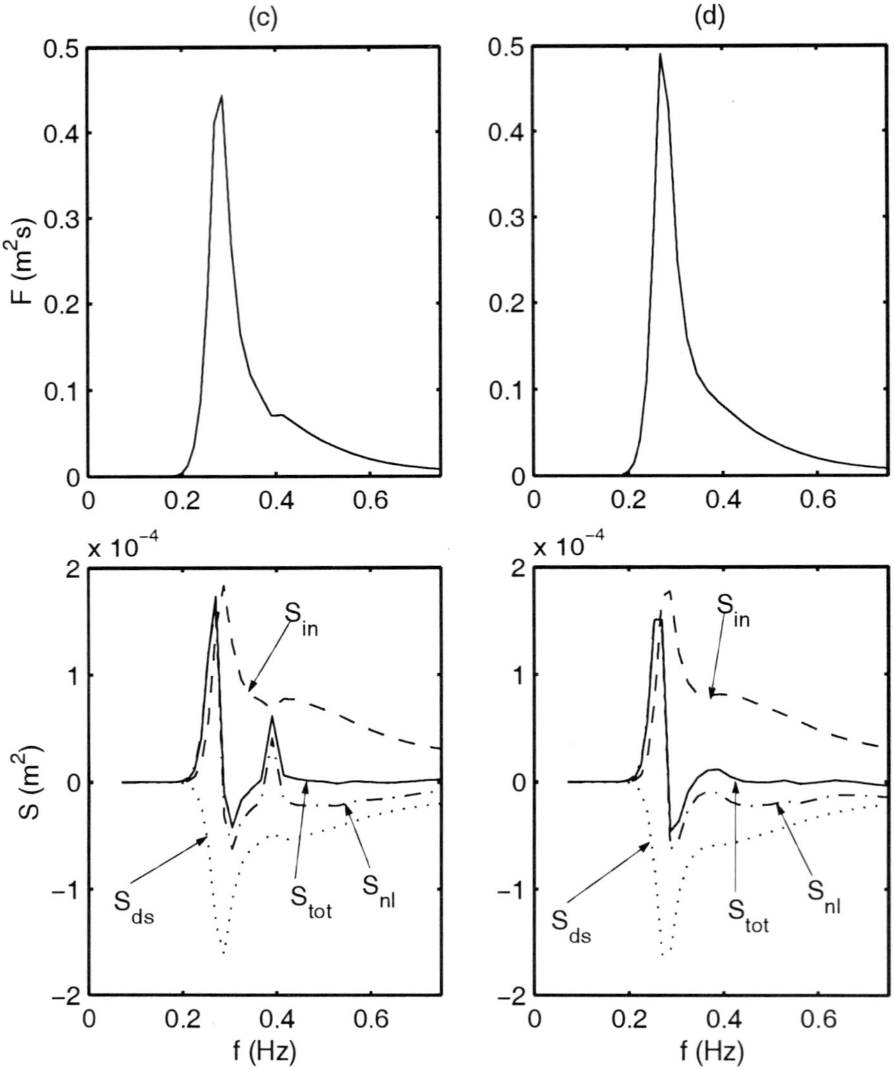

Figure 4.13: The evolution of the spectrum following a perturbation of the spectral level. The spectrum together with the source terms are shown at four times following the introduction of the perturbation. The figures are shown at times of (a) 0 s, (b) 180 s, (c) 300 s and (d) 900 s after the introduction of the perturbation, respectively [after Young and Van Vledder (1993)].

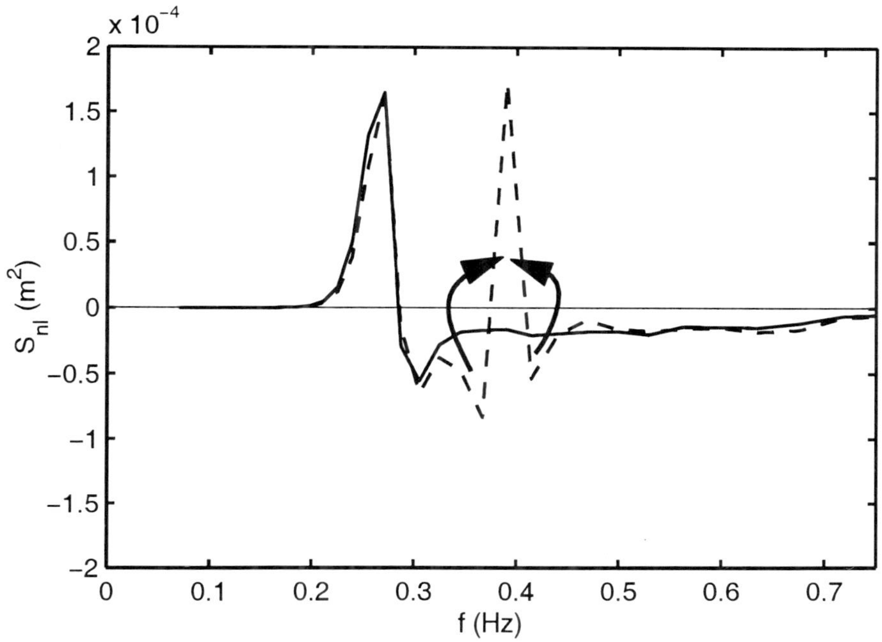

Figure 4.14: The nonlinear source term, S_{nl} directly before (solid line) and directly after (dashed line) the perturbation of the spectrum shown in Figure 4.13. As shown by the arrows, energy flows from frequencies immediately above and below the perturbed frequency to smooth the discontinuity [after Young and Van Vledder (1993)].

Figure 4.15 shows contour plots of $F(f, \theta)$ and $S_{nl}(f, \theta)$ as produced by the fully nonlinear model described by Young and Van Vledder (1993) (4.62). As expected, the spectrum has a directional spread about the mean wind direction. The nonlinear term shows the characteristic narrow (in frequency) positive lobe at frequencies less than the spectral peak with a broader negative lobe above the peak. The positive lobe has a relatively narrow directional distribution. The higher frequency negative lobe decreases in magnitude with angle to the wind direction and becomes positive at large angles. Therefore, at frequencies above the spectral peak, S_{nl} transfers energy from the wind direction to larger angles. Thus, S_{nl} acts to broaden the spectrum at higher frequencies.

Hence, S_{nl} continually forces the spectrum to a shape with a broader directional spread at high frequencies and a relatively narrow spread around the peak. This is consistent with observed directional data (Mitsuyasu et al, 1975; Hasselmann et al, 1980; Holthuijsen, 1983; Donelan et al, 1985; Young et al, 1996) [see Section (5.5)].

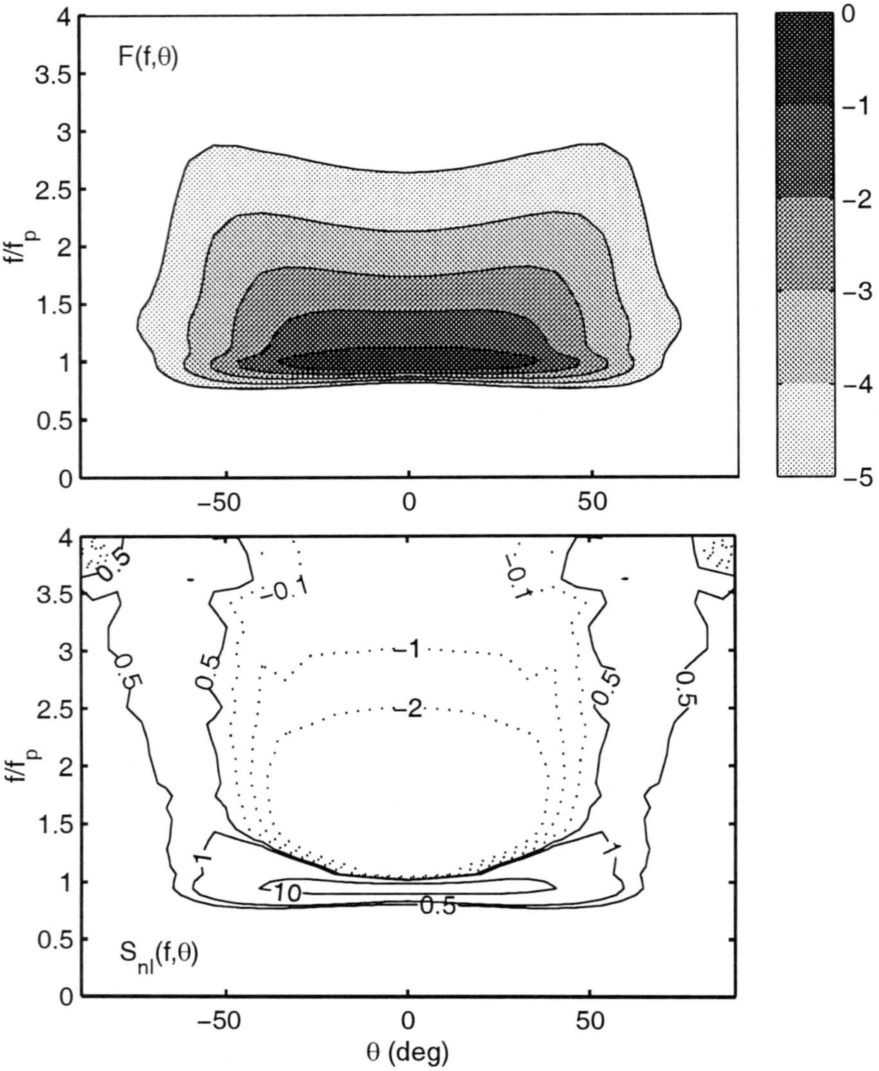

Figure 4.15: A Contour plot of the directional spectrum, $F(f,\theta)$ (top) and the corresponding quadruplet nonlinear interaction term, $S_{nl}(f,\theta)$ (bottom). Values of $F(f,\theta)$ have been normalized to have a maximum of 1 and contours have been drawn with a logarithmic spacing. The contour levels are shown on the colour bar and represent the exponent of the level (i.e. 10^{-5} ... 10^{-3} ... 10^{0}). The contour values shown on S_{nl} are in units of $[10^5\ \mathrm{m}^2]$. Contours representing negative values are shown as dotted lines.

4.5 White-Cap Dissipation, $\mathbf{S_{ds}}$

The energy transfer from the wind to the waves results in an increase in the wave amplitude. This process continues until the wave eventually becomes unstable and breaks. This form of breaking is generally termed *white-capping* in contrast to depth limited breaking, which might occur on a beach [see Section (7.2.6)]. A long standing problem in wind-wave research has been the specification of a source term to represent the energy loss due to this *white-cap dissipation*, S_{ds}. A number of experimental studies have attempted to investigate the dynamics of a single breaking wave (Longuet-Higgins and Cokelet, 1978; Duncan, 1981; Longuet-Higgins, 1988; Rapp and Melville, 1990). In the present context, however, a spectral representation of the energy loss is required.

White-capping represents a strongly nonlinear process which cannot be treated by standard perturbation techniques which are typically applied to weak interactions [e.g. S_{nl}]. Hasselmann (1974) overcame this difficulty by assuming that white-capping was a process which was *weak-in-the-mean*. Such a process induces a change in the wave spectrum per unit wavelength or period which is small, and the Gaussian property of the linear wave field applies to lowest order.

In keeping with the weak-in-the-mean assumption, Hasselmann (1974) treated white-caps as a random distribution of pressure pulses on the water surface, the scale of which was small compared to the wavelength of the waves. Rather than follow the original derivation of Hasselmann (1974), the description of Donelan and Yuan (1994) is adopted here.

It has already been shown by (4.24) that a pressure field applied to the water surface can result in a flux of energy to or from the waves. As white-capping results in a loss of energy, the white-cap pressure pulses must be preferentially situated on the forward faces of the waves [see Section (4.3.1)]. Indeed, it is common knowledge that white-caps form on the forward or down-wind faces of the waves. Donelan and Yuan (1994) assumed that the white-cap and the underlying wave are geometrically similar (Duncan, 1981) as shown in Figure 4.16.

Under this assumption of geometric similarity, it follows that

$$L_w/L = \text{constant},$$

$$h_w/a = \text{constant} \tag{4.63}$$

where h_w is the height and L_w is the length of the white-cap region. If the pressure exerted on the wave by the white-cap is p_w, it follows that

$$p_w \sim \rho_w g h_w \sim \rho_w g a \tag{4.64}$$

Since, from (2.11), $\eta = a \sin \omega t$, it follows from (4.24) that

$$\frac{\partial E}{\partial t} = \frac{1}{\rho_w g} \overline{p_w \frac{\partial \eta}{\partial t}} \sim -\omega a^2 \tag{4.65}$$

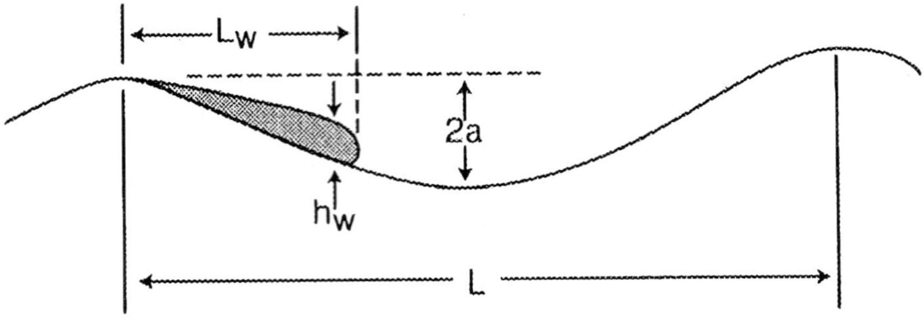

Figure 4.16: The geometric similarity of the white-cap and the underlying wave [after Donelan and Yuan (1994)].

Hence,

$$S_{ds}(f,\theta) = -\gamma_{ds}\omega F(f,\theta) \tag{4.66}$$

where γ_{ds} is a damping coefficient which is a function of non-local properties of the spectrum. The negative sign in (4.66) occurs because of the phase of the white-cap pressure pulse relative to the wave [i.e. it is on the forward face]. Hence, S_{ds} is quasi-linear in F, some non-linearity being introduced through the dependence of γ_{ds} on non-local properties of the spectrum. Hasselmann (1974) shows that such a quasi-linear form is a property of any weak-in-the-mean process. Indeed, during the WAM Group (Komen, 1985) scientific meetings held over the 10 year period 1984-1994, Klaus Hasselmann argued strongly against the validity of models which assumed S_{ds} to be locally nonlinear [ie. $S_{ds} \propto F^n(f,\theta)$]. A number of formulations of this type have been proposed, including Phillips (1985).

In addition to the pressure pulse mechanism described by (4.66), other processes also occur during white-capping. For instance, Banner et al (1989) have shown that the breaking of waves near the peak of the spectrum will attenuate high frequency waves in the breaker's wake. Such a mechanism might reasonably be assumed to be a function of wave frequency relative to that of the spectral peak. In addition, the magnitude of the decay coefficient could also be expected to depend on some measure of the steepness of the waves.

Komen et al (1984) attempted to include such effects by proposing a form for γ_{ds}, such that

$$S_{ds}(f,\theta) = -C_{ds}\left(\frac{\hat{\alpha}}{\hat{\alpha}_{PM}}\right)^m \left(\frac{\omega}{\overline{\omega}}\right)^n \omega F(f,\theta) \tag{4.67}$$

where $\overline{\omega}$ is the mean frequency

$$\overline{\omega} = \left(\sigma^2\right)^{-1} \int F(f,\theta)\omega \mathrm{d}f\mathrm{d}\theta \tag{4.68}$$

and $\hat{\alpha}$ is an integral steepness parameter

$$\hat{\alpha} = \frac{\sigma^2 \overline{\omega}^4}{g^2} \tag{4.69}$$

The total energy, or variance, is represented by $\sigma^2 = \int F(f,\theta) \mathrm{d}f \mathrm{d}\theta$ and $\hat{\alpha}_{PM} = 4.57 \times 10^{-3}$ is the theoretical value of $\hat{\alpha}$ for the Pierson-Moskowitz spectrum (5.51).

Komen et al (1984) optimized the values of C_{ds}, m and n in (4.67) by fitting the output of the fetch-limited EXACT-NL model (Hasselmann and Hasselmann, 1985a) (4.62) to the Pierson-Moskowitz fully developed state (Pierson and Moskowitz, 1964) for long fetch. Although they made some comparisons with the fetch limited development predicted by JONSWAP (Hasselmann et al, 1973), these comparisons were not exhaustive. They found optimum values of $C_{ds} = 3.33 \times 10^5$, $m = 2$ and $n = 2$. In a later set of numerical experiments with a similar model, Banner and Young (1994) found that no variation of (4.67) could match all the commonly reported spectral parameters over the full range of non-dimensional fetches. One of the principal difficulties encountered by (Banner and Young, 1994) was that $\hat{\alpha}$ typically varies by only a small amount from young to fully developed waves. Therefore, (4.67) is rather insensitive to the selection of m. Despite these difficulties, they could propose no more appropriate form for S_{ds}. Although (4.67) is largely empirical, its general form has proved successful in operational wave models such as WAM (WAMDI, 1988).

Alternative mechanisms to the pressure pulse model have also been proposed to account for the energy loss due to white-capping. Phillips (1985) and Donelan and Pierson (1987) have considered white-capping as a process which is highly nonlinear in the wave steepness. In this model, white-capping has no influence until the local wave steepness reaches some limit where the wave spills or plunges forward losing significant energy. Such a model may include local changes to the steepness induced by the presence of wave groups.

The pressure pulse or steepness models are deterministic. Observations of wave, however, indicate that white-capping is highly variable. There can be two waves which, for all practical purposes appear identical with the same height, period and steepness. One will break and the other will not. Thus, it may be appropriate to represent white-capping as a stochastic process where each wave is assigned a probability of white-capping. A source term of this form has been proposed by Yuan et al (1986).

The discussion above clearly shows that white-cap dissipation is a process which is poorly understood. No rigorous theory exists for either the onset of white-capping or the resulting energy loss. In the absence of theory, the usual approach is to resort to direct measurement. The problem then arises : How does one measure energy decay due to white-capping in a directional sea state experiencing active wind forcing? The major problem is that the forcing mechanisms cannot be separated. Active white-capping only exists when there is also energy input for the wind. Also, as a wind sea spectrum is present, quadruplet non-linear interactions will also be active. Measurements need to isolate just the influence of S_{ds}. There are presently two instrumentation approaches which show promise. The first of these involves measurements of the kinetic energy dissipation in the water column beneath breaking waves (Drennan et al, 1996; Terray et al, 1996). The second approach involves the measurement of the acoustic signature of the breaking waves which appears to be related to the energy loss (Melville et al, 1991). These approaches show promise but are yet to deliver conclusive results.

An excellent summary of the process of white-capping and how this is represented by existing theories is provided by Donelan and Yuan (1994): "*Let us examine the sequence of events leading up to and during breaking to see how the various approaches fit into nature's scheme of things. Wave energy is continually being drained by viscosity. This process is on a sound theoretical footing and is appropriately included in the model of Donelan and Pierson (1987) among others. When a wave becomes very steep, but not yet in the process of breaking, it radiates energy via parasitic capillaries which themselves lose energy quickly to viscosity. This process is ignored in all the models. Once a wave starts breaking it loses energy in two ways: fluid is injected into the whitecap to become turbulent and the turbulence itself interacts with the orbital flow on the forward face of the wave. The presence of the whitecap on the forward face of the wave provides a downward pressure in the right phase to extract energy from the wave. This last effect is the basis of the Hasselmann (1974) model.*"

4.6 The Spectral Balance

In the preceding discussion, each of the source terms has been considered in isolation. It is, however, the sum of the individual terms, as represented by S_{tot} (4.23), which is responsible for the evolution of the wave spectrum. The source term balance for both "young" and "fully developed" spectra is shown in Figure 4.12. The spectral evolution shown in Figure 4.12 is a consequence of the balance between the various source terms and has been achieved without any external constraints on the spectral shape. At high frequencies the spectrum takes the characteristic f^{-n} form. In this region the three source terms sum to yield zero, $S_{tot} = 0$. This is largely a result of S_{nl}. Banner and Young (1994) have shown that changes in the magnitude of either S_{in} or S_{ds} has a small influence on the amount of energy in the spectral tail, but little influence on the decay exponent n. The nonlinear term, S_{nl} continually forces the spectrum to conform to this shape. The reason for this behaviour by S_{nl} is yet to be explained.

As the spectrum becomes more mature, the spectral peak gradually shifts to lower frequencies and the spectrum broadens. As a result the balance achieved at high frequencies extends over a wider frequency band. In addition, S_{in} and S_{nl} decrease in magnitude. The atmospheric input, S_{in} decreases because the ratio U_{10}/C approaches 1. The nonlinear term, S_{nl} decreases because the peak frequency, f_p decreases and because of the broadening of the spectral shape. In addition, the positive lobe of S_{nl} moves from frequencies less than f_p to be equal to f_p. As a result, the migration of f_p to lower frequencies slows. The region where the source terms completely balance, $S_{tot} = 0$ increases and growth almost ceases at all frequencies. This state is commonly referred to as *full development*.

The total source term, S_{tot} is typically negative at frequencies slightly above the spectral peak. As a result, the energy at a fixed frequency decreases slightly once the peak migrates past the point to lower frequencies. This feature has been observed in recorded data and was termed the *overshoot effect* by Barnett and Sutherland (1968). As shown by Banner and Young (1994), the source term balance also influences the directional spread which varies slightly as the spectrum becomes more mature.

As is seen by the discussion above, the source terms used to produce results such as those in Figure 4.12 are capable of reproducing remarkable agreement with observational data. As a result, one might be led to believe that our understanding of the physics of wind wave generation is almost complete. Impressive as the results are, a number of questions still remain. Firstly, the process used by Komen et al (1984) to determine S_{ds}, implicitly forced this term to include all influences required to balance input and nonlinear transfer. Whatever was required to balance these terms was attributed to white-cap dissipation. There could be other processes, unknown at this time, which are also include in this term. Banner and Young (1994) have also shown that many of the detailed features of spectral development (e.g. energy in the spectral tail and directional spreading) are not particularly well modelled by such source terms.

Although knowledge of the source terms has improved markedly over the last 30 years, a number of questions still remain. A complete knowledge of the physics of wind wave generation still eludes modern science.

Chapter 5

Fetch and Duration Limited Growth

5.1 Introduction

There are many factors which can influence the development of a wave field. Some of these include: wind speed and its variation, position and geometry of the coastline, wind direction and water depth. The combinations of these variables leads to a vast number of situations which cannot be simply characterized. Two idealized cases are, however, commonly examined: *fetch limited growth* and *duration limited growth*. Although highly idealized, these cases provide valuable insight into many of the physical processes responsible for wind wave evolution. In addition, they form valuable "test beds" for evaluating the performance of more sophisticated wind wave prediction techniques. These cases also provide estimates of wave conditions which could be expected at a site and are often used in preliminary engineering design.

Fetch limited growth occurs when a wind of constant magnitude and direction blows perpendicular to a long and straight coastline. This case is shown diagrammatically in Figure 5.1. The water is assumed deep and the wind blows for a sufficiently long time that the wave field reaches steady state (independent of time). Hence, for the given wind speed, the wave field becomes a function of the distance from the shoreline, which is termed the *fetch*, x.

The case of fetch limited growth has been extensively investigated both in the laboratory and the field. The conclusions of a number of these studies are considered below.

A related problem to fetch limited growth is *duration limited growth*. This case considers the development of the wave field from an initially calm sea. All land boundaries are assumed sufficiently distant that there is no fetch limitation

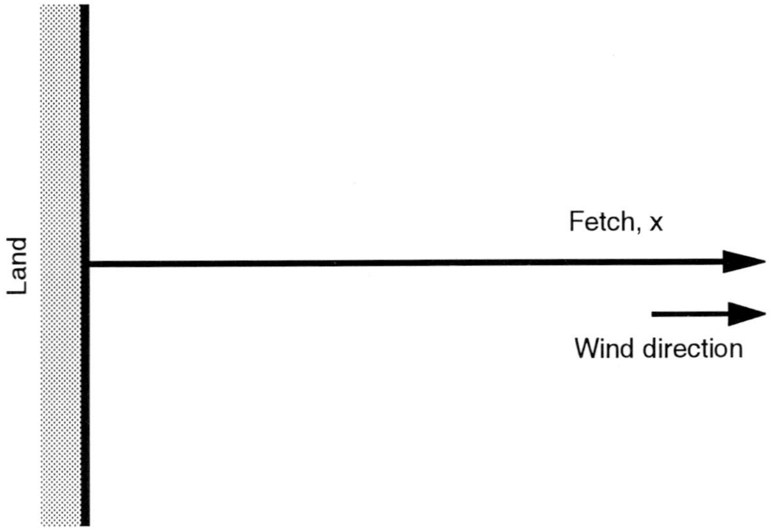

Figure 5.1: Schematic diagram showing fetch limited growth. The wind blows perpendicular to the infinitely long coastline shown to the left. The fetch, x is measured offshore in the direction of the wind. For a constant wind speed, the wave field develops as a function of fetch.

to growth. The wind field is of constant speed and direction and spatially homogeneous. The water is also assumed to be infinitely deep. For the given wind speed, the resulting wave field will be spatially homogeneous and only a function of the duration or time that the wind has been blowing, t. As indicated above, reasonable approximations to fetch limited condition commonly occur, particularly at short fetch. Duration limited conditions are less common and hence the available data base is significantly smaller. The available data will, however, be considered below.

5.2 Similarity Theory and Dimensionless Scaling

5.2.1 Integral Parameters, ε and ν

Based on an unpublished suggestion by M.P. Obrien (1943), Sverdrup and Munk (1947) and Kitaigorodskii (1962, 1970) considered the variables which may be active in fetch and duration limited growth. They assumed the following quantities to be of importance: σ^2 - the variance of the water surface elevation (2.49), $H_s = 4\sigma$, U_a - the wind speed measured at a reference height a, x - the fetch [see Figure 5.1], g - gravitational acceleration, t - the duration for which the wind blows, f_p - the frequency of the spectral peak. Dimensional analysis indicates that there should be four non-dimensional grouping of these parameters

$$\varepsilon = \frac{\sigma^2 g^2}{U_a^4} \quad \text{- the non-dimensional energy} \tag{5.1}$$

$$\nu = \frac{f_p U_a}{g} \quad \text{- the non-dimensional frequency} \tag{5.2}$$

$$\chi = \frac{g x}{U_a^2} \quad \text{- the non-dimensional fetch} \tag{5.3}$$

$$\varsigma = \frac{g t}{U_a} \quad \text{- the non-dimensional duration} \tag{5.4}$$

Therefore for fetch limited conditions

$$\varepsilon = f_1(\chi) \quad ; \quad \nu = f_2(\chi) \tag{5.5}$$

and for duration limited conditions

$$\varepsilon = f_3(\varsigma) \quad ; \quad \nu = f_4(\varsigma) \tag{5.6}$$

where $f_1 \cdots f_4$ are functions to be determined.

5.2.2 Wave Spectrum, F(f)

Phillips (1958) considered the structure of the high frequency face of the one-dimensional spectrum, $F(f)$. He assumed that this region was controlled by gravity, such that, $F = f_5(g, f)$. Under this assumption dimensional analysis yields the result

$$F \propto g^2 f^{-5} \tag{5.7}$$

Toba (1973) argued that the wind speed, as represented by the friction velocity, u_* was also important in this region of the spectrum, $F = f_6(g, u_*, f)$. Again, on dimensional grounds they obtained the form

$$F \propto g u_* f^{-4} \tag{5.8}$$

5.3 Growth Curves for Energy and Peak Frequency

In this section the experimental evidence for the development of the non-dimensional energy, ε and the non-dimensional peak frequency, ν as a function of the non-dimensional fetch, χ and the non-dimensional duration, ς will be reviewed. As there is a relatively small amount of data pertaining to duration limited growth, this situation will receive only passing consideration, with interest concentrated on fetch limited conditions.

Although the cases of fetch and duration limited growth appear highly idealized, there are still a number of largely unresolved scientific question concerning such situations. The most fundamental of these is the selection of the scaling wind speed, U_a to be used in the non-dimensional variables.

5.3.1 The Scaling Wind Speed

A measure of the wind speed, U_a appears in all of the non-dimensional quantities (5.2) to (5.4). The choice of this value is, hence, central to the definition of fetch and duration limited growth. In order to discuss the various choices for U_a which have been proposed, it is first necessary to review the structure of the marine atmospheric boundary layer. The description below largely follows the development by Makin et al (1995).

Under neutral conditions (i.e. air and water temperature approximately equal), the atmospheric boundary layer is generally assumed to have a logarithmic profile

$$\frac{u(z)}{u_*} = \frac{1}{\kappa} \ln\left(\frac{z}{z_0}\right) \tag{5.9}$$

where u is the wind speed measured at a height z above the surface, z_0 is a surface roughness length, u_* is the friction velocity and $\kappa \approx 0.4$ is the von Kármán constant. The shear stress within the boundary layer can be represented as [e.g. Geernaert (1990)]

$$\tau = \rho_a u_*^2 = \rho_a C_d U_{10}^2 \tag{5.10}$$

where τ is the shear stress, ρ_a the density of air, U_{10} the wind speed measured at a reference height of 10m and C_d is the drag coefficient (dimensionless). From (5.9) and (5.10) it follows that

$$C_d = \frac{u_*^2}{U_{10}^2} = \frac{\kappa^2}{\ln^2(10/z_0)} \tag{5.11}$$

At low wind speeds (less than approximately 3 m/s) the sea surface is calm and the air flow can be characterized as aerodynamically smooth (Phillips, 1977; Geernaert, 1990; Makin et al, 1995). In such cases, the surface roughness is determined by the scale of the molecular sublayer (Monin and Yaglom, 1971)

$$z_0 = c_1 \frac{\nu_a}{u_*} \tag{5.12}$$

where ν_a is the kinematic viscosity of air and c_1 is a constant of $O(10^{-1})$. At the other limit of very strong winds, the sea surface can be described as aerodynamically rough. In such cases the surface roughness is determined by the mean squared surface displacement, which can be represented in terms of u_* and g

$$z_0 = c_h \frac{u_*^2}{g} \tag{5.13}$$

Equation (5.13) is termed the Charnock relationship as it was originally developed on dimensional grounds by Charnock (1955). The Charnock parameter, c_h is generally assumed constant, although a range of values have been reported from measurements: 0.014 Garratt (1977), 0.018 Wu (1980), 0.035 (Kitaigorodskii, 1970).

With the surface roughness defined, (5.9), (5.11) and (5.13) can be solved to determine C_d as a function of U_{10}. Assuming $c_h = 0.014$, Garratt (1977) obtained

$$C_d = (0.75 + 0.067U_{10})\, 10^{-3} \tag{5.14}$$

where U_{10} has units of m/s. Equation (5.14) is only one of many parameterizations of the drag coefficient. A comparison has been provided by Geernaert (1990).

The Charnock relationship (5.13) indicates that the surface roughness is completely determined by the wind. This result is, however, limited to the case where the mean squared surface displacement can be determined in terms of u_* and g alone. In fetch, duration or transient (frontal systems etc.) conditions a wide range of different sea states can exist for the same wind speed. Hence, it is reasonable to assume that, in such cases, a measure of the sea state may also be important. For pure wind sea conditions (ie. no swell), this sea state dependence has been represented in terms of the inverse wave age, U_{10}/C_p (or u_*/C_p) where C_p is the phase speed of components at the spectral peak frequency.

Experimental data suggests that in such cases the Charnock parameter becomes a function of the inverse wave age

$$z_0 \frac{g}{u_*^2} = a_1 \left(\frac{u_*}{C_p} \right)^m \tag{5.15}$$

Smith et al (1992) have evaluated the parameters in (5.15) from the HEXOS experimental data set, obtaining $a_1 = 0.48$ and $m \approx 1$. Hence, z_0 is larger for "young" seas than for more developed cases. Toba et al (1990) combined field and laboratory data to argue the opposite trend. Donelan et al (1993) argues that laboratory data should be disregarded, accepting the parameters of Smith et al (1992). Models developed by Janssen (1989), Chalikov and Makin (1991) and Chalikov and Belevich (1993) support the HEXOS trend, showing that for high winds, the drag coefficient is more than twice as large for young seas as it is for old seas.

For given values of u_*/C_p, the dependence of the drag coefficient, C_d on the wind speed, U_{10} can be determined from (5.9), (5.11) and (5.15). This is shown in Figure 5.2 for the cases of young waves ($u_*/C_p = 0.1$) and mature waves near full development ($u_*/C_p = 0.03$). Equation (5.14) is also shown. The influence of sea state is clear, with the drag coefficient being a factor of approximately 1.4 higher for the young waves, compared to the mature case. Equation (5.14) agrees well with the situation for a mature sea state.

The comments above assume that a neutral boundary layer exists (ie. no temperature difference between air and water). Should the air and water temperatures differ there will be a temperature gradient in the vertical and hence the air density will vary with height. The resulting stratification of the atmospheric boundary layer may be important because density differences in conjunction with gravity will affect the magnitude of turbulent mixing [e.g. Geernaert and Plant (1990)].

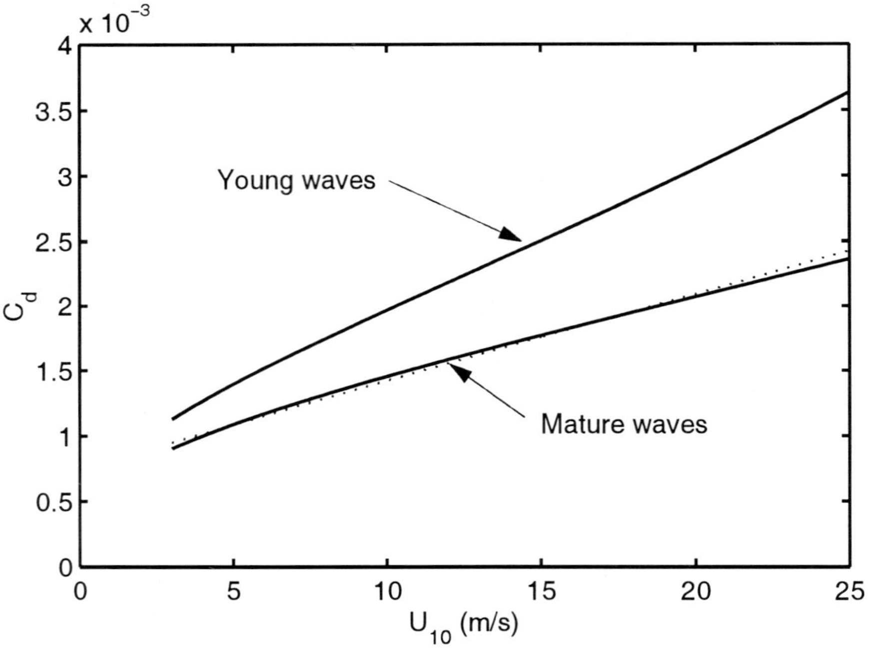

Figure 5.2: The variation in the drag coefficient, C_d as a function of the wind speed, U_{10}. The two extreme cases of young waves ($u_*/Cp = 0.1$) and mature waves ($u_*/C_p = 0.03$) are shown. The dotted line represents the formulation of Garratt (1977) (5.14), which includes no sea state dependence.

Measurements over the ocean by Large and Pond (1982) have shown that the effect can be represented in terms of the Obukhov scale height (Arya, 1988)

$$L_o = -\frac{u_*^3 T_v}{\kappa g < w' T_v' >} \tag{5.16}$$

which defines the height at which the production of turbulence by shear equals the production by buoyancy. In (5.16), T_v is the mean virtual temperature at a height of 10 m, g is the gravitational acceleration, w is the vertical component of velocity and primed quantities denote fluctuations.

When the water is warmer than the air there is an upward flux of heat, an atmospheric density gradient with denser air overlying less dense air and the value of L_o is negative. Such cases are described as unstable. Conversely when the water is colder than the air, L_o is positive and conditions are described as stable. Neutral stratification occurs for large $|L_o|$. An alternative indicator of stability is the Bulk Richardson Number (Kahma and Calkoen, 1992)

$$R_b = \frac{g\,(T_a - T_w)}{z_t T_a \,(u/z)^2} \tag{5.17}$$

where T_a and T_w are the air and water temperatures (°K) respectively, z_t is the height at which the temperature is measured and z is the height at which the wind speed is measured. Again negative values of R_b represent unstable conditions.

The vertical heat flux and associated buoyancy modifies the logarithmic profile (5.9) [e.g. Priestly (1959), Lumley and Panofsky (1964), Webb (1970)] yielding

$$\frac{u(z)}{u_*} = \frac{1}{\kappa}\left[\ln\left(\frac{z}{z_0}\right) - \psi\left(z/L_o\right)\right] \tag{5.18}$$

where ψ is called the integrated universal function and is based on the Monin-Obukhov formalism (Monin and Obukhov, 1954; Monin and Yaglom, 1971). It has been determined by Webb (1970), Dyer and Hicks (1970) and Businger et al (1971). For stable conditions ($L_o > 0$), ψ becomes

$$\psi = -\frac{5z}{L_o} \tag{5.19}$$

and for unstable conditions ($L_o < 0$)

$$\psi = 2\ln\left(\frac{1+x}{2}\right) + \ln\left(\frac{1+x^2}{2}\right) - 2\tan^{-1}x + \pi/2 \tag{5.20}$$

where $x = (1 - 16z/L_o)^{1/4}$.

The effect of atmospheric stability on the mean velocity profile is shown in Figure 5.3. This figure shows the mean velocity profile normalized in terms of the wind speed measured at a reference height of 10m (U_{10}) for values of L_o between −100m and +100m. As the magnitude of the stability effect increases (i.e. $|L_o|$ decreases), the profiles progressively deviate from the neutral logarithmic profile predicted by (5.9). The enhanced vertical mixing which occurs for unstable conditions results in a more uniform profile with height. Thus, in unstable conditions higher velocities are expected closer to the ground than in stable conditions.

Hence, both the sea state and the atmospheric stability influence the shape of the atmospheric boundary layer. The lack of a universal atmospheric boundary layer profile complicates the choice of a scaling velocity to use in the non-dimensional relationships (5.2) to (5.4). The wind speed at a reference height of 10 m, U_{10}, the friction velocity, u_* and the wind speed at a height of half the ocean wave length, $U_{L/2}$ have all been proposed. The example shown in Figure 5.3 illustrates the influence of stability on each of these quantities. In all cases shown, the value of U_{10} will be the same. The shear stress and hence u_* are related to the curvature of the boundary layer. Due to the higher values of mean wind speed closer to the water surface which result in unstable conditions, the gradient of wind velocity in such cases must be larger than in stable conditions. Hence, the surface shear stress and u_* will be larger in unstable than in stable conditions. The relative values of $U_{L/2}$ will depend on the value of L under consideration. Similarly, the same value of U_{10} measured above young and mature waves will yield values of u_* which differ widely.

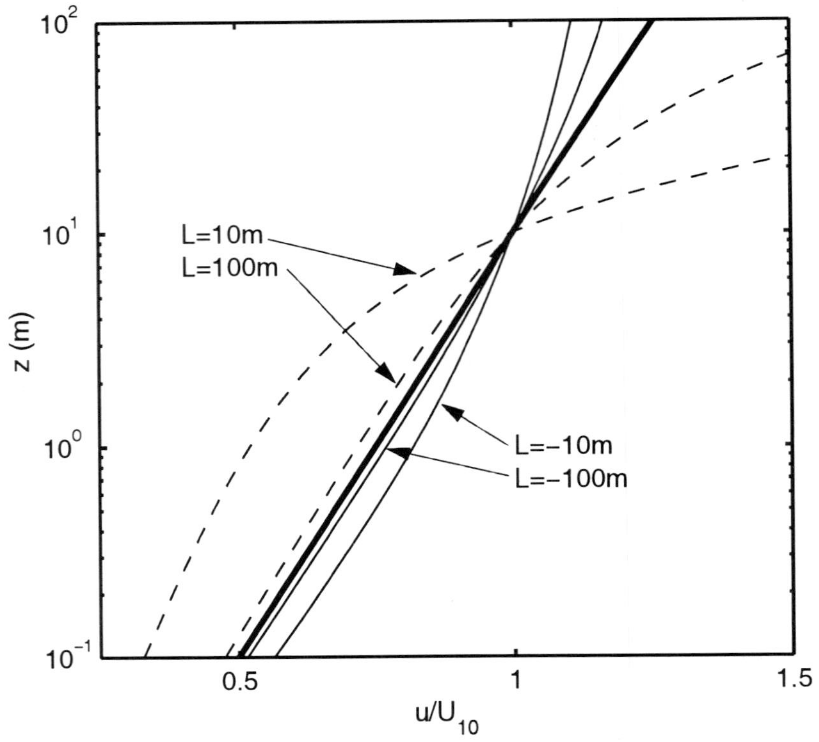

Figure 5.3: Shape of the atmospheric boundary layer for different values of the Obukhov scale height, L_o as calculated from (5.18). Stable conditions are represented by positive values of L_o (dashed lines) and unstable conditions by negative values of L_o (solid lines). The thick solid line represents the logarithmic relationship (5.9) which occurs for neutral conditions.

The Kitaigorodskii (1962) similarity analysis adopted, U_∞, the wind speed above the boundary layer as the scaling value. In practice, U_∞ must be replaced by a more easily measurable quantity. The quantity most commonly adopted is the wind speed at a fixed reference height, usually 10m, U_{10}. Janssen et al (1987) argue that scaling with the friction velocity, u_* rather than U_{10} reduces the scatter in experimental data and hence recommend its adoption. In contrast, Kahma and Calkoen (1992) concluded that: "*... we were not able to improve the results by replacing U_{10} by u_* calculated using the wind speed dependent drag coefficient ...*". It is important to note here that Kahma and Calkoen (1992) indicated that a significant problem in the use of u_* is that it is seldom directly measured. As seen above, the indirect determination of u_* from measured quantities is an involved task and this step potentially introduces error. If u_* was measured directly, an advantage may result.

Donelan et al (1985) favour the use of U_{10} over u_* and raise the question of the physical basis for the use of u_*. As seen by (5.10) and (5.11), u_* is related to the surface shear stress τ. A significant proportion of this stress is supported by waves with frequencies much higher than those of interest in typical wind wave evolution studies. They therefore argue that u_* is not an appropriate scaling parameter.

Chapter 4 indicated that energy is transferred from the air to the water due to wave induced surface pressure. In addition, it was shown that these wave induced pressure fluctuations decay exponentially with height [i.e. $p \propto e^{-kz}$, (4.33)]. Hence, the height at which these pressure fluctuations become negligible scales with k or L. Based on studies by Al-Zanaidi and Hui (1984), Donelan and Pierson (1987) have proposed that the wind speed measured at a height equal to half the wave length may be an appropriate scaling parameter, $U_{L/2}$. Hence, the height at which the scaling wind was measured would decrease with increasing frequency. Although, the arguments in favour of $U_{L/2}$ have merit, the practicalities of its measurement preclude its routine use.

Of the various proposals, only U_{10} is routinely and directly measured and hence in the analysis to follow, it has been adopted as the scaling wind speed.

5.3.2 Fetch Limited Observations

In this section, a number of the significant field studies of fetch limited growth will be reviewed. Interest will focus on the differences between these studies and whether a consensus view exists concerning the evolution of the non-dimensional energy, ε and non-dimensional frequency, ν as a function of non-dimensional fetch, χ. The following studies will be considered:

- The original pioneering work of Sverdrup and Munk (1947) and Bretschneider (1952a, 1958); the so-called SMB result.
- The studies of the fully developed state by Pierson and Moskowitz (1964)
- The JONSWAP studies of Hasselmann et al (1973)
- Measurements in the Bothnian Sea by Kahma (1981)
- The detailed measurements in Lake Ontario by Donelan et al (1985)
- The measurements off the North Atlantic coast of Canada by Dobson et al (1989)
- The multi-station measurements in Lake St. Clair by Donelan et al (1992)

5.3.2.1 The SMB Curves

The first significant attempts to predict fetch limited waves occured during World War II in connection with amphibious landings (Sverdrup and Munk, 1944a,b, 1946, 1947). Sverdrup and Munk compiled existing data from a number of sources, including both field and laboratory. This data set was further augmented and the results refined in a number of subsequent contributions (Bretschneider, 1952a,b, 1958; Wiegel, 1961). Due to the significant contributions of Sverdrup, Munk and Bretschneider, the resulting non-dimensional growth curves became known as the SMB Curves. The results have been summarized in CERC (1977) as

$$\varepsilon = 5.0 \times 10^{-3} \tanh^2 \left[0.0125\chi^{0.42}\right] \tag{5.21}$$

$$\nu = 0.133 \tanh^{-1} \left[0.077\chi^{0.25}\right] \tag{5.22}$$

Due to the properties of the hyperbolic tanh functions in (5.21) and (5.22), the following asymptotic forms result

$$\varepsilon = \begin{cases} 7.82 \times 10^{-7}\chi^{0.84} & \text{for small } \chi \\ 5.00 \times 10^{-3} & \text{for large } \chi \end{cases} \tag{5.23}$$

$$\nu = \begin{cases} 1.722\chi^{-0.25} & \text{for small } \chi \\ 0.133 & \text{for large } \chi \end{cases} \tag{5.24}$$

5.3.2.2 The Pierson-Moskowitz Limit

At large non-dimensional fetch, the SMB curves reach asymptotic limits defined by (5.23) and (5.24). These limits indicate that all further development ceases. This limit is commonly called the *fully developed state*. Pierson and Moskowitz (1964) examined spectra obtained from North Atlantic weather ships. Based on these data, they proposed a limiting form for the one-dimensional spectrum [see Section (5.4)]. Correction of the wind speed from the originally measured height of 19.5 m to 10 m and integration of the P-M (Pierson-Moskowitz) spectral form yields the limits

$$\varepsilon = 3.64 \times 10^{-3} \tag{5.25}$$

$$\nu = 0.13 \tag{5.26}$$

Assuming the phase speed of linear waves in deep water is given by $C_p = g/\omega$ [see Table 2.3], (5.26) becomes $U_{10}/C_p = 0.82$. That is, at full development, the phase speed of the waves can actually exceed the speed of the generating wind. As indicated in Section (4.3.1) it is generally believed that direct atmospheric input cannot transfer energy to waves propagating faster than the wind. Nonlinear interactions [see Section (4.4.2)] can however transfer energy to such waves and this is believed to be the mechanism responsible for this limit to growth being such that $U_{10}/C_p < 1$.

5.3.2.3 JONSWAP

The Joint North Sea Wave Project (JONSWAP) (Hasselmann et al, 1973) was a major international attempt to investigate fetch limited wave evolution. The measurement array consisted of 13 stations spaced along a 160 km transect extending westward from the island of Sylt, northern Germany, into the North Sea. Data were collected for four weeks during July 1969. Although the full data set consisted of approximately 2,000 spectra, only 121 corresponded to "ideal" fetch limited conditions. Hasselmann et al (1973) augmented the JONSWAP data with data from a variety of other studies, both field and laboratory. These composite data sets are shown in Figures 5.4 and 5.5.

Based on these data, Hasselmann et al (1973) determined the fetch limited relationships

$$\varepsilon = 1.6 \times 10^{-7}\chi \tag{5.27}$$

$$\nu = 3.5\chi^{-0.33} \tag{5.28}$$

The JONSWAP relationships (5.27) and (5.28) do not attempt to accommodate any transition to full development as predicted by Pierson and Moskowitz (1964) in a manner similar to the corresponding SMB relationships (5.21) and (5.22). Hence, (5.27) and (5.28) are not applicable at large values of χ. The JONSWAP data were confined to $\chi < 1 \times 10^4$ which seems a reasonable upper limit to the applicability of (5.27) and (5.28). JONSWAP is widely recognized as a landmark study with data of significantly higher quality than previous studies.

5.3.2.4 Bothnian Sea

Kahma (1981) investigated the fetch limited growth of waves using four waverider buoys deployed over a 70 km transect in the Bothnian Sea between Finland and Sweden. They found the development of ν with χ to be comparable to JONSWAP but growth rates for ε were larger than JONSWAP by a factor of approximately 2. Their data yielded the growth curves

$$\varepsilon = 3.60 \times 10^{-7}\chi \tag{5.29}$$

$$\nu = 3.18\chi^{-0.33} \tag{5.30}$$

Kahma (1981) examined his data carefully and concluded the large growth rates measured were real. He speculated that the enhanced growth rate may be associated with atmospheric stability, the data being recorded in unstable atmospheric conditions. Kahma and Calkoen (1992) later confirmed the significant influence of atmospheric stability on energy growth rates, this influence being considered further in Section (5.3.3). Because of the strongly unstable conditions, (5.29) should be considered as atypical.

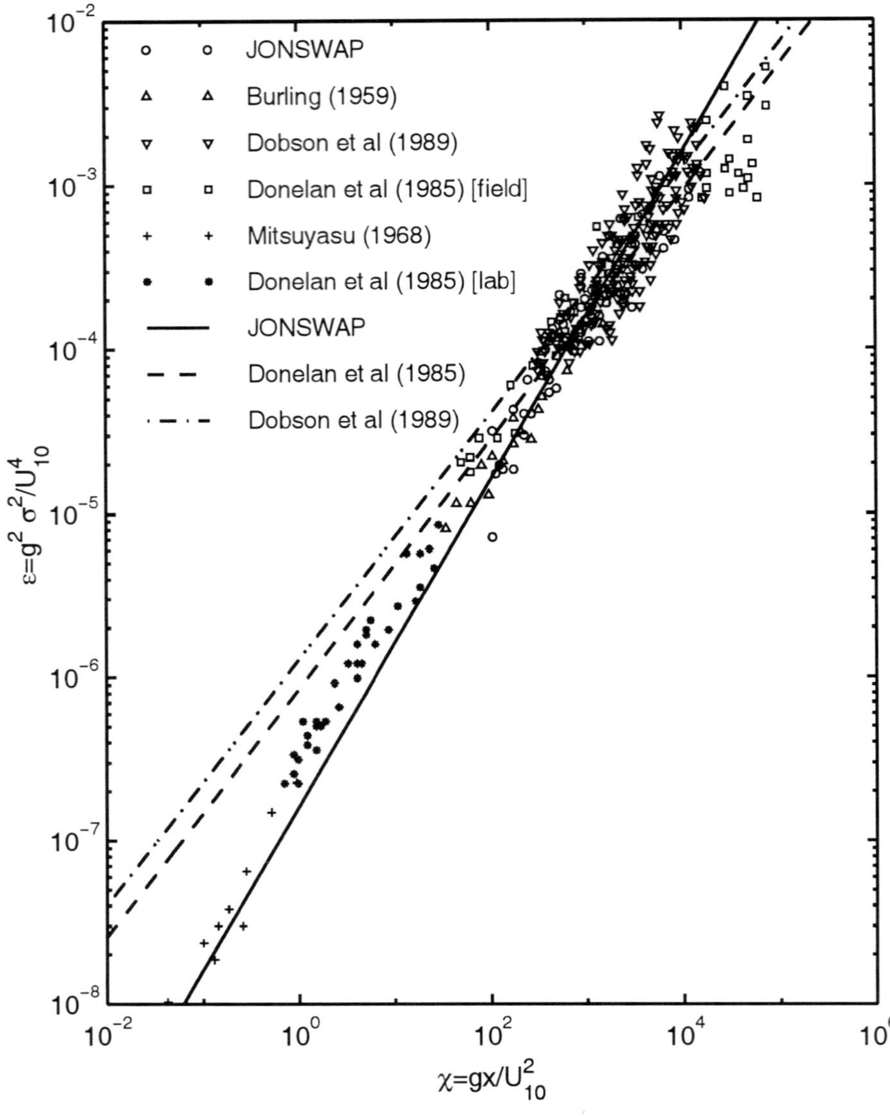

Figure 5.4: A composite of data from a variety of studies showing the development of the non-dimensional energy, ε as a function of non-dimensional fetch, χ. The original JONSWAP study (Hasselmann et al, 1973) used the data marked, JONSWAP, together with that of Burling (1959) and Mitsuyasu (1968). Also shown are a number of growth curves obtained from the various data sets. These include: JONSWAP Eq.(5.27), Donelan et al (1985) Eq.(5.33) and Dobson et al (1989) Eq.(5.38).

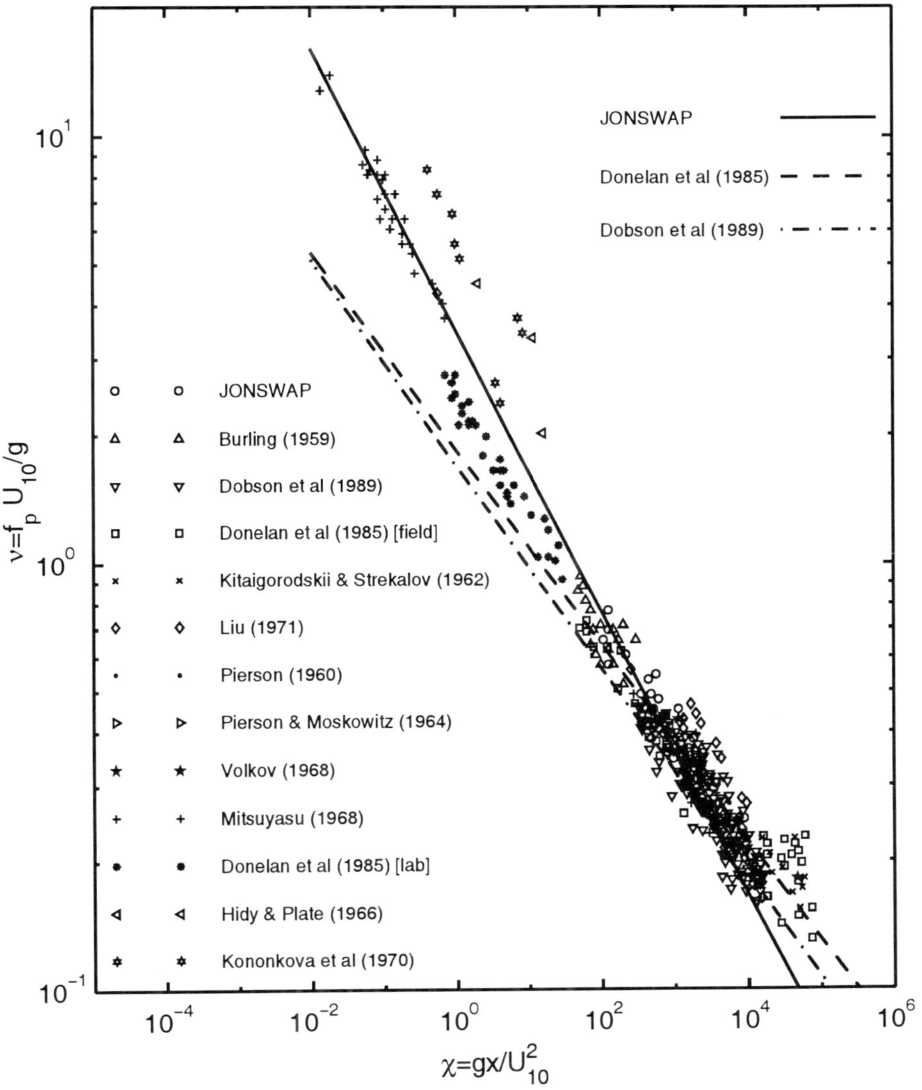

Figure 5.5: A composite of data from a variety of studies showing the development of the non-dimensional peak frequency, ν as a function of non-dimensional fetch, χ. The original JONSWAP study (Hasselmann et al, 1973) used all the data shown with the exception of that marked Donelan et al (1985) and Dobson et al (1989). Also shown are a number of growth curves obtained from the various data sets. These include: JONSWAP Eq.(5.28), Donelan et al (1985) Eq.(5.34) and Dobson et al (1989) Eq.(5.39).

5.3.2.5 Lake Ontario

Donelan et al (1985) examined fetch limited spectral development at one location in Lake Ontario, Canada. In contrast to previous measurements, they did not present their data directly as a function of non-dimensional fetch, χ. Rather, they adopted the inverse wave age, U_{10}/C_p as the independent variable and expressed both ε and χ as functions of this quantity. Their results are shown in Figure 5.6. In an attempt to include younger wave ages (high U_{10}/C_p) than measured in the field, data from a large wind-wave flume was also included. As can be seen from Figure 5.6, these laboratory data appear inconsistent with the field data and hence Donelan et al (1985) excluded them from further analysis. Based on the field data, they obtained the relationships

$$\varepsilon = 0.00274 \left(\frac{U_{10}}{C_p} \right)^{-3.3} \tag{5.31}$$

$$\frac{U_{10}}{C_p} = 11.6\chi^{-0.23} \tag{5.32}$$

Combining (5.31) and (5.32) ε and ν can be expressed as functions of χ

$$\varepsilon = 8.415 \times 10^{-7}\chi^{0.76} \tag{5.33}$$

$$\nu = 1.85\chi^{-0.23} \tag{5.34}$$

For comparison with other data sets, the data shown in Figure 5.6 have also been plotted on Figures 5.4 and 5.5.

The Donelan et al (1985) field data are generally consistent with the JONSWAP data. The quality of the data does, however, appear superior and the scatter within the data is reduced, compared to earlier studies.

5.3.2.6 North Atlantic Open Ocean

The well controlled field experiments of JONSWAP (Hasselmann et al, 1973), Kahma (1981) and Donelan et al (1985) were all in relatively enclosed waters free of swell. In order to extend the data set to conditions which included background swell, Dobson et al (1989) staged a fetch limited study off the North Atlantic coast of Canada. The measurement array consisted of 9 buoys deployed along a 29 km transect.

One of the primary assumptions adopted in all previous fetch limited studies was that the wind speed is constant over the fetch. Dobson et al (1989) recognized that, even in the most stable of meteorological situations, such conditions do not occur. At the shoreline there is a discontinuity in the surface roughness, from the aerodynamically "rough" land surface to the relatively smooth water surface. As a result, an internal marine boundary layer begins to grow within the thicker terrestrial boundary layer. Hence, the wind speed measured at a constant reference

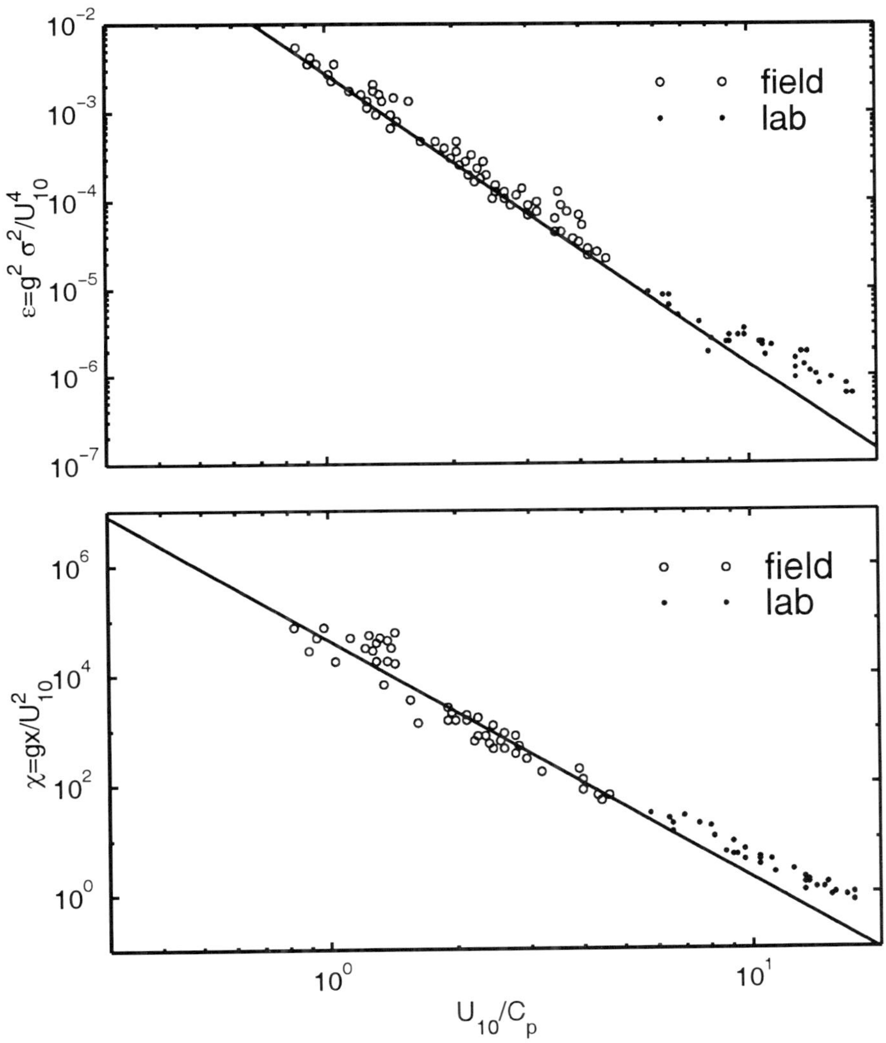

Figure 5.6: The data of Donelan et al (1985): ε vs U_{10}/C_p (top) and χ vs U_{10}/C_p (bottom). The line in the top panel is (5.31) and in the bottom (5.32) [after Donelan et al (1985)].

height will gradually increase moving down the fetch. The development of such internal boundary layers has been previously investigated by Taylor and Lee (1984) and Smith and MacPherson (1987). The data of Smith and MacPherson (1987), as presented by Dobson et al (1989) [Figure 5.7], clearly show the increase in wind speed with fetch.

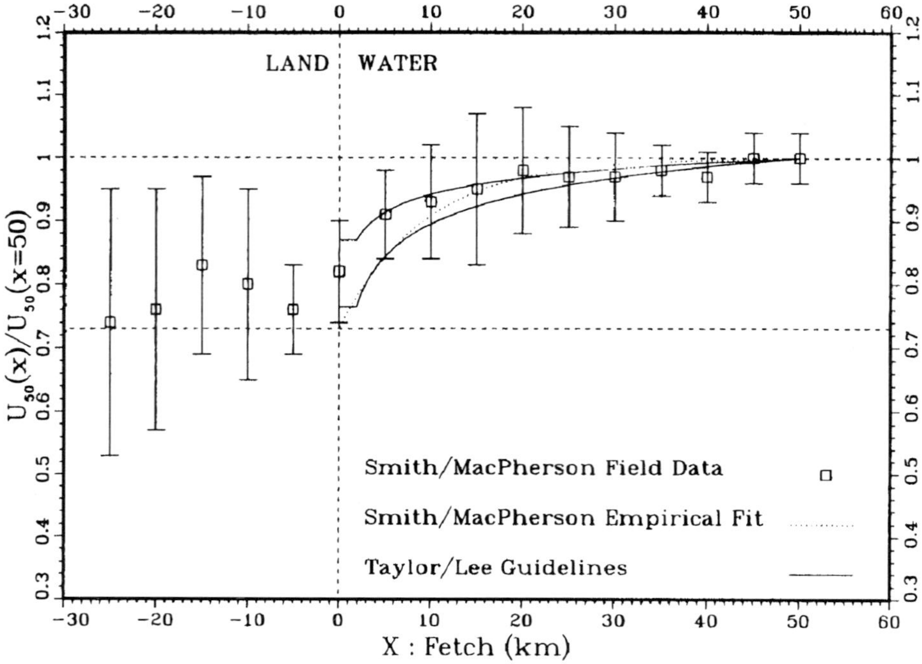

Figure 5.7: A comparison between measured aircraft winds at 50 m elevation, U_{50}, as measured by Smith and MacPherson (1987) with the predictions of Taylor and Lee (1984). The squares represent the ratio of the wind as measured at fetch, x, $U_{50}(x)$ to the wind at a fetch of 50 km, $U_{50}(x = 50)$. The error bars represent standard deviations on the measurements. An empirical fit by Smith and MacPherson (1987) is shown by the dotted line. The solid lines represent the predictions of Taylor and Lee (1984) with upwind roughness lengths of 0.07 m (upper line) and 1m (lower line). [after Dobson et al (1989)].

Taylor and Lee (1984) found that the variation of the wind speed in such cases can be modelled by the assumption that the thickness of the internal boundary layer, δ_i can be represented by

$$\delta_i = 0.75 z_0 \left(\frac{x}{z_0} \right)^{0.8} \tag{5.35}$$

where x is the fetch length and z_0 is the roughness length. With this assumption, it follows that the fractional change in wind speed, $\Delta R = [U(x) - U_u] / U_u$ (where the subscript $_u$ refers to quantities at a point "upwind" of x) is given by (Taylor and Lee, 1984)

$$\Delta R = \begin{cases} 0 & \text{for } z \geq \delta_i \\ \frac{\ln(z/z_0)\ln(\delta_i/z_{0u})}{\ln(z/z_{0u})\ln(\delta_i/z_0)} - 1 & \text{for } z < \delta_i \end{cases} \tag{5.36}$$

where z is the reference measurement height. In the present case, U_u is taken as the wind speed at the shoreline and z_{0u} as the roughness length for the upwind land surface. Equation (5.36) assumes that both the upwind boundary layer is logarithmic and that the developing internal boundary layer is logarithmic for $z < \delta_i$.

To apply this model requires estimates of both the upwind roughness, z_{0u} and the roughness at the fetch x, z_0. Dobson et al (1989) calculated z_0 from the Charnock relationship (5.13) with $c_h = 0.01$. They adopted an upstream roughness, $z_{0u} \approx 0.07$ m, as shown in Figure 5.7.

Dobson et al (1989) argued that the wave field at a given fetch is the space and time history of the growth conditions along the fetch. Therefore, the scaling wind velocity should be an average over the down wind fetch, x

$$\overline{U}_{10} = \frac{1}{x} \int_0^x U_{10}(x) \mathrm{d}x \tag{5.37}$$

Adopting a scaling wind speed as defined above, Dobson et al (1989) obtained the growth relationships

$$\varepsilon = 12.7 \times 10^{-7} \chi^{0.75} \tag{5.38}$$

$$\nu = 1.7 \chi^{-0.24} \tag{5.39}$$

5.3.2.7 Lake St. Clair Growth Rate Measurements

Donelan et al (1992) adopted an innovative approach to the problem of fetch limited growth by examining the differential growth between adjacent measurement stations along the fetch. Although a number of the studies cited in previous sections involved measurements at multiple stations along the fetch, no attempts were made to investigate the rate of growth between these stations. Donelan et al (1992) argued that this approach avoids the problems of inhomogeneity of the wind field over long fetches as highlighted by Dobson et al (1989). Also, difficulties associated with accurately defining the fetch when the upwind land boundary is not straight are avoided. The approach does, however, place significant demands on the accuracy of the wave measurements. The change in the spectrum between adjacent stations can be small and hence measurement errors can become significant.

Despite these difficulties, Donelan et al (1992) were able to investigate the fractional energy increase per radian, $\Gamma = C_g/(\omega_p \sigma^2) \frac{\Delta \sigma^2}{\Delta x}$ as a function of the inverse wave age U_{10}/C_p where C_g and C_p are the group velocity and phase speed respectively of components at the spectral peak frequency $\omega_p = 2\pi f_p$. They found that their data could be adequately represented by the relationship

$$\Gamma = \frac{C_g}{\omega_p \sigma^2} \frac{\Delta \sigma^2}{\Delta x} = A \left(\frac{U_{10}}{C_p} - 0.83 \right) \tag{5.40}$$

where A is a constant between 6.5×10^{-5} and 7.2×10^{-5}. This relationship indicates that the growth rate becomes zero at the point of full development ($U_{10}/C_p = 0.83$) determined by Pierson and Moskowitz (1964).

In order to integrate (5.40) and determine the fetch dependent growth $\sigma^2(x)$, a relationship between σ^2 and U_{10}/C_p is required. Based on data recorded at 5 stations along a 25 km transect in Lake St. Clair, Canada, Donelan et al (1992) adopted the relationship

$$\varepsilon = 0.0023 \left(\frac{U_{10}}{C_p} \right)^{-3.2} \tag{5.41}$$

which is in close agreement with the Lake Ontario result of Donelan et al (1985), (5.31). By integrating (5.40) and (5.41) for the case of a constant wind speed, U_{10} along the fetch, Donelan et al (1992) obtained the analytical relationships

$$\chi = 4.09 \times 10^4 \ln \left[\frac{1}{1 - 5.54\varepsilon^{0.313}} \right] - 2.27 \times 10^5 \left(1 + 2.77\varepsilon^{0.313} \right) \varepsilon^{0.313} \tag{5.42}$$

$$\chi = 4.09 \times 10^4 \ln \left[\frac{\nu}{\nu - 0.132} \right] - 5.411 \times 10^3 \nu^{-2} \left(\nu + 0.066 \right) \tag{5.43}$$

Donelan et al (1992) originally expressed (5.43) in terms of U_{10}/C_p, the present form has been obtained by applying the deep water relationship $U_{10}/C_p = 2\pi\nu$, [Table 2.3]. Equations (5.42) and (5.43) are compared with other proposed growth law relationships in Figures 5.8 and 5.9, respectively. At small values of χ, (5.42) and (5.43) approximate power laws and at large values of χ they asymptote to the Pierson-Moskowitz limit (Pierson and Moskowitz, 1964). This upper limit to growth is to be expected as it was an assumption in the Donelan et al (1992) analysis. It should be pointed out that (5.42) and (5.43) are not well defined for $\chi < 10^1$.

The transcendental nature of (5.42) and (5.43) makes solving for ε or ν for given χ a cumbersome process. The real potential of the differential approach to the problem comes in its ability to handle more realistic situations. The differential equation (5.40) can be numerically integrated for cases where the wind speed, U_{10} is not constant along the fetch. Donelan et al (1992) have shown examples of this for fluctuating wind speeds and for boundary layer development along the fetch [see Dobson et al (1989)]. The influences of these processes on fetch limited growth will be considered in more detail in Section (5.3.3).

5.3.2.8 Comparison of Fetch Limited Relationships

The studies cited in the preceeding sections present a myriad of growth law relationships for the apparently well defined case of fetch limited growth. In addition, the relationships presented are not exhaustive, with other proposals, including Wilson (1965) and Ebuchi et al (1992) having been ignored. The above relationships are compared in Figures 5.8 and 5.9.

A further comparison appears in Table 5.1 which shows the asymptotic limits of the various proposals at both short and long fetch.

Study	Eq. Nos.	Asymptotic Limits	
		Small χ	Large χ
SMB	ε - (5.21)	$\varepsilon = 7.82 \times 10^{-7}\chi^{0.84}$	$\varepsilon = 5.00 \times 10^{-3}$
CERC (1977)	ν - (5.22)	$\nu = 1.722\chi^{-0.25}$	$\nu = 0.133$
PM Limit	ε - (5.25)	N/A	$\varepsilon = 3.64 \times 10^{-3}$
Pierson & Moskowitz (1964)	ν - (5.26)	N/A	$\nu = 0.13$
JONSWAP	ε - (5.27)	$\varepsilon = 1.6 \times 10^{-7}\chi$	N/A
Hasselmann et al (1973)	ν - (5.28)	$\nu = 3.5\chi^{-0.33}$	N/A
Bothnian Sea	ε - (5.29)	$\varepsilon = 3.60 \times 10^{-7}\chi$	N/A
Kahma (1981)	ν - (5.30)	$\nu = 3.18\chi^{-0.33}$	N/A
Lake Ontario	ε - (5.33)	$\varepsilon = 8.415 \times 10^{-7}\chi^{0.76}$	N/A
Donelan et al (1985)	ν - (5.34)	$\nu = 1.85\chi^{-0.23}$	N/A
North Atlantic	ε - (5.38)	$\varepsilon = 12.7 \times 10^{-7}\chi^{0.75}$	N/A
Dobson et al (1989)	ν - (5.39)	$\nu = 1.7\chi^{-0.24}$	N/A
Lake St. Clair	ε - (5.42)	$\varepsilon \approx 1.7 \times 10^{-7}\chi$	$\varepsilon = 4.07 \times 10^{-3}$
Donelan et al (1992)	ν - (5.43)	$\nu \approx 3.6\chi^{-0.33}$	$\nu = 0.132$

Table 5.1: A summary of various proposed growth law relationships for fetch limited growth. References are given to the equation numbers for each relationship, together with the asymptotic forms of the relationships at both short (small χ) and long (large χ) fetch.

Although the comparisons in Figures 5.8 and 5.9 present a qualitatively consistent picture, the logarithmic axes mask significant quantitative differences. The JONSWAP relationship for ε (5.27) has a more rapid increase as a function of χ than most other proposals (power of 1 compared to other values typically near 0.8).

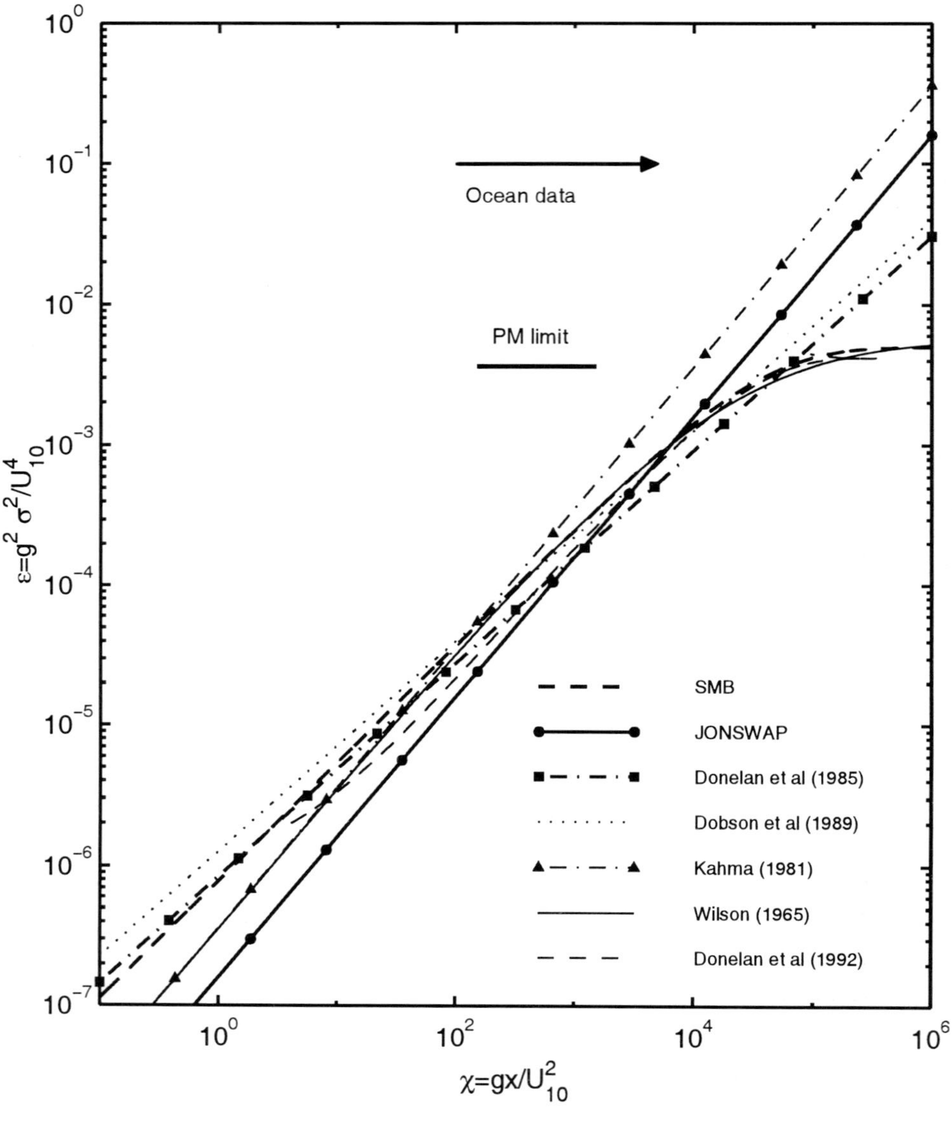

Figure 5.8: A comparison of proposed relationships between the non-dimensional energy, ε and non-dimensional fetch, χ. The arrow marked 'Ocean data', indicates that typical field data exists at values of χ larger than approximately 100.

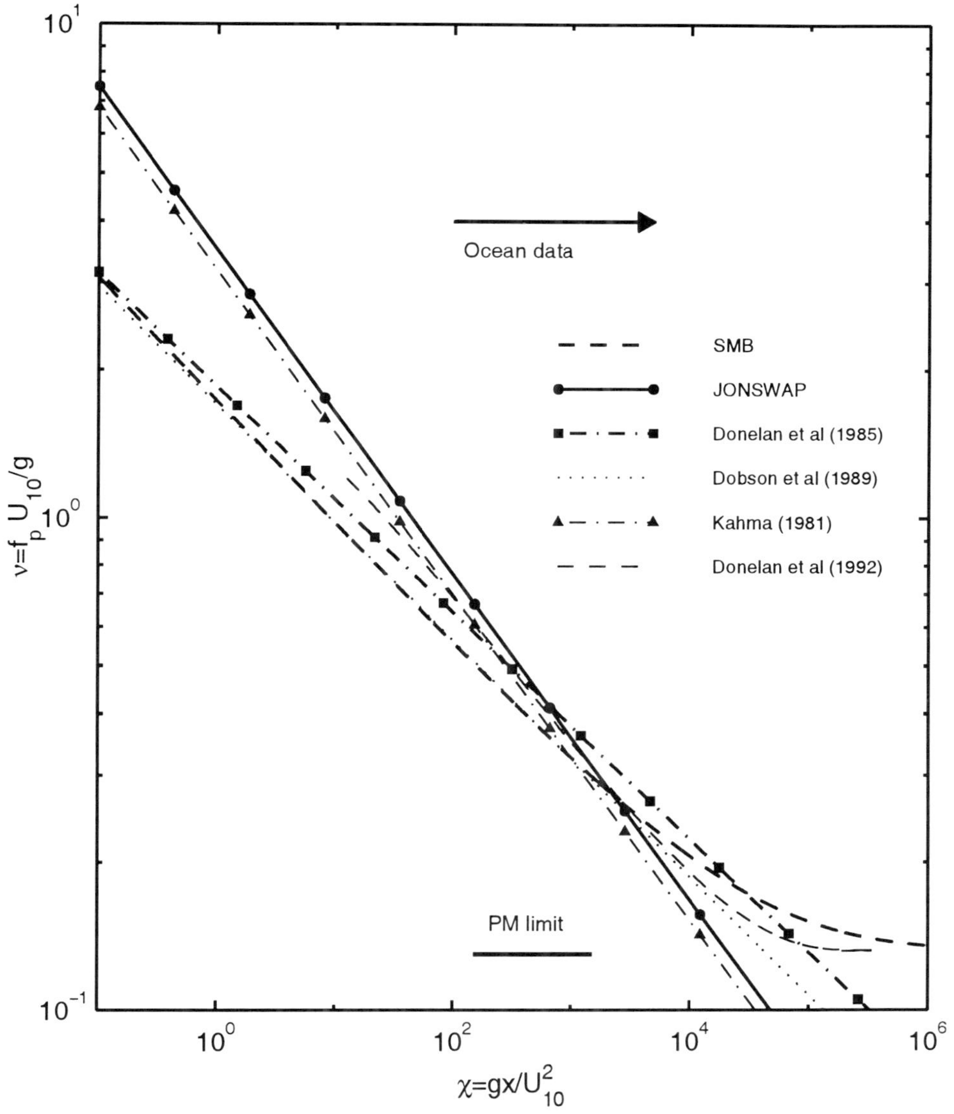

Figure 5.9: A comparison of proposed relationships between the non-dimensional frequency, ν and non-dimensional fetch, χ. The arrow marked 'Ocean data', indicates that typical field data exists at values of χ larger than approximately 100.

As a result, at small values of χ the JONSWAP result is significantly smaller than other proposals. In the development of the JONSWAP relationship, Hasselmann et al (1973) augmented their field data with laboratory data. As shown in Figures 5.8 and 5.9, field data is generally confined to $\chi > 100$. As shown in Figure 5.4, laboratory data appears inconsistent with field data. This feature was also found by Donelan et al (1985) who ignored their own laboratory data in their analysis. The use of these laboratory data by Hasselmann et al (1973) appears to have biased the resulting JONSWAP relationship (5.27), an exponent closer to 0.8 appearing more consistent with the various field data sets shown in Figure 5.8. As mentioned earlier, the data of Kahma (1981) should also be treated with caution as they were collected in unstable atmospheric conditions and show significantly higher growth rates than other data sets.

The bias introduced by the inclusion of laboratory data is even clearer in Figure 5.9 which shows the development of ν. Again JONSWAP is inconsistent with most other data sets with a power law exponent of –0.33 compared with typical values of –0.25.

It is interesting to note that, despite the fact that Dobson et al (1989) used the wind speed averaged over the down wind fetch rather that the local wind speed, their results are consistent with the other studies.

If the JONSWAP and Kahma (1981) results are excluded for the reasons outlined above, and comparisons are confined to $\chi > 100$ (field data), a generally consistent picture develops. A reasonable representation of the composite data set, including a fully developed limit, can be given by

$$\varepsilon = \max \begin{cases} (7.5 \pm 2.0) \times 10^{-7} \chi^{0.8} \\ (3.6 \pm 0.9) \times 10^{-3} \end{cases} \tag{5.44}$$

$$\nu = \min \begin{cases} (2.0 \pm 0.3) \chi^{-0.25} \\ (0.13 \pm 0.02) \end{cases} \tag{5.45}$$

Equations (5.44) and (5.45) are shown in Figure 5.10

5.3.3 Secondary Influences

The fetch limited growth analysis considered in Section (5.3.2) assumes that the development of both ε and ν depend on only one quantity, χ. The scatter inherent within individual data sets and differences between data sets suggest that other parameters may also play a role, all be it secondary. To date, a number of such processes have been identified including: atmospheric stability, boundary layer development along the fetch, gustiness of the wind, the geometry of the down wind fetch and the presence of swell.

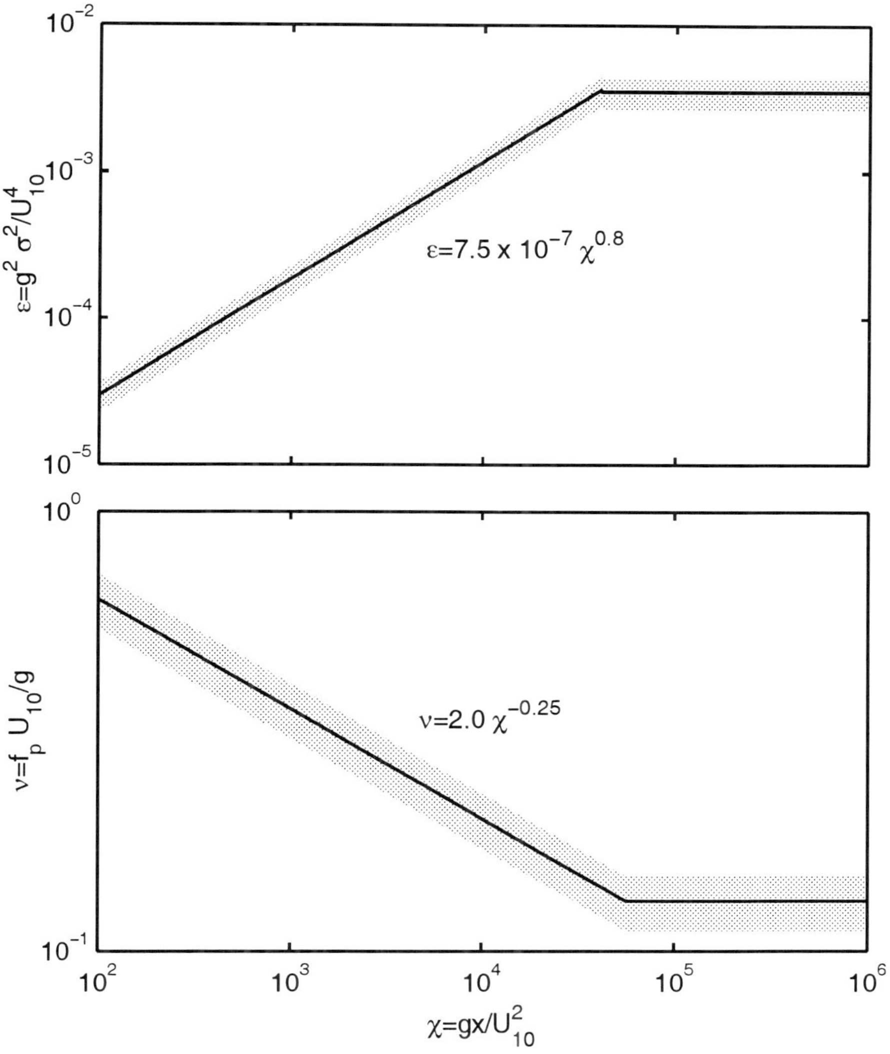

Figure 5.10: Average growth laws as given by (5.44) (top) and (5.45) (bottom). The shaded region shows the typical variation between the various experimental data sets and provides a representation of the accuracy of the relationships.

5.3.3.1 Atmospheric Stability

As indicated in Section (5.3.1), the difference in temperature between the air and the water influences the atmospheric stability and hence the structure of the atmospheric boundary layer. That atmospheric stability has an influence on wind wave development is clear from a comparison of the Bothnian Sea data of Kahma (1981) with other data sets. Waves appear to grow more rapidly in unstable conditions (water warmer than air). In a careful study of a number of data sets Kahma and Calkoen (1992) confirmed that enhanced growth in unstable conditions was a feature of a number of data sets. Their combined data set was, however, insufficient for them to accurately parameterize the effect. They concluded that: *"Unfortunately, the existing data are too scanty for this purpose. The scatter is far too large to reveal anything more than that the growth is faster in unstable than stable stratification."* Despite the data scatter, they speculated that the effect of stability was concentrated to a small region near neutral stability. Based on this conclusion, one would not expect the growth rate to continue to increase as the air-water temperature difference increased.

More recently, Young (1998a) has considered the effect of stability on the fetch limited development of ε using data collected at Lake George, Australia. The data were presented in terms of the observed deviation of energy from that which could be expected in neutral conditions. This deviation, ξ, was defined as

$$\xi = \frac{\varepsilon - \varepsilon_m}{\varepsilon_m} \tag{5.46}$$

where ε is the non-dimensional energy calculated for each of the data and ε_m is the corresponding value of non-dimensional energy predicted by a relationship such as (5.44) which was obtained for neutral conditions. Hence, (5.46) is a measure of the deviation of the measured non-dimensional energy from the mean (neutral) result.

Figure 5.11 shows ξ as a function of the Richardson number, R_b (5.17). Positive values of ξ indicate growth rates larger than the mean and negative values of ξ indicate growth rates smaller than the mean. Although there is significant scatter in the data a clear trend showing enhanced growth for unstable conditions and reduced growth for stable conditions is apparent. In contrast to the findings of Kahma and Calkoen (1992) this effect is not confined to a small region near neutral stability. The deviation from the mean result continues to increase in magnitude as the magnitude of R_b increases.

Although there is significant scatter in the data, a linear regression yields (Young, 1998a)

$$\xi = -1.22R_b + 0.01 \tag{5.47}$$

Also shown on Figure 5.11 are 95% confidence limited on the linear regression (5.47). Although there is significant scatter within the data, these confidence limits indicate that (5.47) is a reasonable representation of the observable trend. The results indicate that significant deviations from the mean result can occur due to the effects of atmospheric stability. As an example, for a relatively large value of $|R_b| = 0.5$, the energy deviates from the mean by approximately 50%.

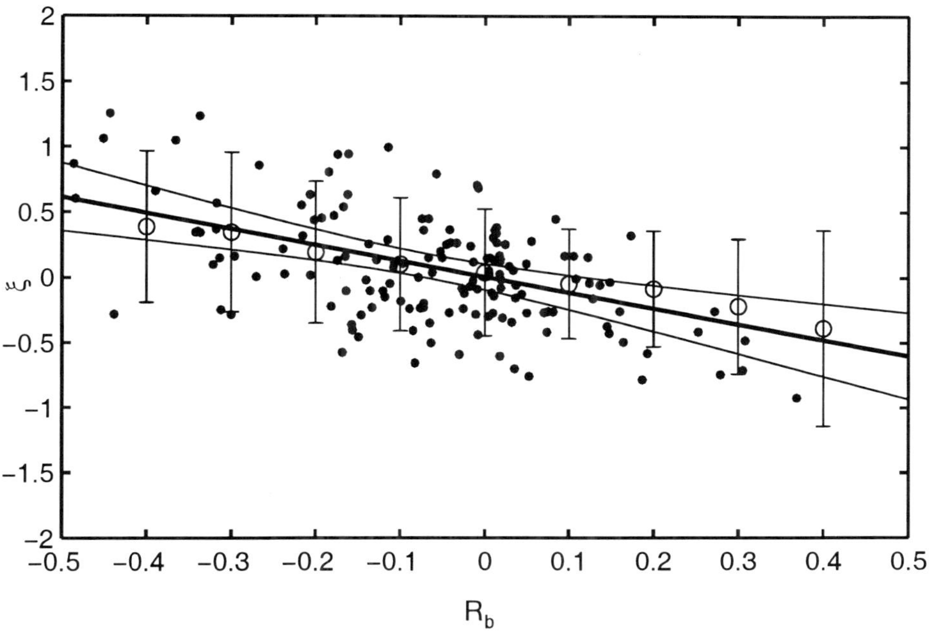

Figure 5.11: Correction to the mean energy, ξ as a function of Bulk Richardson Number, R_b. The solid thick line represents the linear regression to the data of (5.47). Also shown are the 95% confidence limits for (5.47). The open circles represent mean values of ξ over a bin of extent ± 0.2 in R_b space. The error bars represent plus and minus one standard deviation for the data in each bin. [after Young (1998a)]

Hence, it appears that atmospheric stability can have a significant effect on the growth rate of waves. The influence of stability on ν is less well defined. The data of Kahma (1981) and the analysis of Kahma and Calkoen (1992) indicates that ν is less sensitive to the effects of stability than ε.

5.3.3.2 Boundary Layer Development

Dobson et al (1989) realized that under typical fetch limited conditions the wind speed, U_{10} is not constant along the fetch. Rather, U_{10} gradually increases along the fetch as an internal boundary layer develops following the transition form the aerodynamically rough land surface to the smoother water [see Section (5.3.2.6)]. They attempted to account for this effect by using the wind speed averaged over the down wind fetch in all non-dimensional variables.

The differential growth rate relationship of Donelan et al (1992) (5.40) can be integrated assuming a wind speed which increases with fetch. Donelan et al (1992) performed such calculations assuming the boundary layer developed as proposed by Taylor and Lee (1984). Their results are shown in Figure 5.12. This figure clearly shows that the inclusion of the variable wind speed yields growth rates significantly smaller than for a constant wind speed. Further, as the development of the boundary layer with fetch is a function of the magnitude of the wind speed, no universal growth profile exists. A similar result was found by Young (1997b) in the context of finite depth wave growth [see Chapter 7]. Young (1997b) found short fetch growth rates significantly lower than would be expected. Inclusion of a realistic represention of the boundary layer development, however, explained the lower growth rates.

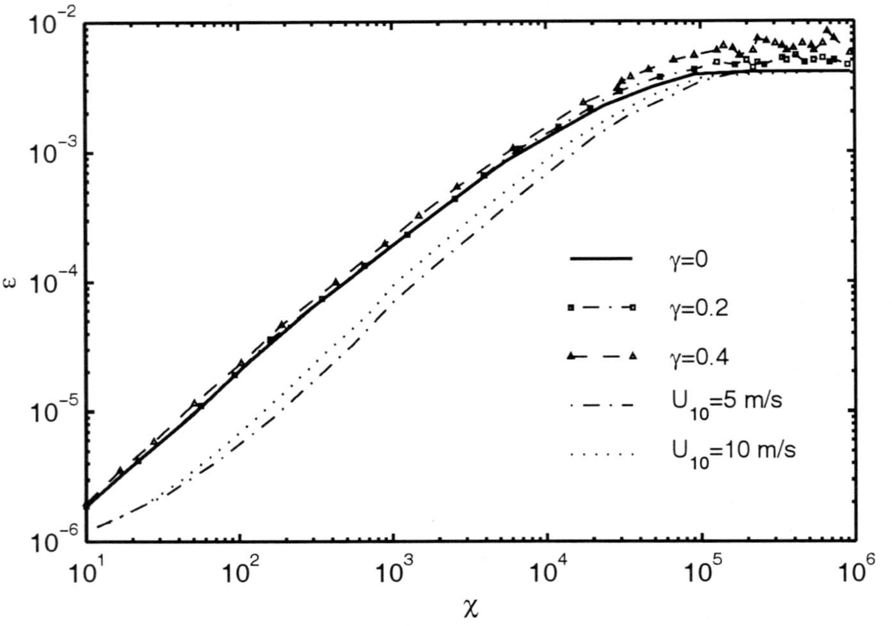

Figure 5.12: Growth rates assuming (a)wind speed which follows a Gaussian distribution with mean U_{10} and standard deviation γU_{10}. The values of γ are marked. (b) wind speed which varies with fetch according to Taylor and Lee (1984). In these cases the up-wind values of U_{10} are shown. Results obtained from the numerical integration of (5.40). [after Donelan et al (1992)]

5.3.3.3 Gustiness of the Wind

The wind speed measured at a fixed point is not constant with time but fluctuates about the mean value. Cavaleri et al (1981) and Smith et al (1990) have shown that U_{10} has a Gaussian distribution. The magnitude of the fluctuations (*gustiness*) can be represented as $\gamma = \sigma_U/U_{10}$, where σ_U is the standard deviation of the oscillations. Values of γ up to 0.3 have been reported (Cavaleri, 1994). The theory of Miles (1957) [see Section (4.3.1.3)] indicates that wind can pump energy to a slower wave, whilst a faster wave remains unaffected. Hence, high wind speed fluctuations transfer extra energy to the waves, which is not compensated during phases of lower speed.

Numerical calculations by Cavaleri and Burgers (1992) show enhanced growth rates with increasing γ. They show that with $\gamma = 0.3$ growth rates can increase by 25% compared to cases with $\gamma = 0$. In contrast, Donelan et al (1992) find that numerical integration of (5.40) is largely insensitive to the value of γ [see Figure 5.12]. The only significant effect of the gustiness occurs at large fetches, where an increase in γ increases the level of the fully developed state.

Measurements to reconcile these differences are lacking and the magnitude of the effects of gustiness is yet to be clearly defined.

5.3.3.4 Fetch Geometry

Under ideal fetch limited conditions, the upwind land boundary is straight, infinitely long and perpendicular to the wind. In practice, such ideal conditions are seldom met. Donelan et al (1985) showed that when the wind is not perpendicular to the land boundary, there is a strong gradient in the fetch about the wind direction. In such cases the reduction in the effective forcing wind, $U_{10}\cos\theta$, is more than compensated for by the extended fetch (θ is the angle the wind makes to the normal at the coastline). In such cases, the wave direction deviates from the wind direction and waves preferentially propagate from a direction with a longer fetch than that measured directly along the wind direction. Donelan et al (1985) showed cases where this deviation, θ, was as large as 50°. They argued that, in such slanting fetch cases, the appropriate scaling values to use are the fetch and wind speed both measured in the wave direction.

Kahma and Pettersson (1996) investigated fetch limited growth in the narrow bays represented by the Gulf of Finland and the Bothnian Sea. Both these bodies of water are approximately 3 times as long as they are wide. Kahma and Pettersson (1996) found that, for the same values of χ, ε was significantly lower when the wind direction was along the long axis of the bay, than when it was directed along the short axis. The difference was as large as a factor of 3. Apparently, the narrow geometry constrained the development of the waves. The non-dimensional frequency, ν was also affected, but not to the same extent as ε. In contrast, Young and Verhagen (1996a) have shown that in shallow Lake George, Australia (aspect ration 2:1), this same influence is small. Apparently, the magnitude of the influence is sensitive the actual aspect ratio of the water body.

5.3.3.5 The Influence of Swell

A number of laboratory experiments have shown that the presence of swell reduces the growth of shorter wind generated waves (Mitsuyasu, 1966; Phillips and Banner, 1974; Hatori et al, 1981; Bliven et al, 1986; Kusaba and Mitsuyasu, 1986; Donelan, 1987). The mechanism by which this might occur is not clear. Young et al (1985) have shown that the non-linear coupling [Section (4.4.2)] between the swell and wind-sea should be weak and hence little interaction would be expected.

Evidence of swell influencing wind wave development in the field is lacking. Kahma and Calkoen (1992) examined data from both JONSWAP (Hasselmann et al, 1973) and the Bothnian Sea (Kahma, 1981) and could find no difference in growth between cases with and without the presence of swell. Similarly, the data of Dobson et al (1989) were collected in the presence of North Atlantic swell and yet are still consistent with other data sets obtained in the absence of swell.

5.3.4 Duration Limited Observations

As stated in Section (5.1), observations of wind wave development under duration limited conditions are far rarer than observations of fetch limited development. Wiegel (1961) compiled data from a number of sources to produce the results shown in Figure 5.13.

An approximation to the data of Figure 5.13 was provided by CERC (1977)

$$\varsigma = K \exp\left\{\left[A\left(\ln\chi\right)^2 - B\ln\chi + C\right]^{1/2} + D\ln\chi\right\} \tag{5.48}$$

where $K = 6.5882$, $A = 0.0161$, $B = 0.3692$, $C = 2.2024$ and $D = 0.8798$. Values of ε and ν can then be calculated from (5.21) and (5.22) using values of χ determined from (5.48) for given values of ς.

Sanders (1976) observed that during a gale in the Norwegian Sea, wave development appeared independent of fetch. Based on this rather limited data set, he proposed a relationship of the form

$$\varepsilon = 3.22 \times 10^{-3}\left[\tanh\left(1.26 \times 10^{-3}\varsigma^{0.75}\right)\right]^2 \tag{5.49}$$

The original relationship proposed by Sanders (1976) was scaled in terms of the shear velocity, u_* rather than U_{10} as used here. Equation (5.49) was obtained by assuming that $U_{10} \approx 21u_*$, which ensures that (5.49) asymptotes close to the Pierson-Moskowitz value (5.25), at large values of ς.

A third representation for duration limited growth has been proposed by CERC (1984)

$$\varsigma = 68.8\chi^{0.67} \tag{5.50}$$

Equation (5.50) can be used in conjunction with the JONSWAP relationships (5.27) and (5.28) to obtain ε as a function of ς and ν as a function of ς, respectively.

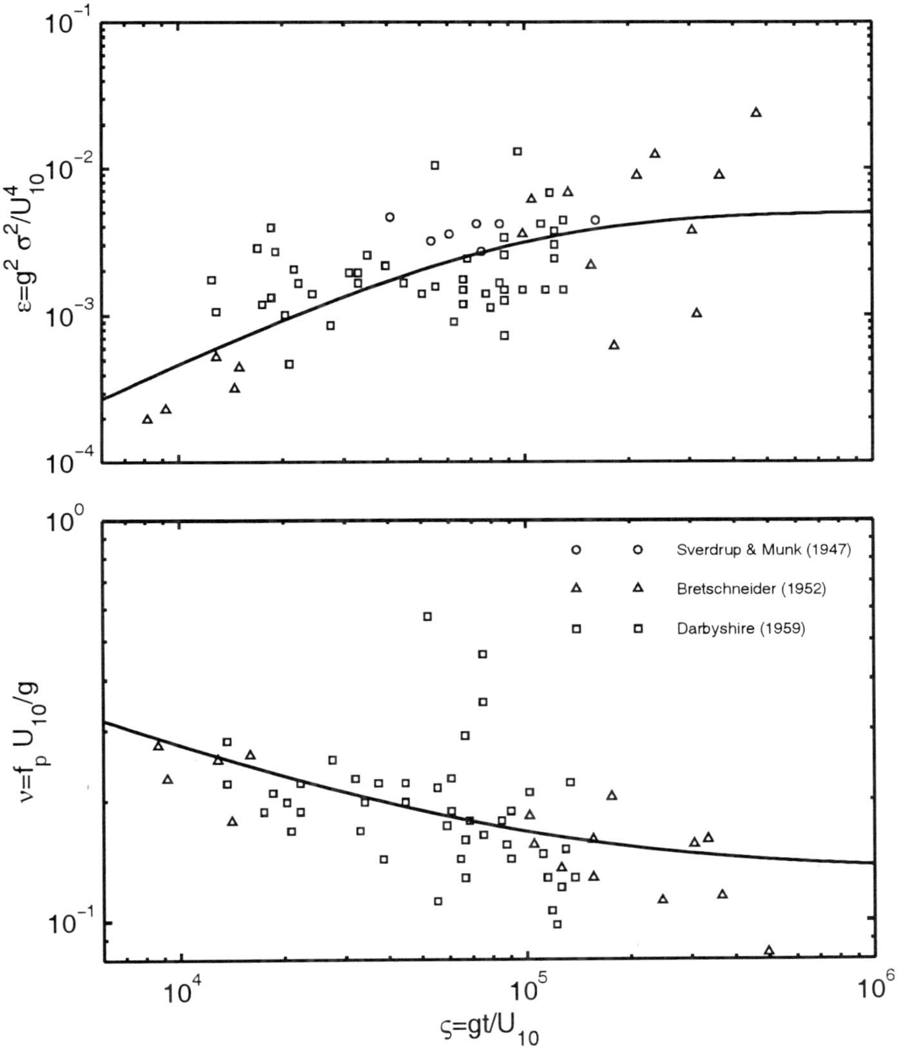

Figure 5.13: Data compiled by Wiegel (1961) showing duration limited growth: non-dimensional energy, ε vs non-dimensional duration, ς (top) and non-dimensional peak frequency, ν verses non-dimensional duration, ς (bottom). The solid lines represent the results of (5.48) together with (5.21) (top) and (5.22) (bottom).

The three proposed representations for duration limited growth described above are compared in Figure 5.14.

The results shown in Figure 5.14 show considerable divergence between the proposed duration limited relationships, reflecting the paucity of available data. The data used to construct the CERC (1977) relationship (5.48), although old, and exhibiting significant scatter [see Figure 5.13], appears to be the most comprehensive.

The wave field can either be limited by the fetch or the duration. At long fetches, the probability that the waves will be duration limited increases. Equation (5.48) provides a means of determining whether the fetch or the duration will be the limiting factor. Figure 5.15 shows the relationship between χ and ς as predicted by (5.48). Cases where the combination of χ and ς lie above the line shown in Figure 5.15 represent duration limited growth whereas cases below the line are fetch limited. Consider a case where: $U_{10} = 15$ m/s and $x = 50$ km, Figure 5.15 indicates that the duration must exceed approximately 3.5 hours for the waves to be fetch limited. Hence, even in coastal regions, duration limited conditions are not uncommon.

5.4 One-Dimensional Spectrum

As indicated in Section (5.2.2), two basic forms have been proposed for the high frequency portion ($f >> f_p$) of the one-dimensional spectrum, $F(f)$: $\propto f^{-5}$ (Phillips, 1958) and $\propto f^{-4}$ (Toba, 1973). Numerous field measurements of wind seas have confirmed a high frequency spectral region, $\propto f^{-n}$, although debate still surrounds the value of the exponent. These measurements have also confirmed that wind wave spectra have a single, well defined, peak and a low frequency face which rises rapidly to this peak. As shown in Section (4.4.2.1), this well defined spectral shape is largely a result of non-linear energy transfer across the spectrum. The non-linear source term continually forces the spectrum to this well defined shape. This process has been called *shape stabilization.* In order to determine the exact form of spectral shape, and how it develops with fetch or duration requires resort to field measurements.

5.4.1 The JONSWAP Form, $\propto f^{-5}$

Based on the high frequency formulation of Phillips (1958), the JONSWAP experiment (Hasselmann et al, 1973) found that the frequency spectrum of fetch-limited waves could be represented by the form

$$F(f) = \underbrace{\alpha g^2 (2\pi)^{-4} f^{-5} \exp\left[-\frac{5}{4}\left(\frac{f}{f_p}\right)^{-4}\right]}_{\text{Pierson-Moskowitz spectrum}} \cdot \gamma^{exp\left[\frac{-(f-f_p)^2}{2\sigma^2 f_p{}^2}\right]} \tag{5.51}$$

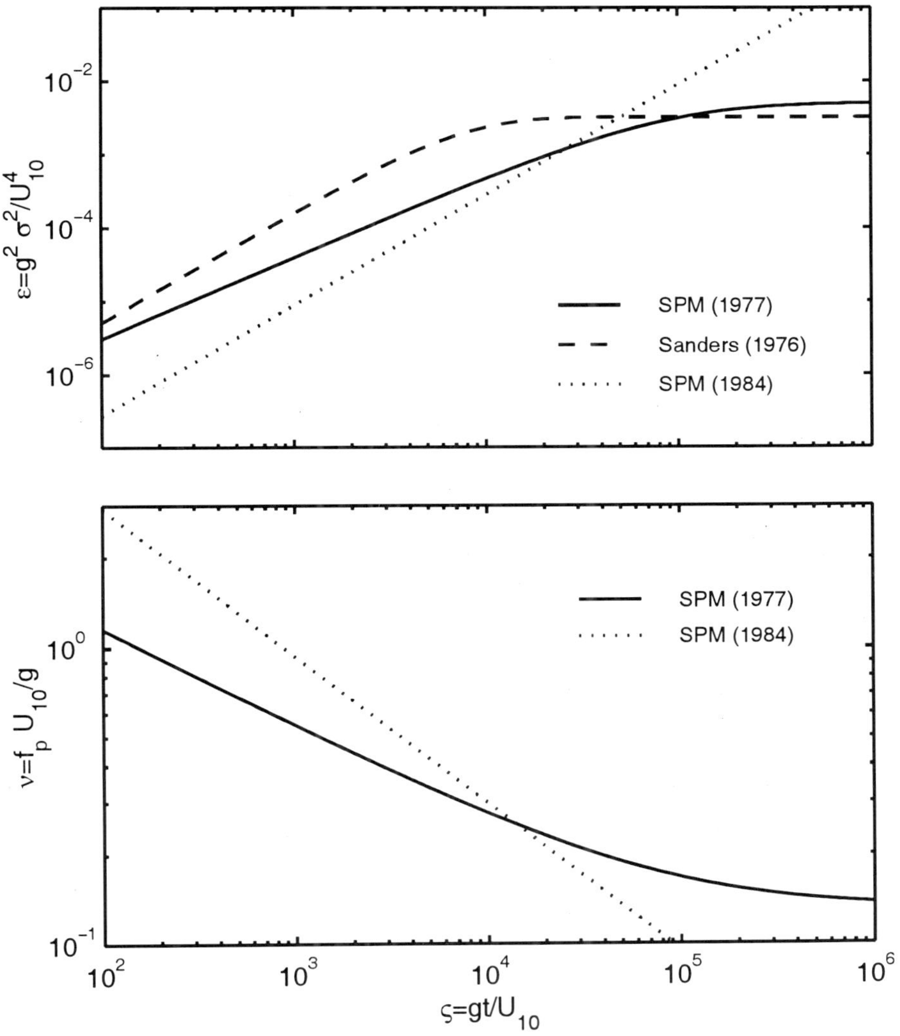

Figure 5.14: A comparison between proposed representations for duration limited growth: non-dimensional energy, ε vs non-dimensional duration, ς (top) and non-dimensional peak frequency, ν verses non-dimensional duration, ς (bottom). The results of CERC (1977), (5.48) together with (5.21) (top) and (5.22) (bottom); Sanders (1976) (5.49); CERC (1984), (5.50) together with (5.27) (top) and (5.28) (bottom).

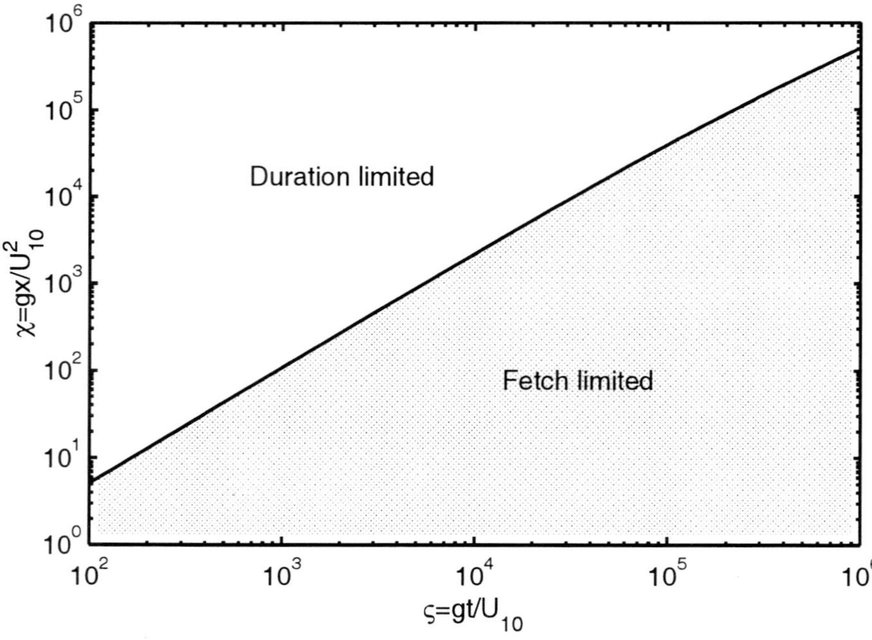

Figure 5.15: Relationship between non-dimensional fetch, χ and equivalent non-dimensional duration, ς. Cases which lie above the line represent duration limited situations, those in the shaded region below the line, fetch limited situations. The curve was evaluated from (5.48).

where

$$\sigma = \begin{cases} \sigma_a & \text{for } f \leq f_p \\ \sigma_b & \text{for } f > f_p \end{cases} \tag{5.52}$$

and $F(f)$ is the energy spectrum and f is frequency. Equation (5.51) contains five parameters which define the spectral shape. The parameters f_p and α are scale parameters; f_p represents the frequency at the maximum of the spectrum and α corresponds to the Phillips proportionality "constant" (Phillips, 1958). The remaining three parameters define the shape of the spectrum: γ, the peak enhancement factor, is the ratio of the maximum spectral energy to the maximum of the corresponding Pierson-Moskowitz (Pierson and Moskowitz, 1964) spectrum with the same value of α and σ_a and σ_b define the left and right side widths respectively of the spectral peak region. The first term of (5.51) is the called the Pierson-Moskowitz spectrum with the second term modifying the shape to make it more peaked. This enhancement is only significant for $f \approx f_p$. At large f/f_p the spectrum reverts to an f^{-5} decay as proposed by Phillips (1958).

Specification of the five spectral parameters in (5.51) completely defines the spectrum. Hasselmann et al (1973) attempted to define these parameters in terms of the non-dimensional variables χ, ν and ε. The JONSWAP field data set was combined with other available measurements, as indicated in Section (5.3.2.3) and it was found that ε, α and χ could be related to ν by power law relationships of the form $a_k = r_k \nu^{s_k}$ ($a_k = \varepsilon, \alpha, \chi$). The values of r_k and s_k are shown in Table 5.2, [see also (5.27), (5.28) and Figures 5.4, 5.5]. The JONSWAP data used to develop these relationships are shown in: Figure 5.16 [α vs ν], Figure 5.17 [γ vs ν] and Figure 5.18 [σ_a, σ_b vs ν].

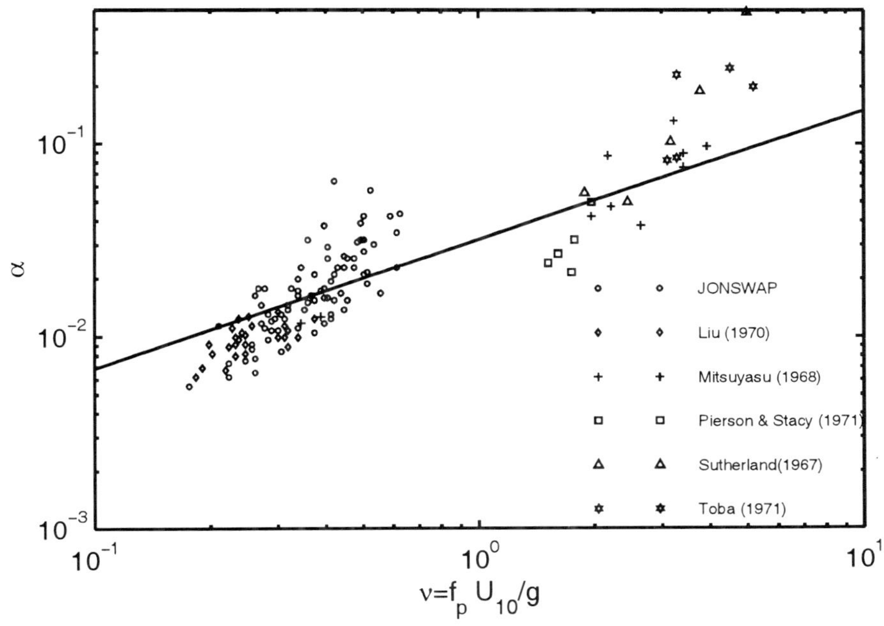

Figure 5.16: JONSWAP data (Hasselmann et al, 1973) showing the relationship between α and non-dimensional frequency, ν. Also shown is the power law relationship of Lewis and Allos (1990) [see Table 5.2].

The data scatter for γ, σ_a and σ_b is large and showed no consistent trends. Hence, Hasselmann et al (1973) represented them by their mean values $\gamma = 3.3$, $\sigma_a = 0.07$ and $\sigma_b = 0.09$. These parameters are extremely difficult to determine from recorded spectral data which are defined on a relatively course spectral grid (Young, 1995). Hence, the lack of correlation shown in Figure 5.17 and Figure 5.18 does not prove that the shape parameters are not related to ν, only that none can be determined from these data.

Hasselmann et al (1976) demonstrated that although the JONSWAP relationships represented a considerable advance in understanding, there were a number of inconsistencies in the power law relationships. They firstly augmented the data sets used in the Hasselmann et al (1973) analysis and obtained modified power law relationships [see Table 5.2]. There was still considerable scatter in the results for γ, σ_a and σ_b. Although these relationships represented the best fit results to

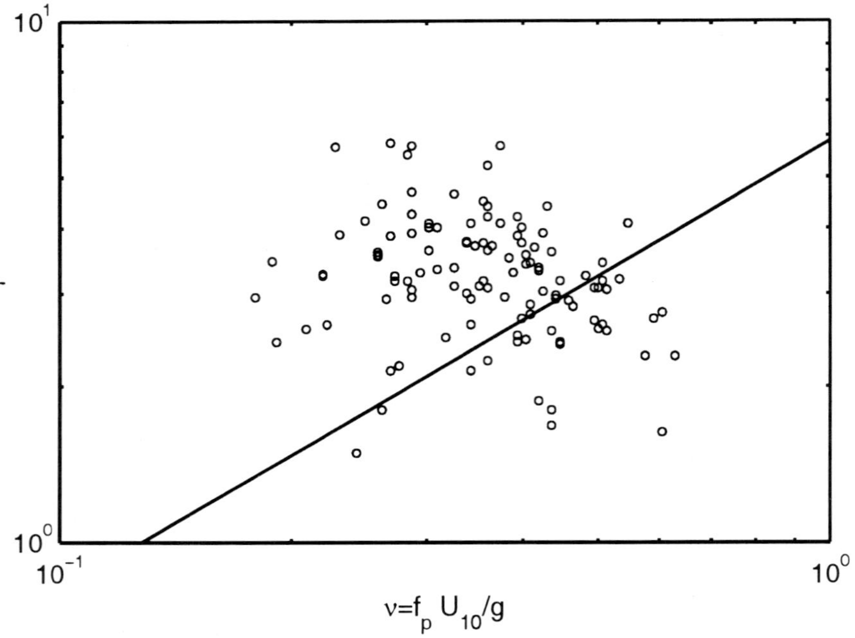

Figure 5.17: JONSWAP data (Hasselmann et al, 1973) showing relationship between γ and non-dimensional frequency, ν. Also shown is the power law relationship of Lewis and Allos (1990) [see Table 5.2].

the available data, Hasselmann et al (1976) further modified them by the use of a number of theoretical arguments to develop slightly modified relationships for χ, α and χ [see Table 5.2].

There still remained a number of apparent inconsistencies in the results. As a result of JONSWAP it had become clear that the development of the spectrum with fetch was the result of the sum of processes representing atmospheric input from the wind, nonlinear interactions amongst spectral components and dissipation due to white-capping [see Section (4.6)]. As a result of these processes the spectral peak frequency moves to lower frequencies and the energy increases with increasing fetch. In addition, the values of both α and γ decrease with fetch. This results in a gradual broadening of the spectrum as it matures. In principle, the JONSWAP spectrum should asymptotically approach the Pierson-Moskowitz form at large fetch. Thus, as $\nu \rightarrow 0.13$, the Pierson-Moskowitz (PM) value, $\alpha \rightarrow 0.0081$ and $\gamma \rightarrow 1.0$. Within the scatter of the data, α appeared to reproduce such a trend. Although there was considerable scatter in the data, γ appeared to have a mean value at $\nu = 0.13$ of approximately 2.0. Consequently, Hasselmann et al (1976) reanalyzed the original data of Pierson and Moskowitz (1964). They found that a little more than half the spectra in this data set were double peaked. When these spectra, which do not conform to purely wind-sea conditions, were excluded a mean value of $\gamma = 1.4$ was

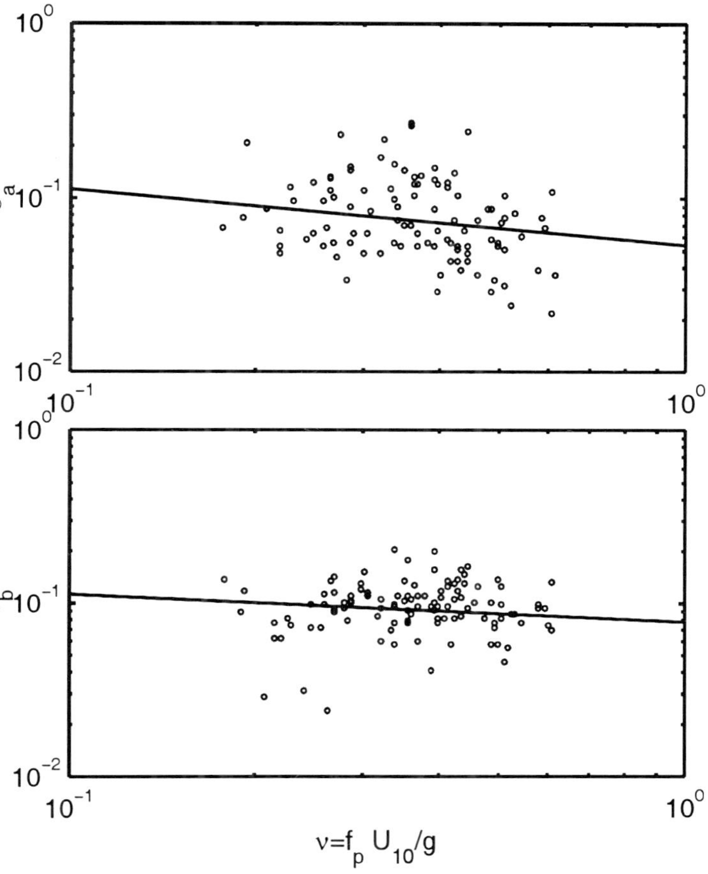

Figure 5.18: JONSWAP data (Hasselmann et al, 1973) showing relationships between σ_a and non-dimensional frequency, ν (top) and σ_b and ν (bottom). Also shown are the power law relationships of Lewis and Allos (1990) [see Table 5.2].

obtained. Based on this result, Hasselmann et al (1976) speculated that the final transition to a value of $\gamma = 1.0$ may occur very rapidly as the spectrum reached full development.

Mitsuyasu et al (1980) considered an independent data set to that of Hasselmann et al (1973, 1976). They assumed that σ_a and σ_b were constant at their mean JONSWAP values of 0.07 and 0.09 respectively and developed power law relationships for α, ε and ξ [see Table 5.2]. Rather than choose γ as a constant, they numerically integrated (5.51) to determine the value of γ which produced σ^2 and hence ε consistent with their observations. The value of γ obtained in this manner agreed well with their observations which had far less scatter than the previous studies. Again, their results showed a value of γ of approximately 1.7 at the Pierson-Moskowitz limit. Although the Mitsuyasu et al (1980) result appears quite compelling, it is interesting to note that almost half their data are for situations where $\nu < 0.13$. Such data would not normally be considered as representing locally generated wind-seas.

An obvious inconsistency in these studies had developed. There was general agreement as to the values of α and ε at the Pierson-Moskowitz limit. However, with constant values of σ_a and σ_b and the observed value of γ near 1.7, integration of (5.51) would not produce consistent values of ε. Lewis and Allos (1990) concluded that this inconsistency resulted from forcing σ_a and σ_b to maintain constant values. They adopted the power law relationships for σ_a and σ_b obtained by Hasselmann et al (1976) from their combined data set. With these values they integrated (5.51) in a similar manner to Mitsuyasu et al (1980) and determined γ required to obtain values of ε consistent with observations. The parameter γ was then represented in terms of ν as a power law which passed through 1.0 at the Pierson-Moskowitz limit. Their results for all the spectral parameters are shown in Table 5.2.

$a_k \rightarrow$	α		γ		σ_a		σ_b	
Author ↓	r_k	s_k	r_k	s_k	r_k	s_k	r_k	s_k
JONSWAP	0.033	0.67						
Hasselmann et al (1976)	0.033	0.67						
Hasselmann et al (1976)(cd)	0.036	0.87	4.12	0.32	0.54	–0.32	0.078	–0.16
Mitsuyasu et al (1980)	0.033	0.86	4.42	0.43				
Lewis & Allos (1990)	0.032	0.67	5.87	0.86	0.054	–0.32	0.078	–0.16

$a_k \rightarrow$	ε		χ	
Author ↓	$r_k \times 10^6$	s_k	r_k	s_k
JONSWAP	7.13	–3.03	44.4	–3.03
Hasselmann et al (1976)	5.30	–3.33	32.5	–3.33
Hasselmann et al (1976)(cd)	7.40	–3.05		
Mitsuyasu et al (1980)	6.84	–3.00	25.8	–3.03
Lewis & Allos (1990)	7.58	–3.03	47.4	–3.03

Table 5.2: Proposed power law relationships for the parameters of the JONSWAP spectrum, (5.51). The equations are all of the form $a_k = r_k \nu^{s_k}$. The rows labeled Hasselmann et al (1976)(cd) are for the curve fit results of Hasselmann et al (1976) with the combined data set they considered.

The proposed power law relationships for the spectral parameters as described in Table 5.2 are compared in Figure 5.19. There is general agreement for the dependence of α and ε on ν but the various proposals for γ vary considerably. In particular, the Lewis and Allos (1990) result diverges markedly from the other proposals near the Pierson-Moskowitz limit. Although Lewis and Allos (1990) argue that this is a result of allowing σ_a and σ_b to vary, their result explicitly assumes that the Pierson-Moskowitz form is correct. As already mentioned, the reanalysis of this data set by Hasselmann et al (1976) casts doubt on the Pierson-Moskowitz form.

For many engineering applications, the detailed behaviour of γ is not important. For such cases, the results of both Mitsuyasu et al (1980) and Lewis and Allos (1990) produce self-consistent relationships. A full understanding of the behaviour of σ and γ awaits more refined data.

A typical example of the development of the one-dimensional spectrum, $F(f)$ as represented by (5.51) with the parameters defined by Lewis and Allos (1990) is shown in Figure 5.20. The case shown is for fetch limited growth with a constant wind speed of $U_{10} = 15$ m/s. At short fetch the peak enhancement, γ is large and this can be clearly seen in the figure. As the fetch increases, γ decreases. In addition, the value of α also decreases with fetch, which can be seen by the decreasing energy levels in the high frequency tail of the spectrum. The longest fetch shown in Figure 5.20 is close to full development where the JONSWAP form reverts to that of Pierson and Moskowitz (1964). In comparison to the "younger" (shorter fetch) cases, this asymptotic form has a significantly broader shape.

5.4.2 The Toba Form, $\propto f^{-4}$

As indicated in Section (5.2.2), Toba (1973) proposed an alternative high frequency form to that of Phillips (1958), upon which the JONSWAP spectrum is based. His proposed form was proportional to f^{-4} rather than f^{-5}. Subsequent field results supporting this form have been presented by Kawai et al (1977), Mitsuyasu et al (1980), Kahma (1981), Forristall (1981) and Donelan et al (1985). This form has also been supported on theoretical grounds by Kitaigorodskii (1983) and Phillips (1985). In a reanalysis of the JONSWAP data Battjes et al (1987) found that the data were better represented by this form than the original f^{-5} proposal.

Based on this high frequency relationship, Donelan et al (1985) proposed a modification to the JONSWAP spectral form

$$F(f) = \beta g^2 (2\pi)^{-4} f_p^{-1} f^{-4} \exp\left[-\left(\frac{f}{f_p}\right)^{-4}\right] \cdot \gamma_d^{\exp\left[\frac{-(f-f_p)^2}{2\sigma^2 f_p{}^2}\right]} \tag{5.53}$$

Note that the notation, β and γ_d have been used to distinguish these parameters from their JONSWAP counterparts, α and γ. At first appearances, the differences between (5.51) and (5.53) may appear insignificant for engineering applications as it is only the high frequency portion of the spectrum which is altered. It is, however, the level of this region of the spectrum which largely influences the value of α [or β for (5.53)]. At the Pierson-Moskowitz limit, for example, it is possible

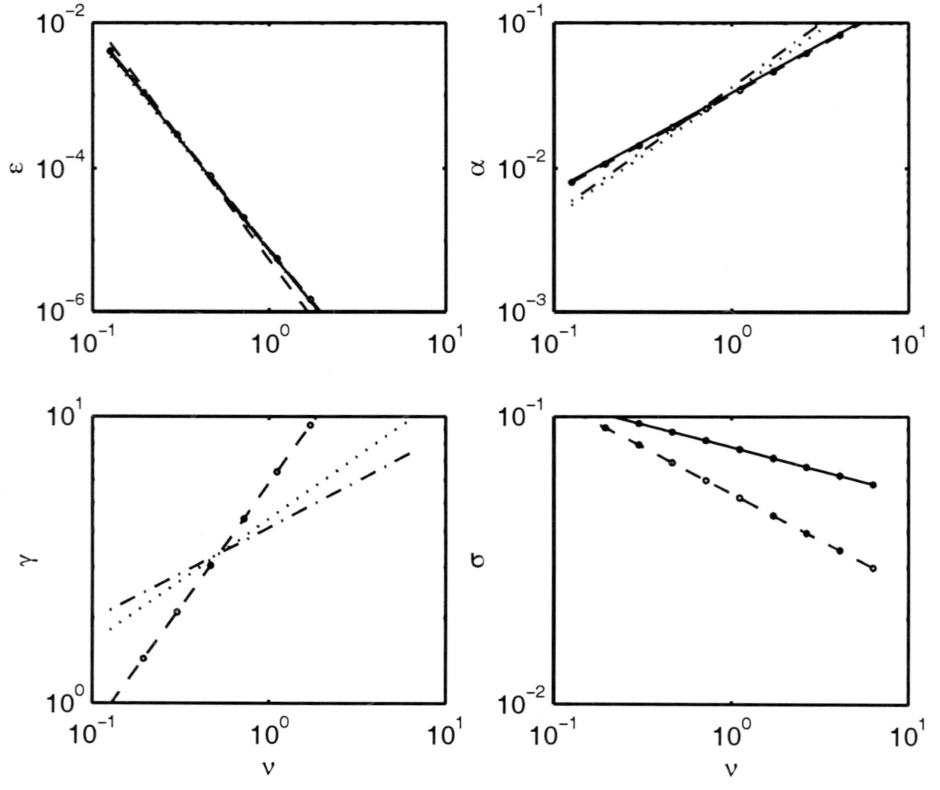

Figure 5.19: Power law relationships for the JONSWAP spectral parameters, ε, α, γ and σ as functions of ν [see Table 5.2]. JONSWAP, (Hasselmann et al, 1973) (solid lines); Hasselmann et al (1976) (dashed lines); Hasselmann et al (1976) (combined data) (dash-dot lines); Mitsuyasu et al (1980) (dotted lines); Lewis and Allos (1990) (dashed with circles lines). For the parameter σ, σ_a (dashed with circles line) and σ_b (solid with circles line). The σ relationships are those proposed by Hasselmann et al (1976) (combined data).

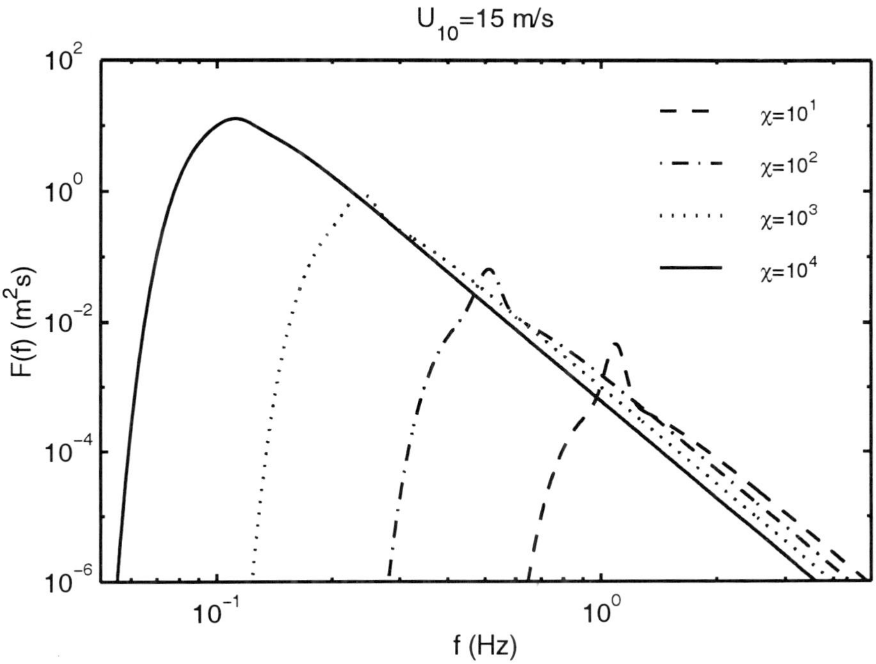

Figure 5.20: An example of the development of the JONSWAP spectrum (5.51) with fetch for a wind speed, $U_{10} = 15$ m/s. The spectral parameters are as defined by Lewis and Allos (1990) [see Table 5.2]. Spectra are shown at four non-dimensional fetches, χ. The largest value of χ is close to full development.

to obtain values of ε equal to the commonly observed value with $\gamma \approx 1.7$, provided $\alpha < 0.0081$. Hence, the apparent inconsistencies at the PM limit may simply be a result of α values which are artificially too high due to forcing the curve fit to follow an f^{-5} high frequency tail.

The data set of Donelan et al (1985) spans a wide range of ν and shows consistent trends in all spectral parameters with little scatter. As examples, the Donelan et al (1985) data for β and γ are shown in Figure 5.21 and Figure 5.22, respectively.

Donelan et al (1985) represented the parameters as

$$\varepsilon = 6.365 \times 10^{-6} \nu^{-3.3} \tag{5.54}$$

$$\beta = 0.0165 \nu^{0.55} \tag{5.55}$$

$$\gamma_d = \begin{cases} 6.489 + 6 \log \nu & ; \nu \geq 0.159 \\ 1.7 & ; \nu < 0.159 \end{cases} \tag{5.56}$$

$$\sigma = 0.08 + 1.29 \times 10^{-3} \nu^{-3} \tag{5.57}$$

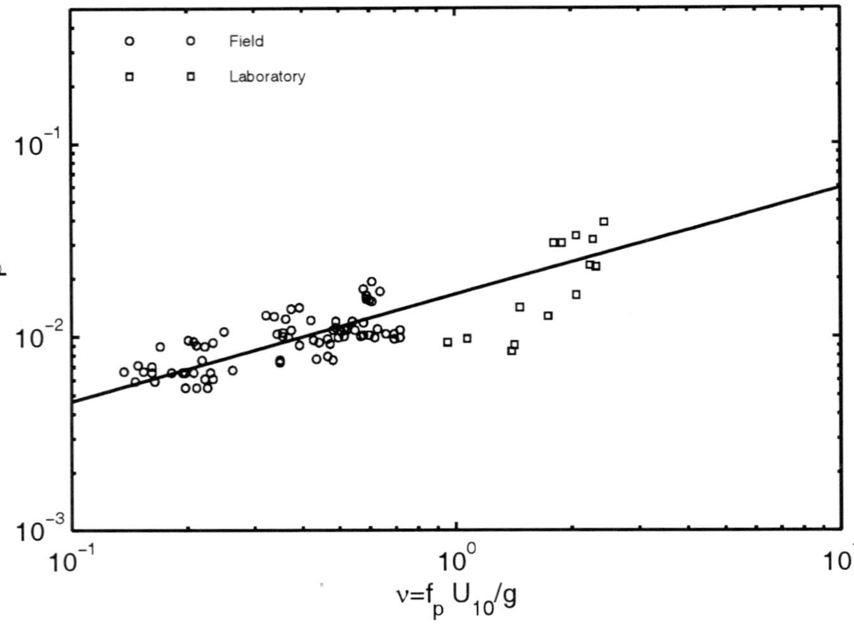

Figure 5.21: The data of Donelan et al (1985) showing the dependence of β on ν. The solid line is (5.55).

Equations (5.54) - (5.57) are shown in Figure 5.23. The differences between this formulation and those based on (5.51) are now clear. As a result of forcing the high frequency components to conform to an f^{-5} form, the resulting α is significantly greater than β. The Donelan et al. (1985) values for σ are similar to JONSWAP for larger values of ν. As the spectrum approaches the Pierson-Moskowitz limit, however, σ increases very rapidly. This results in a rapid broadening in the spectral shape as full development is reached. As a result of the relatively low values of β, the resulting values of γ_d are generally higher than those obtained with (5.51). In particular, a value of 1.7 occurs at the Pierson-Moskowitz limit.

The consequences of selecting an f^{-5} formulation over an f^{-4} becomes apparent in Figure 5.24. This figure shows a comparison between a JONSWAP spectrum, (5.51) and a Donelan et al (1985) form, (5.53). The two spectra have very similar shapes but quite different spectral parameters. For the JONSWAP form, $f_p = 0.1$ Hz, $\alpha = 0.016$, $\gamma = 2.0$ and $\sigma = \sigma_a = \sigma_b = 0.08$ whereas the Donelan et al (1985) form has, $f_p = 0.1$ Hz, $\beta = 0.008$, $\gamma_d = 3.1$ and $\sigma = 0.12$. These parameters were selected from the appropriate relationships proposed by Lewis and Allos (1990) and Donelan et al (1985) for a value of $\nu = 0.3$. The total energy of the two spectra differ by less than 1%. It is quite possible that much of the scatter in the values of γ and σ reported from studies in which JONSWAP spectra were fitted to the data is a result of an inappropriate choice of the spectral shape.

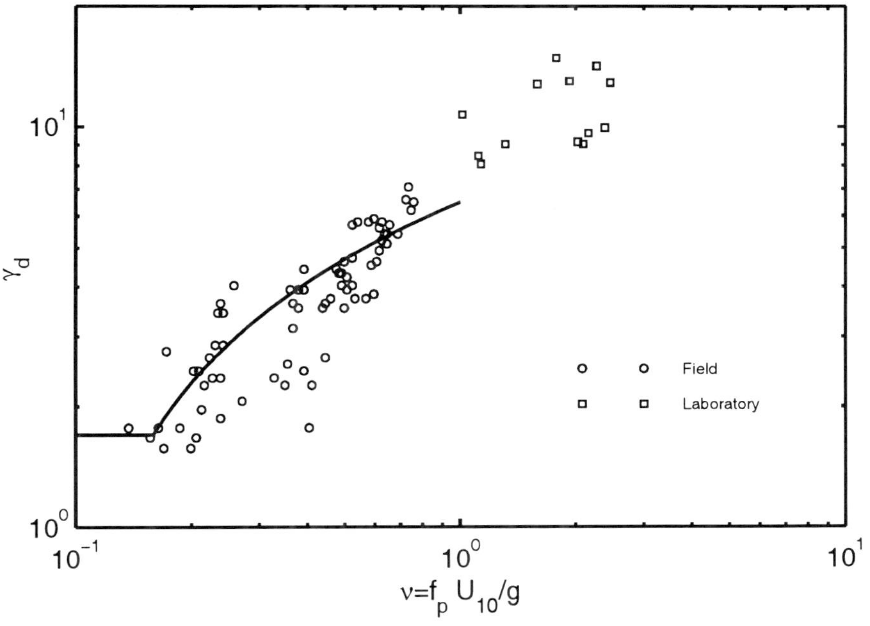

Figure 5.22: The data of Donelan et al (1985) showing the dependence of γ on ν. The solid line is (5.56).

The Donelan et al (1985) results reveal consistent trends in the development of the spectrum. As the spectrum evolves, both β and γ_d progressively decrease. The parameter σ remains relatively constant until near full development, where it increases rapidly. Because of the relatively low values of β and the rapid broadening of the spectrum represented by the increase in σ, γ_d does not reach a value of 1.0 at the Pierson-Moskowitz limit, but plateaus at 1.7.

5.4.3 The Value of the Exponent in f^{-n}

The sections above have presented observational evidence to support a high frequency region of the spectrum ($f >> f_p$) of the form f^{-n}. The actual value of n, however, appears to be in dispute, with values of both 4 and 5 having been proposed. As shown in Section (4.4.2), the detailed structure of the high frequency region of the spectrum is a result of the balance between the processes of atmospheric input, S_{in}, non-linear interaction, S_{nl} and white-cap dissipation, S_{ds}. In particular, S_{nl} seems to have a controlling influence, forcing the spectrum to conform to the f^{-n} form. The dimensional analysis arguments developed in Section (5.2.2) to support both $n = 4$ and $n = 5$, although interesting, have no relevance to the physics which controls the formation of this region of the spectrum. This fact was recognized by Phillips (1985) who revised his earlier proposal of $n = 5$.

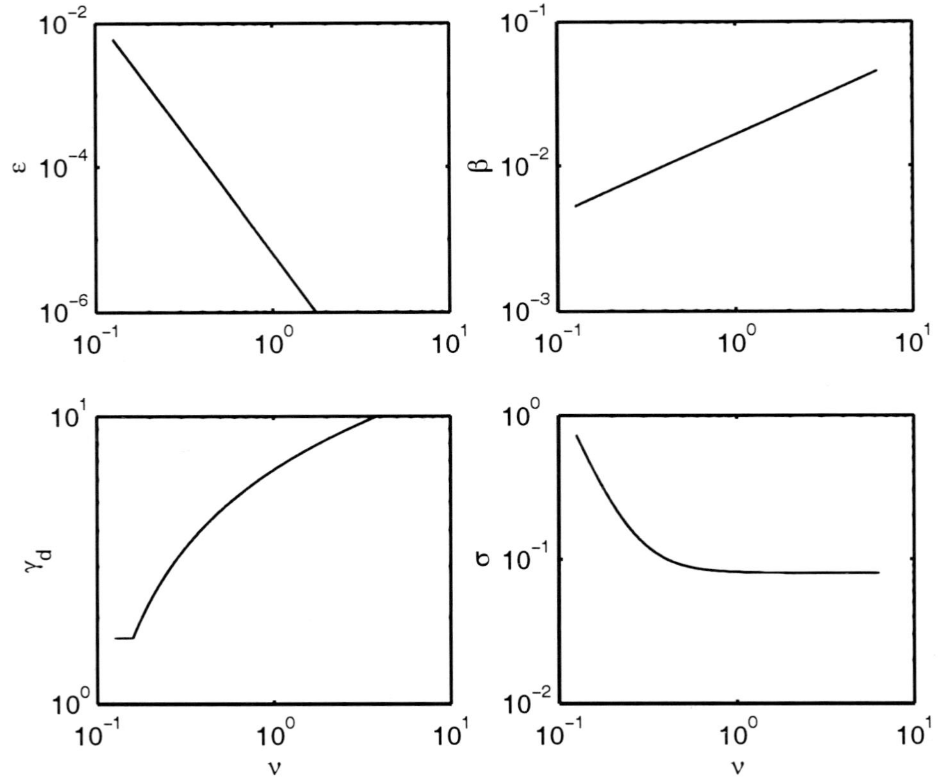

Figure 5.23: The dependence of the parameters for the Donelan et al (1985) spectral representation, (5.53). The relationships show are defined by (5.54) - (5.57).

The observational evidence is even more confusing than the representation of Sections (5.4.1) and (5.4.2) would suggest. Liu (1989) compared more than 2200 spectra obtained from data buoys in the Great Lakes. As shown in Figure 5.25 he found a wide range of values of n, although most values clustered between 3.5 and 5.5. He concluded that: "*... a representation of f^{-4} is perhaps a useful approximation for practical applications. What the correct slope is for the equilibrium range, or even whether or not a unique slope emerges, remains elusive and has yet to be satisfactorily substantiated.*"

Using stereo-photographic data, Banner (1990) has speculated that the equilibrium level really exists in the wind direction "slice" of the wave number spectrum. The one-dimensional frequency spectrum, $F(f)$ is the integral of the wave number spectrum. Therefore, the value of n will depend on the detailed directional structure of the spectrum. In addition, high frequency components of $F(f)$ are Doppler shifted by the longer wavelength waves near the peak of the spectrum. The magnitude of these components is determined by the peak enhancement factor, γ. The

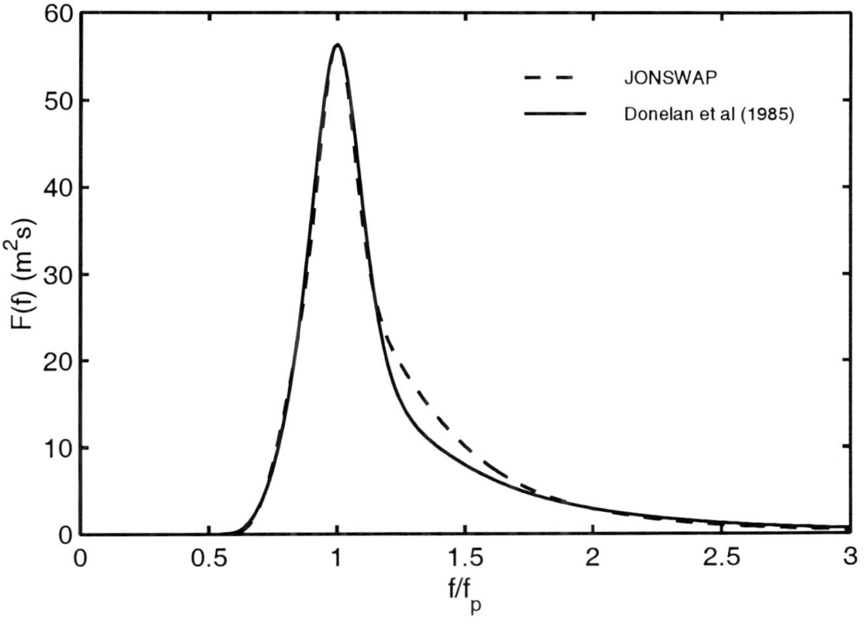

Figure 5.24: Comparison between spectra represented by the JONSWAP form, (5.51) and the Donelan et al (1985) form, (5.53). The JONSWAP spectrum has parameters: $f_p = 0.1$ Hz, $\alpha = 0.016$, $\gamma = 2.0$ and $\sigma = \sigma_a = \sigma_b = 0.08$ whereas the Donelan et al (1985) form has, $f_p = 0.1$ Hz, $\beta = 0.008$, $\gamma_d = 3.1$ and $\sigma = 0.12$.

result of the Doppler shifting is to change the value of the exponent n. Therefore, the existence of a universal value for n may be questioned.

In summary, the observational evidence is not conclusive. It is likely that there is no universal value of n. For practical engineering applications, the choice of n generally makes little difference. For other application, such as remote sensing of the ocean, the detailed structure of the high frequency region of the spectrum is important. In these cases, a choice should be made with knowledge of the potential variability.

5.5 Directional Spreading

The directional frequency spectrum, $F(f, \theta)$, as defined by (2.52), is commonly represented in terms of the one-dimensional spectrum, $F(f)$, as (Longuet-Higgins et al, 1963)

$$F(f, \theta) = F(f)D(f, \theta) \tag{5.58}$$

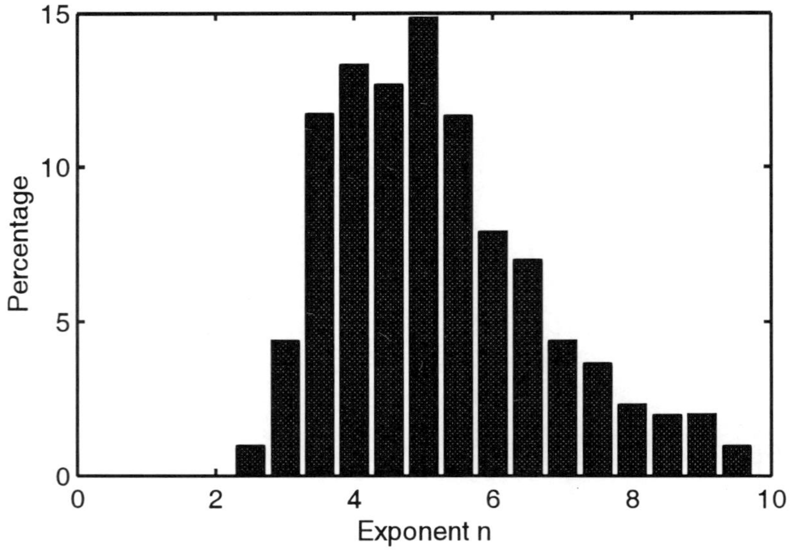

Figure 5.25: Probability distribution of the exponent, n in the relationship $F(f) \propto f^{-n}$ for the high frequency portion of the spectrum [after Liu (1989)].

The directional spreading function, $D(f, \theta)$ must satisfy the condition

$$\int D(f, \theta) \mathrm{d}\theta = 1 \tag{5.59}$$

Based on field data, analytical forms for $D(f, \theta)$ have been proposed by Mitsuyasu et al (1975), Hasselmann et al (1980) and Donelan et al (1985). Mitsuyasu et al (1975) considered data collected with a cloverleaf buoy which measures six quantities related to the surface wave field (the vertical acceleration of the water surface, η_{tt}, the wave slope, η_x, η_y and the surface curvature, η_{xx}, η_{yy}, η_{xy}). Their analysis procedure, considered only the surface acceleration and slope information, reducing the data to that which would be collected with a pitch/roll buoy [see Section (9.3.2)]. Following Longuet-Higgins et al (1963) they represented $D(f, \theta)$ in the form

$$D(f, \theta) = Q(s) \cos^{2s} \left\{ \frac{[\theta - \theta_m(f)]}{2} \right\} \tag{5.60}$$

where $Q(s)$ is a normalization factor required to satisfy (5.59) and θ_m is the mean wave direction at frequency f. Based on their data, Mitsuyasu et al (1975) parameterized s as

$$s = \begin{cases} s_p \left(\frac{f}{f_p}\right)^5 & \text{for } f < f_p \\ s_p \left(\frac{f}{f_p}\right)^{-2.5} & \text{for } f \geq f_p \end{cases} \tag{5.61}$$

where s_p is the value of s at the frequency of the spectral peak, f_p, given by

$$s_p = 11.5 \left(\frac{U_{10}}{C_p}\right)^{-2.5} \tag{5.62}$$

$C_p = g/\omega_p$ is the deep water phase speed of components at the spectral peak and U_{10} the wind speed at a reference height of 10 m.

Hasselmann et al (1980) considered pitch/roll buoy data, also representing their data in the form of (5.60) but with a different parameterization for s

$$s = \begin{cases} 6.97 \left(\frac{f}{f_p}\right)^{4.06} & \text{for } f < 1.05 f_p \\ 9.77 \left(\frac{f}{f_p}\right)^{\mu} & \text{for } f \geq 1.05 f_p \end{cases} \tag{5.63}$$

where μ has a weak dependence on wave age

$$\mu = -2.33 - 1.45 \left(\frac{U_{10}}{C_p} - 1.17\right) \tag{5.64}$$

Based on data from an array of 14 wave gauges, Donelan et al (1985) found that the form represented by (5.60) did not adequately represent their data and adopted the alternative form

$$D(f,\theta) = 0.5\beta_s \mathrm{sech}^2 \beta \left[\theta - \theta_m(f)\right] \tag{5.65}$$

They found that β_s varied as a function of non-dimensional frequency, f/f_p. Their data, however, extended only to $f/f_p = 1.6$ and beyond this point they assumed β_s was constant. Banner (1990) reviewed this conclusion in the context of high frequency stereo photography data and concluded that the assumption that $\beta_s =$ constant for $f/f_p > 1.6$ was unreasonable. He proposed a formulation for β_s beyond $1.6 f_p$ which is combined with the Donelan et al (1985) parameterizations for $f < 1.6 f_p$ in (5.66)

$$\beta_s = \begin{cases} 2.61 \left(\frac{f}{f_p}\right)^{1.3} & \text{for } 0.56 < f/f_p < 0.95 \\ 2.28 \left(\frac{f}{f_p}\right)^{-1.3} & \text{for } 0.95 < f/f_p < 1.6 \\ 10^{\left\{-0.4+0.8393\exp\left[-0.567\ln\left((f/f_p)^2\right)\right]\right\}} & f/f_p > 1.6 \end{cases} \tag{5.66}$$

5.5.1 Comparison of Spreading Functions

Figure 5.26 shows the directional spreading functions calculated using (5.61), (5.63) and (5.66) at frequencies of $f/f_p = 1$, 2 and 3 and values of the inverse wave age $U_{10}/C_p = 1$ and 2. The differences between these proposed formulations are significant. It is clear that the spreading functions derived from pitch/roll buoy data yield broader spreading functions than the spatial array data (Donelan et al, 1985). A clearer representation of the differences is evident if the mean spectral width, $\overline{\theta}$ is examined

$$\overline{\theta}(f) = \frac{\int_0^{2\pi} F(f,\theta)(\theta - \theta_m)\mathrm{d}\theta}{\int_0^{2\pi} F(f,\theta)\mathrm{d}\theta} \tag{5.67}$$

Figure 5.27 shows the mean spectral width as a function of f/f_p and inverse wave age, U_{10}/C_p for each of the above spreading parameterizations. All representations indicate that the spreading is narrowest in the region of the spectral peak, becoming broader at frequencies both above and below f_p. The Donelan et al (1985) formulation is not a function of wave age whereas the representations of both Mitsuyasu et al (1975) and Hasselmann et al (1980) indicate increased broadening as a function of wave age (decreasing U_{10}/C_p). This dependence on wave age is considerably stronger in the Mitsuyasu et al (1975) formulation than that of Hasselmann et al (1980). Whether the directional spreading should be a function of wave age has been discussed by both Hasselmann et al (1980) and Donelan et al (1985). The answer to this question depends on which physical process is controlling the directional spreading. If it is nonlinear interactions within the spectrum, as proposed by Hasselmann (1962), then wave age should play no role. If, however, atmospheric input plays a role, the spreading should be dependent on wave age. Young and Van Vledder (1993) and Banner and Young (1994) have investigated the development of the directional spectrum under fetch limited conditions using a spectral model with a full solution to the nonlinear source term. They find that the directional spreading is almost completely controlled by nonlinear interactions, input playing only a very minor role. Hence, it would appear that either no dependence on wave age (Donelan et al, 1985) or a weak dependence (Hasselmann et al, 1980) is most appropriate.

In interpreting these proposed spreading functions, the parameter range of the data upon which the various formulations are based should be noted. The regions of data validity have been marked on Figure 5.27. It is clear that the data upon which the formulation of Mitsuyasu et al (1975) is based, covers a very small range of wind-sea wave ages and hence the accuracy of the strong dependence on wave age in this formulation must be questioned. Over the parameter range for which the data of Mitsuyasu et al (1975) and Hasselmann et al (1980) overlap, there is in fact quite good agreement between these formulations. These formulations are however considerably broader than that of Donelan et al (1985).

Young (1994a) has investigated the analysis techniques used by the three studies and concluded that the combination of buoy data together with the analysis technique [see Chapter 9] used by Mitsuyasu et al (1975) and Hasselmann et al (1980)

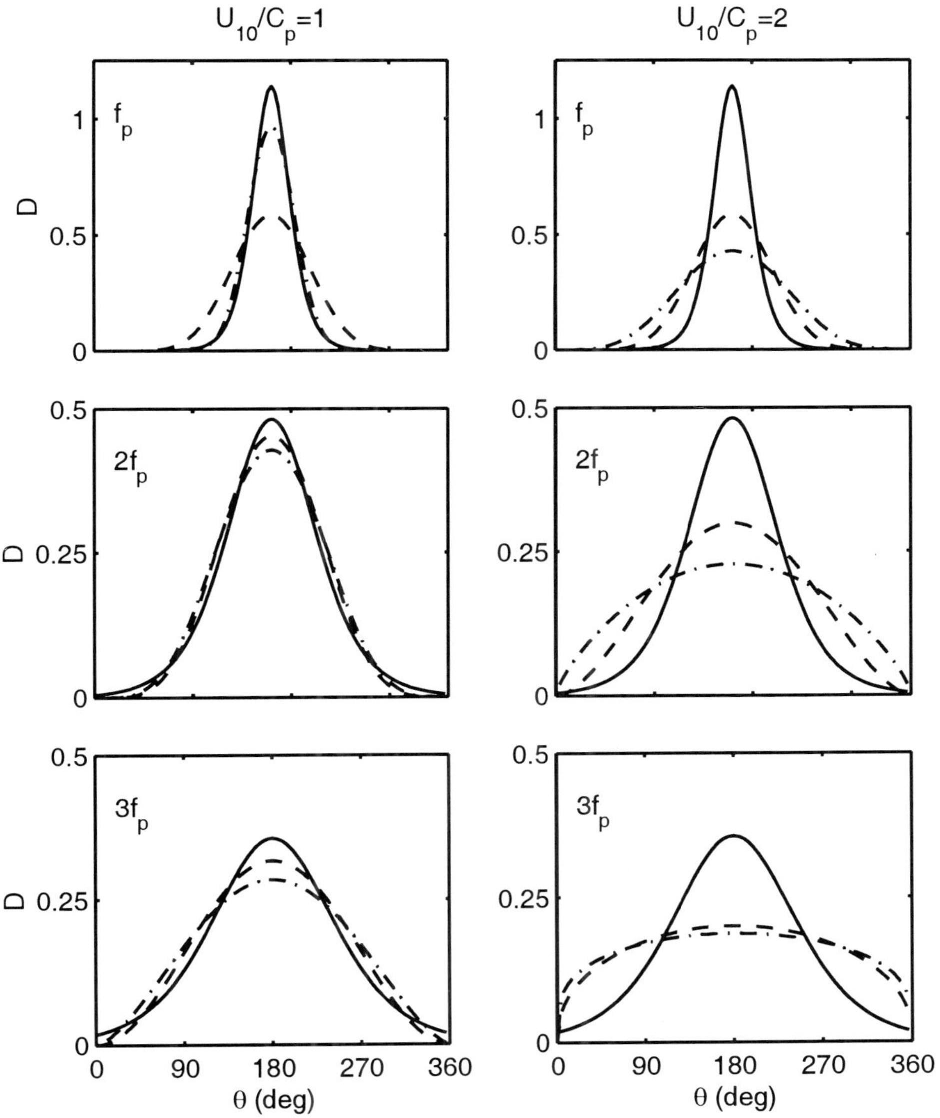

Figure 5.26: The directional spreading function, D as proposed by Mitsuyasu et al (1975) (dash-dot line), Hasselmann et al (1980) (dashed line) and Donelan et al (1985) (solid line). Spreading functions are shown at frequencies of f_p, $2f_p$ and $3f_p$. In addition, two wave ages are shown: $U_{10}/C_p = 1$ (mature waves) and $U_{10}/C_p = 2$ (younger waves).

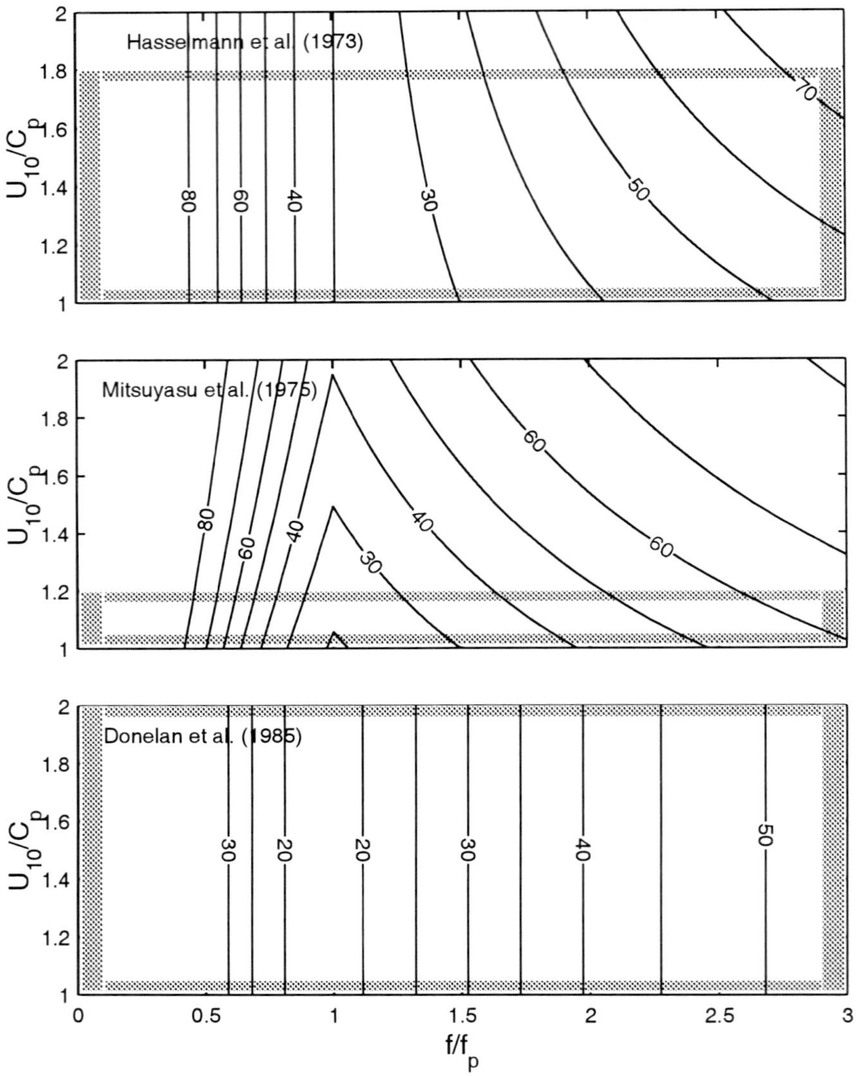

Figure 5.27: Contours of the mean directional width, $\overline{\theta}$ (5.67) as a function of the inverse wave age, U_{10}/C_p and non-dimensional frequency, f/f_p: Hasselmann et al (1980) (top), Mitsuyasu et al (1975) (middle), Donelan et al (1985) (bottom). The parameter range for which data were collected is outlined by the shading.

result in artificially broad spectra. Therefore, it is reasonable to conclude that the narrower distributions reported by Donelan et al (1985) are a better approximation to the directional spreading of wind generated waves.

Chapter 6

Non-stationary Wind Fields

6.1 Introduction

Cases such as fetch or duration limited growth [see Chapter 5] represent highly idealized situations. In the vast majority of wind wave generation cases, the wind speed is not constant in speed or direction. In such cases, the prediction of the wave spectrum is generally achieved through the use of numerical models [see Chapter 8]. Before such models are considered in detail, a number of situations of non-stationary wind fields will be considered in this chapter. Four specific cases will be investigated: the interaction of swell and wind-sea, a rapid change in wind direction, a rapid change in wind speed and the vortex wind field of the hurricane. These cases all provide additional insight into the physical processes responsible for the evolution of the wave spectrum.

6.2 The Interaction of Swell and Wind Sea

Swell is ubiquitous in the open ocean. Once generated, swell can propagate huge distances with apparently little attenuation. Snodgrass et al (1966) have observed that swell originally generated south of New Zealand can propagate across the Pacific Ocean before breaking on the shores of Alaska with no measurable change in amplitude. In addition, as indicated in Section (5.3.3.5), it is not clear whether fetch limited wind waves are influenced in their development by background swell.

There are presumably two mechanisms by which swell and wind sea may interact. The first would be by the presence of the swell in some manner influencing the air flow over the wind waves and hence altering the energy flux from the wind to the water. Whether such influences are significant is unknown and no detailed studies of this type have been conducted. The second mechanism is by non-linear quadruplet interaction between the swell and wind sea. This mechanism has been investigated theoretically by Young et al (1985) and Masson (1993).

Young et al (1985) investigated the magnitude of the nonlinear source term, S_{nl} for spectra which consisted of the summation of two JONSWAP spectra (5.51). The spectra had peak frequencies, f_{m1}, f_{m2} and mean directions θ_1, θ_2, respectively. The magnitude of the nonlinear transfer of the composite spectrum was then investigated as both f_{m2}/f_{m1} and $\Delta\theta = \theta_2 - \theta_1$ were increased. Figure 6.1 shows the magnitude of the positive lobe of S_{nl} for the first of these sub-spectra (f_{m1}, θ_1). The results are presented as the ratio of the non-linear transfer which would be expected if the two sub-spectra did not interact (i.e. simple summation of S_{nl} associated with each sub-spectrum calculated independently) to that calculated for the composite spectrum. As the peaks of the two spectra are increasingly separated in frequency and direction, the magnitude of the coupling between them decreases. From this figure it can be concluded that wave trains separated by more than 90° do not interact significantly, irrespective of their frequencies. Similarly, waves with peak frequencies whose ratios are larger than approximately 3 do not interact significantly.

When the spectra do interact, the lower frequency spectrum is generally parasitic. An energy transfer occurs from the higher frequency spectrum to the lower frequency spectrum. This is an example of the property of *shape stabilization* discussed in Section (4.4.2.2). The nonlinear source term acts to force the spectrum back to a unimodal shape. As the spectra are increasingly separated in frequency-direction space, the coupling however decreases.

6.3 Rapid Change in Wind Speed

The problem to be considered here is where a wind sea has been developing under the action of a steady wind. The magnitude of the wind then instantaneously increases or decreases. There is only one published field investigation of the response of the wave spectrum to changes in the magnitude of the wind. Toba et al (1988) investigated field data collected at a number of sites for the case of a sudden change in the wind speed. They described their results in terms of the Toba (1973) equilibrium spectral form (5.8), $F(f) = \alpha_s g u_* f^{-m}$. For stable fetch limited conditions α_s is a universal constant approximately equal to 0.062, g is the gravitational acceleration, u_* is the friction velocity and m is an exponent equal to 4. The exponent m describes the rate of decay of the high frequency spectral face.

The Toba (1973) relationship predicts that an increase in the magnitude of the wind speed should result in an increase in the energy of the high frequency spectral components, an elevation of the spectral tail. Conversely, a decrease in wind speed should lead to a reduction in the energy of the spectral tail. Toba et al (1988) found, however, that the spectrum could not respond instantly to rapid changes in the wind speed. An example of this is shown in Figure 6.2. In response to the rapid increase in wind speed which occurs at approximately 10–15 min in Figure 6.2 the value of α_s decreases, indicating that the high frequency spectral

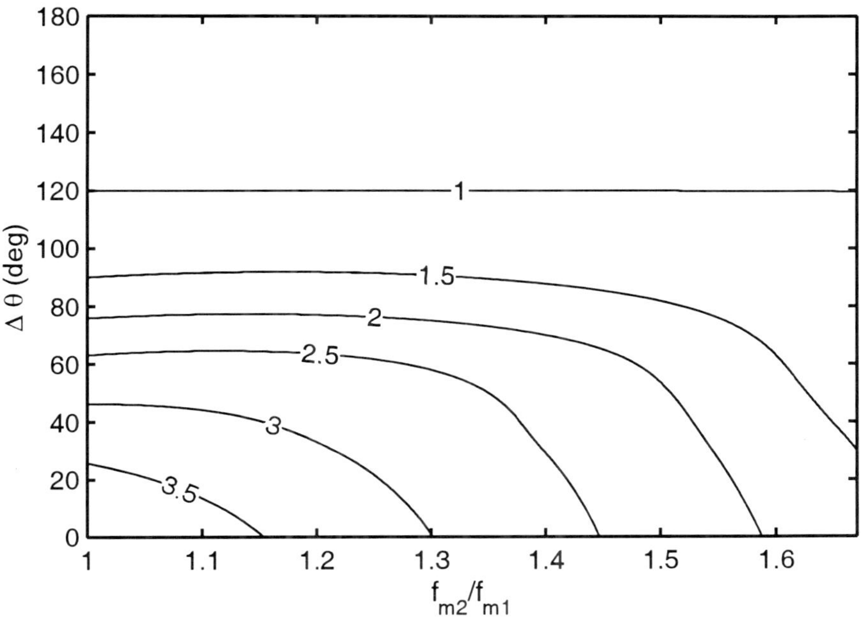

Figure 6.1: Contour plot showing the magnitude of the nonlinear coupling between two JONSWAP spectra. The value f_{m2}/f_{m1} is the ratio of the peak frequencies of the two spectra. The difference in the mean direction of propagation is represented by $\Delta\theta$. The contours show the magnitude of the coupling at the positive lobe of the nonlinear transfer. A contour value of 1 indicates that the spectra are independent and do not interact. Values greater than 1 indicate the ratio by which the nonlinear transfer exceeds that which would occur if the spectra did not interact. For example, a value of 3 indicates that the nonlinear transfer is 3 times larger than if the spectra were independent and did not interact with each other.

energy level is lower than could be expected at the new wind speed. With time, the energy of the high frequency components increase, reflected by a gradual increase in α_s. Although not explicitly explored in their study, there is some suggestion in Figure 6.2 that high frequency components respond more rapidly than lower frequencies. This is exhibited by a change in the value of m. With the increase in wind speed, there is a tendency for m to decrease slightly in magnitude. This would indicate that the high frequency components have increased in energy more rapidly than their lower frequency counterparts, resulting in a less rapid decay of the high frequency spectral face. Similarly, the rapid decrease in wind speed which occurs at approximately 20 min in Figure 6.2 is associated with an increase in α_s and an increase in the magnitude of m.

The data of Toba et al (1988) present a consistent picture which indicates that high frequency components of the spectrum respond more rapidly than lower frequencies. A sudden increase in wind speed will result in an elevation of the high

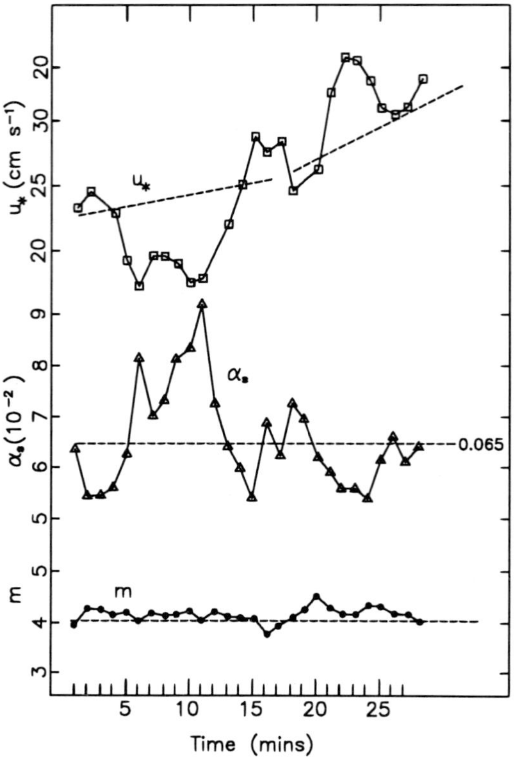

Figure 6.2: Values of the shear velocity, u_*, the Toba "constant", α_s and the high frequency exponent, m, as reported by Toba et al (1988). Note that the value of α_s responds in an inverse manner to the wind speed as represented by u_*. This indicates that the high frequency spectral tail takes a finite time to respond to changes in the wind speed.

frequency components first, this influence gradually "feeding" down to lower frequencies. The converse happens with a decrease in wind speed. The data indicate that the response time is of order a few minutes, although response times as a function of frequency cannot be determined from the data. The spectra presented by Toba et al (1988) (not shown here) show no tendency for a second high frequency peak to develop in the case of an increase in wind speed. The unimodal structure of the spectrum is maintained.

The case of a rapid change in wind speed has also been investigated numerically by Young and Van Agthoven (1998) who considered a duration limited spectral model similar to (4.62), $dF/dt = S_{tot}$. The total source term, S_{tot} included a full solution to the non-linear term, S_{nl} [see Section (4.4.1)]. Figure 6.3 shows the response of the one-dimensional spectrum, $F(f)$ to a sudden increase in the wind speed. In the case shown, the spectrum was allowed to develop under a constant wind speed of $U_{10} = 10$ m/s until f_p attained a value of 0.24 Hz. At this point, the wind speed was doubled to $U_{10} = 20$ m/s.

Figure 6.3 shows the response of the spectrum for the case described above. As expected, there is an almost immediate response in the high frequency regions of the spectrum. At $f = 3f_p$ the energy increases by 35% in the first 50 s following the wind increase and by 55% in the first 200 s. There is little further increase in energy at these frequencies after 200 s. The spectral response becomes progressively slower with decreasing frequency. By $t = 400$ s, frequencies above 1.5 f_p have stabilized at a new equilibrium level with little further increase in energy. Spectral response near the spectral peak is much slower than at higher frequencies. The energy at the peak of the spectrum does, however, continue to increase whilst the re-adjustment of the tail is occurring. By $t = 600$ s the spectrum has fully adjusted to the increased wind speed and continues to develop as for duration limited growth at the increased wind speed. During the re-adjustment process there was no indication of the development of a second high frequency peak, despite the fact that the energy input at high frequencies had increased substantially.

Young and Van Agthoven (1998) examined, in detail, the source term balance for this case. Following the increase in wind speed a natural increase in the atmospheric input occurred, particularly at high frequencies. This increase in input accounts for the elevation of the spectral tail. As shown in Figure 6.3, there is no tendency for the creation of a second high frequency peak. Again, this is due to the *shape stabilizing* effect of the non-linear term which prevents development of such a peak.

Opposite results occurred for a decrease in wind speed, with a decrease in the level of the spectral tail. The response of the tail in such cases is, however, much slower than for an increase in wind speed. The decrease in energy of the tail is now controlled by the imbalance which results between the dissipation and the reduced atmospheric input.

Young and Van Agthoven (1998) summarized the response of the spectrum in terms of a parametric response model. The one-dimensional spectrum was normalized in terms of the spectrum prior to the change in wind speed

$$F'(f,t) = \frac{F(f,t)}{F_0(f)} \tag{6.1}$$

where $F_0(f) = F(f, t = 0)$. The response of $F'(f,t)$ being represented by

$$F'(f,t) = (1 - F'_f) \exp\left[\pm\left(\frac{a}{1 - F'_f}\right) t\right] + F'_f \tag{6.2}$$

where $F'_f = F'(f, t = \infty)$ and $a = a(f, \Delta U)$. The "plus" sign applies for an increase in wind speed and the "minus" sign for a decrease. The parameter a provides a non-dimensional measure of the rate of response as a function of f and ΔU. Figure 6.4 shows the results for an increase in wind speed and Figure 6.5 for a decrease in wind speed.

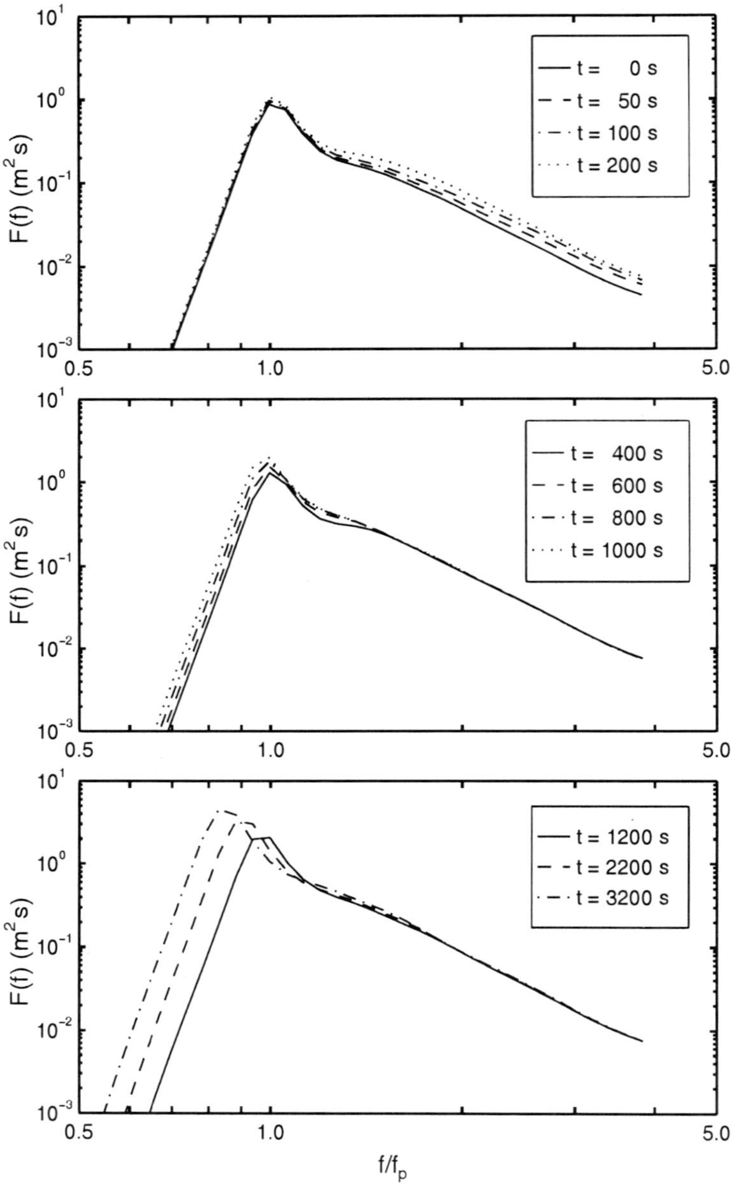

Figure 6.3: Response of the spectrum as a function of time (shown on a logarithmic scale) for a wind speed increase of $\Delta U = 10$ m/s. The initial wind speed was also $U_{10} = 10$ m/s. [after Young and Van Agthoven (1998)]

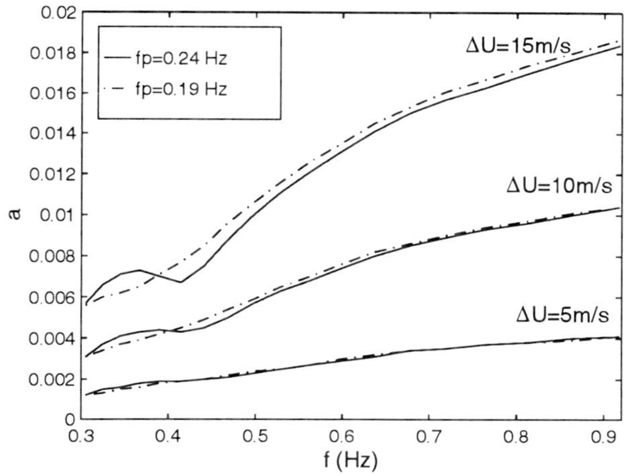

Figure 6.4: The time response coefficient, a, (6.2) as a function of frequency, f, change in wind speed, ΔU and peak spectral frequency, f_p for a sudden *increase* in wind speed.

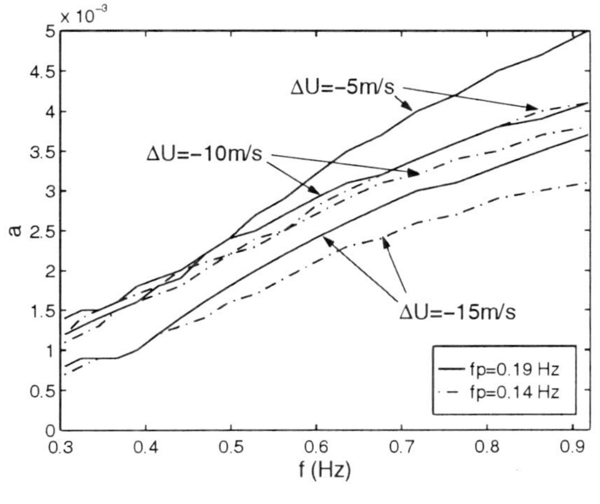

Figure 6.5: The time response coefficient, a, (6.2) as a function of frequency, f, change in wind speed, ΔU and peak spectral frequency, f_p for a sudden *decrease* in wind speed.

The value of a increases as a function of frequency, indicating the more rapid response of the high frequency components to a change in wind speed. In addition, a is an order of magnitude larger for an increase in wind speed than for a corresponding decrease, indicating the more rapid response of the spectrum to an increase in wind speed.

The results for an increase in wind speed are much more ordered, partly because of the insensitivity of the spectral response to f_p but also due to the fact that (6.2) is a better approximation to the numerical model results for an increase in wind speed. Nevertheless, (6.2) and Figures 6.4 and 6.5 represent a useful predictive tool to describe spectral response in such situations.

6.4 Rapid Change in Wind Direction

A closely related problem to that discussed in Section (6.3) is that of a rapid change in the wind direction. Observations of spectral response under such conditions have been reported by Hasselmann et al (1980), Günther et al (1981), Allender et al (1983), Young et al (1985), Masson (1990) and Van Vledder and Holthuijsen (1993). These data have commonly been summarized in terms of a directional relaxation model of the form

$$\frac{\partial \theta_0(f)}{\partial t} = \frac{1}{\tau(f)} \sin\left[\theta_w - \theta_0(f)\right] \tag{6.3}$$

where $\theta_0(f)$ is the mean wave direction and θ_w is the wind direction. The frequency dependent time scale, $\tau(f)$ is often expressed in the form $\tau(f) = 1/(\omega b)$, where b is a non-dimensional directional relaxation coefficient.

Introducing the non-dimensional variables, $\tilde{\tau} = g\tau/U_{10}$ and $\nu = f_p U_{10}/g$, Van Vledder and Holthuijsen (1993) found that the composite field data set yields

$$\tilde{\tau} = 312\nu^{-1.7} \tag{6.4}$$

Van Vledder and Holthuijsen (1993) represented their non-dimensional variables in terms of u_* rather than U_{10} and (6.4) was obtained assuming $U_{10} = 21u_*$ as in Section (5.3.4). The data upon which (6.4) is based are shown in Figure 6.6. Although the trend represented by (6.4) is clear, there is appreciable scatter about this result.

The physical processes responsible for the directional turning of the spectrum were investigated by Young et al (1987) using a duration limited spectral model with a full solution to the non-linear source term, as described in Section (6.3). Figures 6.7 and 6.8 show the response of the directional spectrum with time to wind shifts of 60° and 90°, respectively. For wind shifts of less than approximately 90°, the high frequency components of the spectrum rapidly rotate to align with the new wind direction. The rate of rotation decreases with decreasing frequency, consistent with the field data as represented by (6.4). It is noticeable in Figure 6.7

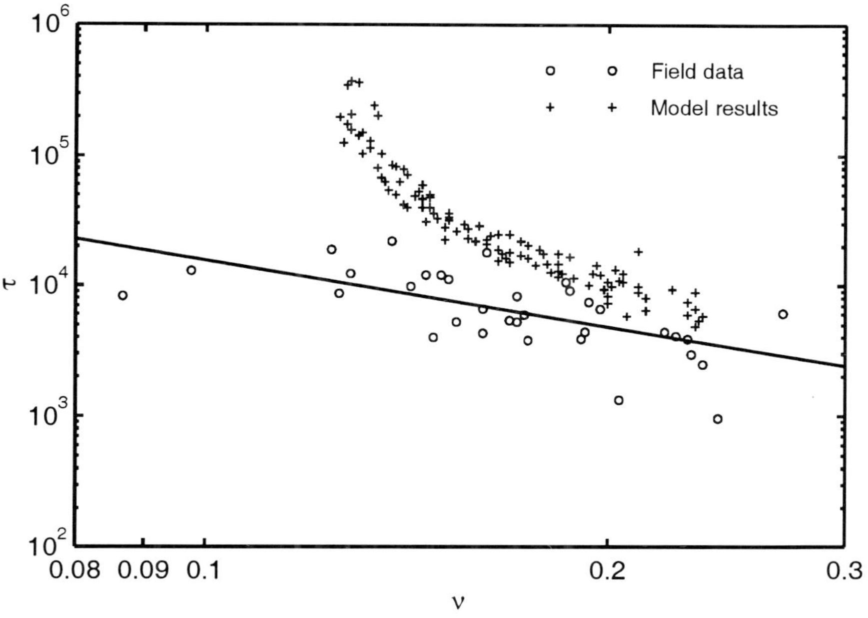

Figure 6.6: Non-dimensional directional relaxation time scale, $\tilde{\tau}$ as a function of the non-dimensional frequency, ν. Both field data and numerical model results are shown. The solid line is (6.4). [after Van Vledder and Holthuijsen (1993)].

that there is no tendency for a new, high frequency, peak to develop in the new wind direction. Young et al (1987) examined the source term balance and found that the rotation of the spectrum was caused by the added atmospheric input in the new wind direction, together with the, now unbalanced, dissipation in the old wind direction. Hence, spectral components in the old wind direction gradually decayed whilst those in the new wind direction developed. The non-linear term aided the smooth rotation of the spectrum by preventing the development of a high frequency peak in the new wind direction. Again, the non-linear terms tend to favour a unimodal spectral structure.

For wind shifts of 90° and larger [see Figure 6.8] the nonlinear coupling between the old wind direction and the new is so weak that the non-linear terms cannot prevent the development of a high frequency peak in the new wind direction. Hence, in such cases a new wind sea develops in the new wind direction whilst the spectrum in the old wind direction gradually decays. For such large wind shifts there is no tendency for the spectrum to rotate to the new wind direction.

Van Vledder and Holthuijsen (1993) used this same numerical model and compared values of $\tilde{\tau}$ with field data. Although the model gives the same qualitative trends as the field data, the rate of spectral turning is larger in the model by a factor of between 2 and 3. This possibly highlights deficiencies in the model source

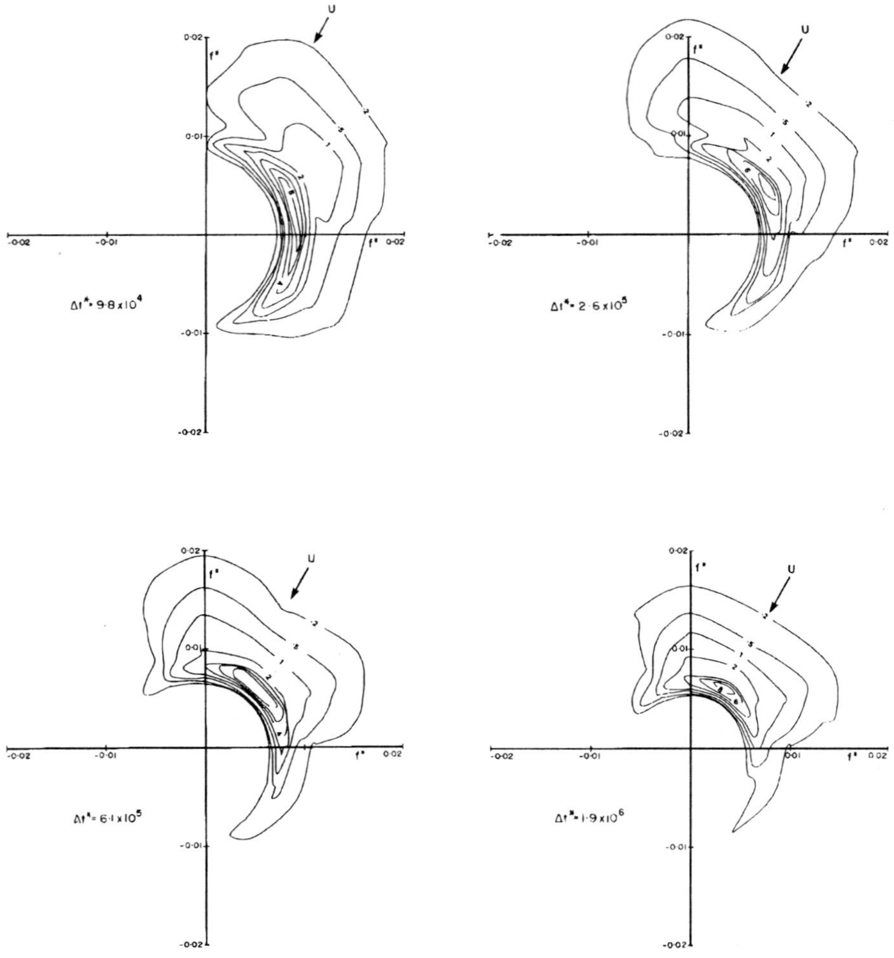

Figure 6.7: Response of the directional wave spectrum to a sudden wind shift of 60°. The spectrum has been normalized such that it has a maximum value of 10. The frequency axes are non-dimensionalized as $f^* = fu_*/g$. The non-dimensional time since the wind shift is shown by Δt^*, where $t^* = tg/u_*$. [after Young et al (1987)].

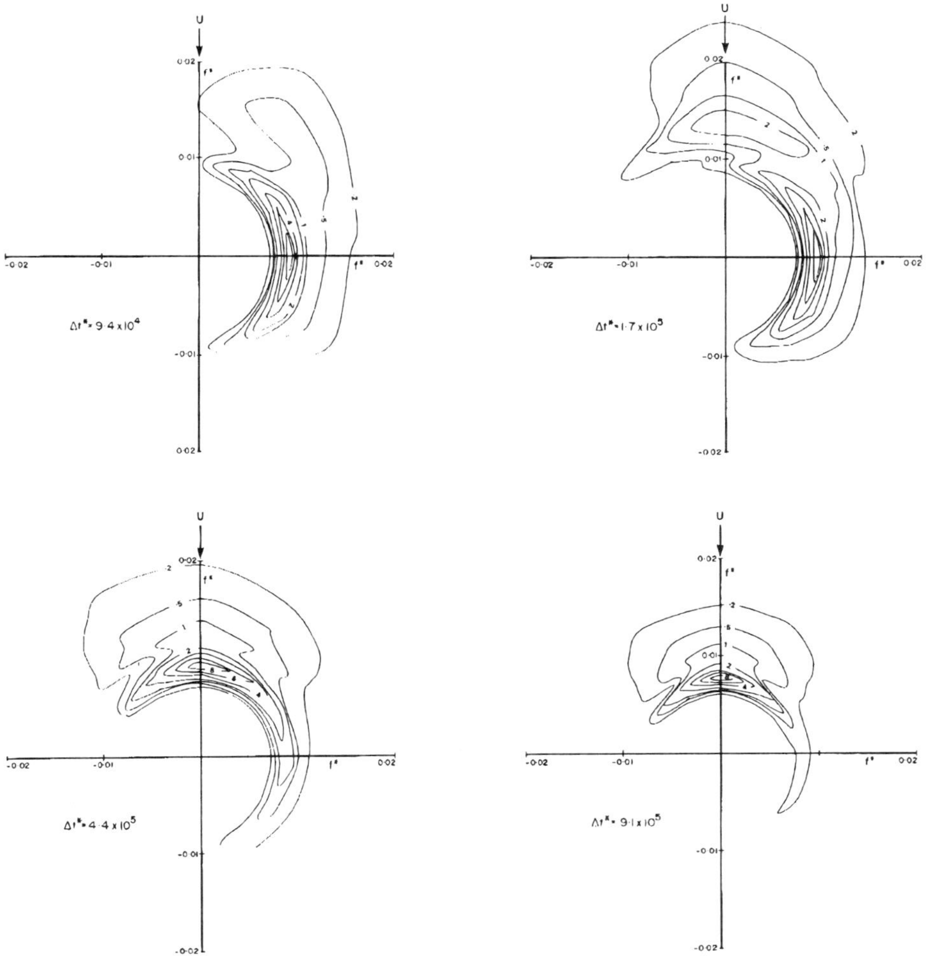

Figure 6.8: Response of the directional wave spectrum to a sudden wind shift of 90°. The spectrum has been normalized such that it has a maximum value of 10. The frequency axes are non-dimensionalized as $f^* = fu_*/g$. The non-dimensional time since the wind shift is shown by Δt^*, where $t^* = tg/u_*$. [after Young et al (1987)].

term representations in such situations.

6.5 Hurricane Wind and Wave Fields

The intense low pressure systems knows as hurricanes, tropical cyclones or typhoons represent the most extreme meteorological event encountered in most tropical areas. For the sake of simplicity the following discussion will focus on northern hemisphere hurricanes, although the discussion holds for both tropical cyclones and typhoons with the appropriate correction for the direction of circulation which occurs in the southern hemisphere.

6.5.1 Hurricane Wind Field

Hurricane wind fields are characteristically intense, spatially inhomogeneous and directionally varying. The large gradients in wind speed and the rapidly varying wind directions of the hurricane vortex generate extremely complex ocean wave fields. Typically, waves generated by hurricanes consist of a combination of both swell and wind sea. In addition, the directional distribution of the wind sea component is often skewed due to the rapid variation in the wind direction. The complexity of the wave generation process within hurricane wind fields provides a very demanding test for wind wave prediction models. Indeed, the complexity of the wind field represents a demanding test for our physical knowledge of wind wave generation.

The structure of the hurricane wind field has been the subject of considerable research (Schloemer, 1954; Graham and Hudson, 1960; Smith, 1968; Gray and Shea, 1973; Shea and Gray, 1973; Simiu et al, 1976; Atkinson and Holliday, 1977; Wang, 1978; Holland, 1980). The vortex-shaped wind fields in these well-formed systems are relatively similar from storm to storm. This similarity makes the prediction of the wind field more reliable than in other less organized meteorological systems. Nevertheless, there is variability between storms, and a full understanding of the hurricane wind field is still lacking.

Holland (1980) has shown that the wind field can be represented in terms of a relatively simple parametric model. Following Holland (1980), the gradient wind can be expressed as

$$U_g = \left[\frac{AB\left(p_n - p_0\right)\exp\left(-A/r^B\right)}{\rho_a r^B} + \frac{r^2 f_c^2}{4} \right]^{0.5} - \frac{r f_c}{2} \tag{6.5}$$

where U_g is the gradient wind at radius r from the center of the storm, f_c is the Coriolis parameter, ρ_a the air density, p_0 the central pressure, and p_n the ambient atmospheric pressure far from the storm. The parameters A and B can be expressed in terms of the radius to maximum winds, R, as

$$R = A^{1/B} \tag{6.6}$$

The dimensionless parameter B defines the shape of the wind field with increasing distance from the center of the hurricane. Increasing B concentrates more of the pressure drop near R. Holland (1980) has shown that B can be related to the central pressure p_0. A linear fit to his data yields

$$B = 1.5 + (980 - p_0)/120 \tag{6.7}$$

where p_0 is expressed in hectopascals (or millibars).

Following Shea and Gray (1973), the radial winds can be assumed to spiral in toward the center of the storm with a constant inflow angle of 25°. Also, a first-order asymmetry has been applied to the wind field by adding the cyclone forward speed V_{fm} to the symmetric flow and rotating the maximum to an angle of 70° to the direction of the cyclone forward motion (to the left in the southern hemisphere and to the right in the northern hemisphere). These modifications are consistent with the results of Shea and Gray (1973) and Shapiro (1983). Naturally, there can be variability in all these parameters, particularly the angle to the maximum winds, which can vary significantly between storms (Shapiro, 1983). The gradient wind velocity U_g can be reduced to U_{10} (1-min. average) by the application of a factor of 0.8 (Powell, 1980). Hence the full wind field is defined by the specification of the position of the center of the storm, the central pressure p_0, radius to maximum winds, R, and the direction and velocity of forward translation.

Figure 6.9 shows a typical example of the two-dimensional wind field generated by this model. The distribution of wind speed along a line through the center of the storm is shown in Figure 6.10. The case shown has a central pressure p_0 of 950 HPa, a velocity of forward movement V_{fm} of 10 m/s, and a radius to maximum winds, R of 50 km. The asymmetric structure of the wind field is clear with the maximum wind occurring in the crescent-shaped region to the right of this northern hemisphere storm. The wind speed profile through the center of the storm [Figure 6.10] shows the relatively calm eye of the hurricane. The wind velocity increases rapidly to a maximum at a distance R from the center before decaying in an approximately exponential manner with a further increase in distance.

The model has been compared with recorded data by Hubbert et al (1991) and extensively validated in a number of detailed studies for the oil industry on Australia's North-West Shelf [B.A. Harper, private communication, 1992]. As part of these studies, data from 24 tropical cyclones that have occurred in this region have been compared with model wind fields. Generally, the model performs remarkably well, particularly for storms with relatively simple tracks and whose parameters do not vary rapidly. Model accuracy will be limited for two reasons. The first, is the model's inherent limitations in representing a possibly complex wind field by a relatively simple parametric form; the second, is the ability to accurately determine the model parameters defining the storm. To give an example of model performance, two cases from this extensive database are considered in Figure 6.11. The two storms considered here have been selected since their respective centres passed directly over anemometer stations and they both had relatively simple tracks. In both cases, the parameters used in the model were those provided by the Australian

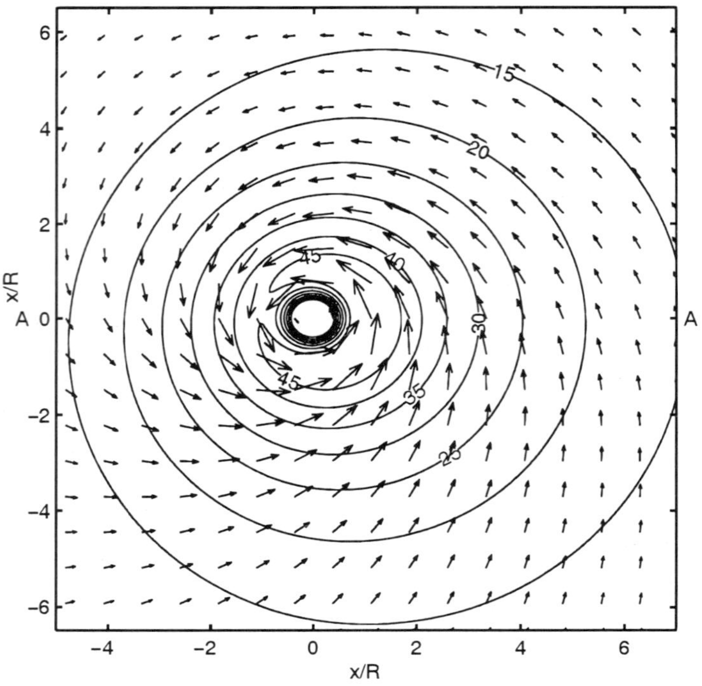

Figure 6.9: An example of the wind field generated by the Holland (1980) hurricane wind field model. The contours are of U_{10} in units of m/s. The vectors indicate the wind direction. The hurricane shown has the following wind field parameters: $p_0 = 950$ HPa, $V_{fm} = 5$ m/s and $R = 30$ km. The storm is propagating up the page with its centre at co-ordinates (0,0).

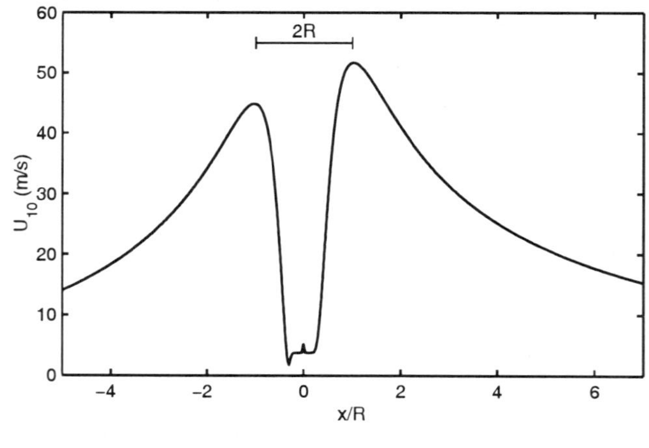

Figure 6.10: The wind speed profile through the centre of the tropical cyclone shown in Figure 6.9 along the line AA.

Bureau of Meteorology, and no attempt has been made to optimize the agreement between model and data by altering these parameters.

Figure 6.11a shows a comparison between the recorded and model winds at Woodside Offshore Petroleum's North Rankin A gas platform (−19.58°S, 116.14°E) during the passage of tropical cyclone Orson in 1989. Tropical cyclone Orson is the most severe tropical cyclone to have occurred in the Australian region since reliable records have been kept (50 years). The storm had a minimum central pressure of 905 HPa at 1030 UT on April 22, 1989, when centered at −18.35°S, 116.35°E. Six hours later at 1630 UT it passed over the North Rankin A platform with a central pressure of 906 HPa. The maximum wind speed recorded was $U_{10} \approx 63$ m/s. Unfortunately, shortly after the passage of the eye of the storm over the measurement location, the anemometer failed. The agreement between model and data at the peak of the storm is very good. During the approach of the storm to the measurement location, the model significantly underestimates the wind speed. This can occur for a variety of reasons. The model considers only the influence of the tropical cyclone vortex, and hence any winds resulting from the background synoptic flow are ignored. As a result, the model could generally be expected to underestimate the wind speed when the storm is distant from the measurement location. In Figure 6.11a, however, it is also likely that the radius to maximum winds, R, has been underestimated during the approach of the storm, and possibly the parameter B was too large.

Figure 6.11b shows the comparison between model and recorded data at Barrow Island (−20.82°S, 115.39°E) during the passage of tropical cyclone Ian in 1992. Tropical cyclone Ian was an intense tropical cyclone which reached a minimum central pressure of 930 HPa at 0700 UT on March 1, 1992, when located at −16.37°S, 117.15°E. Its intensity gradually decreased as it approached Barrow Island, passing over the anemometer site at 1830 UT on March 2, 1992, with a central pressure of 961 HPa. Both model and recorded data indicate a maximum wind velocity of $U_{10} \approx 38$ m/s. The data show a more rapid rise and fall of wind speed during the passage of the storm, indicating the model storm is spatially larger than indicated by the data. This is probably due to an overestimation of the radius to maximum winds, R, which is generally difficult to estimate without additional information.

These results are indicative of the quality of the wind field predictions expected from the model. Differences between model and recored data occur due to limitations of the relatively simple structure of the model and the difficulty of accurately defining the wind field parameters of the model.

6.5.2 Hurricane Wave Height Distribution

Our present understanding of the hurricane wave field has been gained from a combination of in situ measurements (both non-directional and directional) at single points during the passage of hurricanes, remote sensing data and the application of numerical models.

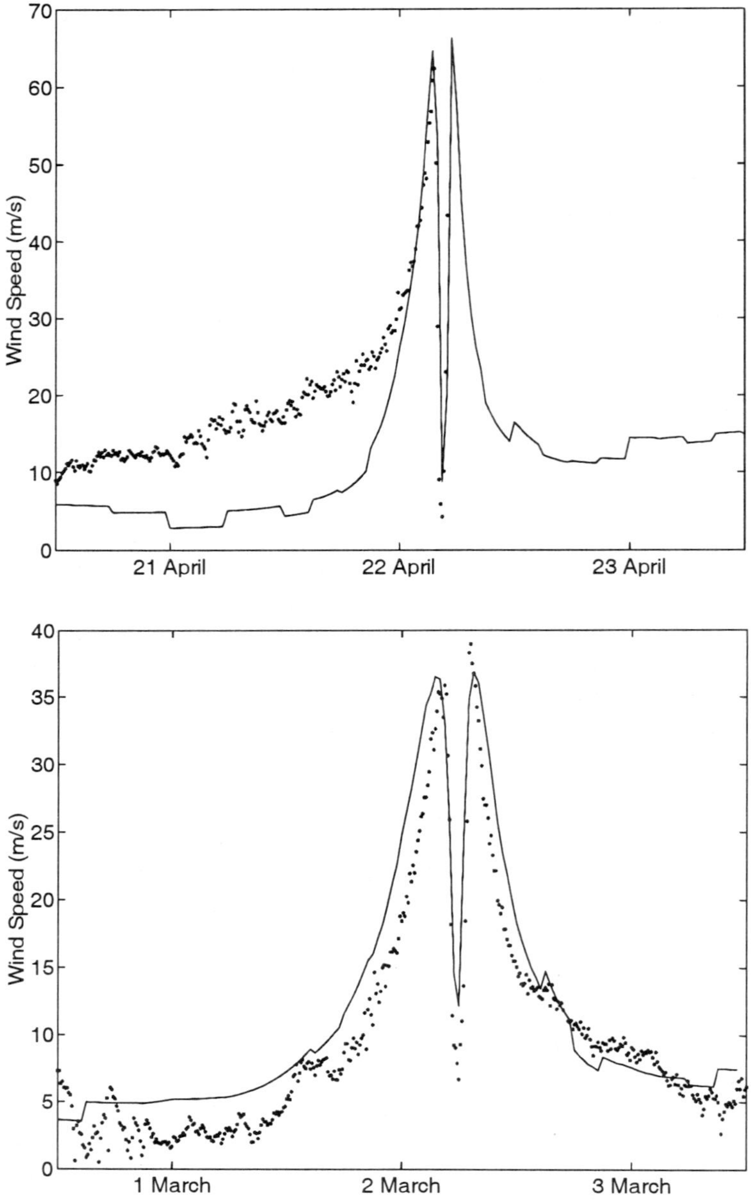

Figure 6.11: (a)[top] Comparison between recorded (dots) and model (solid line) values of U_{10} at North Rankin A platform (−19.58°S, 116.14°E) during the passage of tropical cyclone Orson in 1989. (b) [bottom] Comparison between recorded (dots) and model (solid line) values of U_{10} at Barrow Island (−20.82°S, 115.39°E) during the passage of tropical cyclone Ian in 1992. [after Young (1993)].

Much of the hurricane wave data collected using in situ instruments has been obtained for commercial purposes and consequently, only a small fraction is in the public domain. The first comprehensive attempt to collect data under hurricane conditions was the Ocean Data Gathering Program (Shemdin, 1977). This program involved instruments on six offshore oil platforms in the Gulf of Mexico. These systems operated between 1968 and 1971, during which data were collected for four hurricanes, including Camille. The program has been described in detail by Ward (1974) and Hamilton and Ward (1974).

Since 1972, the US NOAA Data Buoy Office has deployed buoys at various locations around the US coast (NOAA Data Buoy Office, 1973). Hurricane data from these buoys have been presented in a number of publications including Withee and Johnson (1975).

In addition to the US data set a significant hurricane (tropical cyclone) wave data base has been established by the offshore oil industry on the north-west coast of Australia. Data have been obtained from more than 25 hurricanes over a period in excess of twenty years (Harper et al, 1993).

Although non-directional observations of hurricane wave conditions are valuable, a more comprehensive understanding can be obtained from directional measurements. Directional spectral observations during the passage of hurricanes have been reported by Forristall et al (1978) and Black (1979). Remotely sensed data using Synthetic Aperture Radar data have also provided insight into the directional properties of waves within hurricanes (Elachi et al, 1977; King and Shemdin, 1978; Shemdin, 1980; Gonzalez et al, 1982; McLeish and Ross, 1983; Beal et al, 1986; Holt and Gonzalez, 1986).

Numerous models have been proposed for the prediction of waves within hurricanes. These range from the largely empirical (Bretschneider, 1959, 1972; Ijima et al, 1968; Ross, 1976; Young, 1988b) to those based on solution of the radiative transfer equation (Patterson , 1972; Bea, 1974; Uji, 1975; Cardone et al, 1977; Young, 1988a).

Based on knowledge of the hurricane wind field [see Section (6.5.1)] and the composite data base of wave observations, a qualitative description of the hurricane wave field emerges. The wind field is asymmetric, with higher winds to the right (northern hemisphere) of the hurricane centre. The wave field has an even greater degree of asymmetry due to the combined influence of the asymmetry of the wind field and the extended fetch which exists within a translating hurricane. The wind vector in the intense wind region to the right of the storm centre (northern hemisphere) is approximately aligned with the direction of forward propagation. Hence, waves generated in this region tend to move forward with the hurricane and therefore remain in high wind regions for an extended period of time [see Figure 6.12]. In contrast, waves generated on the low wind side of the storm (left side in northern hemisphere) propagate in the opposite direction to the hurricane translation and rapidly move away from the high wind areas. Low frequency waves generated in the intense wind regions to the right of the storm centre will have group velocities which are typically greater than the velocity of forward movement of the storm, V_{fm}. Hence, these components will "out run" the storm and appear as swell ahead of the storm centre. This swell radiates out from the spatially localized generation area. Hence, the spectrum ahead of the storm often consists of locally generated wind sea together with swell radiating from the centre of the storm. The two wave systems often intersect at angles near 90°. As they are significantly separated in both frequency and direction, they can co-exist without any significant non-linear interaction between the two systems [see Section (6.2)].

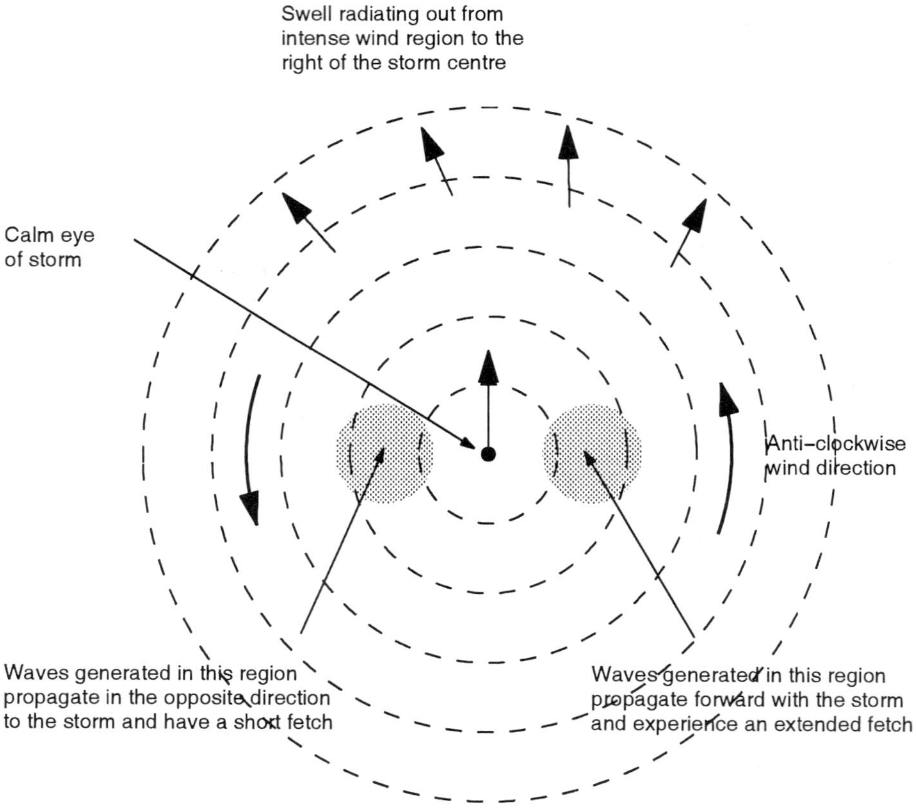

Figure 6.12: Schematic diagram showing the generation of waves within a translating hurricane. The hurricane shown is translating "up the page" as shown by the arrow at the centre of the storm. The wave field is characterized by: (a) swell ahead of the storm, radiating out from the intense wind region to the right of the storm centre and (b) significant asymmetry caused by the higher winds and extended translating fetch to the right of the storm centre.

The concept of an "extended fetch" which occurs due to the forward motion of the storm is also important in determining the wave height within a hurricane. Not only is the wave height determined by the maximum wind speed, V_{max}, but also by the period of time the waves remain within the intense wind region (Shemdin, 1977). As V_{max} increases, the period of the waves generated and hence the group velocity at which they propagate also increases. Consequently, V_{fm} must also increase for the most severe wave conditions to occur. Should the hurricane move too slowly, the waves will "outrun" the hurricane, whereas should the hurricane move too rapidly, the waves will be left behind. Consequently, for a given value of V_{max}, the wave height could be expected to increase with increasing V_{fm} until a maximum is reached. A further increase in V_{fm} will result in a decrease in wave height. Similarly, the spatial distribution of wave height (asymmetry) will also depend on both V_{max} and V_{fm}.

An attempt to incorporate this qualitative understanding of the hurricane wave field into a quantitative model was made by Young (1988b). Young (1988b) assumed that the JONSWAP (Hasselmann et al, 1973) relationships, originally developed for fetch limited conditions [see Section (5.3.2.3)], could also be applied in hurricane wind fields with the specification of a suitable "equivalent fetch". Based on a synthetic data base generated by a numerical model, the equivalent fetch was determined in terms of V_{max}, V_{fm} and R. Although the model is empirical, it implicitly incorporates the intuitively consistent non-dimensional scaling of JONSWAP, together with the parameters observed to be important in determining the wave field within a hurricane.

Given values of V_{max}, V_{fm} and R, Young (1988b) found that an equivalent fetch, F, could be represented by the relationship

$$\frac{x}{R'} = \psi\left[aV_{max}^2 + bV_{max}V_{fm} + cV_{fm}^2 + dV_{max} + eV_{fm} + f\right] \tag{6.8}$$

where $a = -2.175 \times 10^{-3}$, $b = 1.506 \times 10^{-2}$, $c = -1.223 \times 10^{-1}$, $d = 2.190 \times 10^{-1}$, $e = 6.737 \times 10^{-1}$ and $f = 7.980 \times 10^{-1}$. The term ψ is a scaling factor which in the representation of Young (1988b) was taken as 1. The term R' can be found in terms of R as

$$R' = 22.5 \times 10^3 \log R - 70.8 \times 10^3 \tag{6.9}$$

In (6.8) and (6.9) all terms are in standard S.I. units (ie. V_{max}, V_{fm} - [m/s]; F, R, R' - [m]). The maximum significant wave height in the hurricane, H_s^{max} can then be determined from a modified form of the JONSWAP Hasselmann et al (1973) relationship

$$\frac{gH_s^{max}}{V_{max}^2} = 0.0016\left(\frac{gx}{V_{max}^2}\right)^{0.5} \tag{6.10}$$

The spatial distributions of H_s could then be determined in terms of H_s^{max} using a series of diagrams for different values of V_{max} and V_{fm}.

Examples of the spatial distribution of H_s are shown in Figure 6.13. The cases shown all have V_{max} = 40 m/s, but with V_{fm} increasing from 2.5 m/s to 10 m/s. For slow moving storms the spatial distribution is similar to the wind speed distribution [see Figure 6.9]. As V_{fm} increases the asymmetry of the wave field increases significantly as both the asymmetry of the wind field and the extended fetch to the right of the storm increases in significance. For the largest values of V_{fm} the region of maximum waves moves from the right forward quadrant to the right rear quadrant. In these cases it is clear in the wave height distribution that the storm is propagating so quickly that it is "out running" the waves.

The accuracy of the results of the approach proposed by Young (1988b) is limited by the fact that the synthetic data base used, was generated by a numerical model. To assess the applicability of this approach Young and Burchell (1996) compared the results of the model with measurements of H_s obtained from overflights of hurricanes by the GEOSAT satellite altimeter. This satellite database was extensive, consisting of more than 100 hurricanes. Based on these data, Young and Burchell (1996) concluded that the model described by (6.8) contains a systematic bias and proposed that the scaling factor ψ should take the form

$$\psi = -0.015V_{max} + 0.0431V_{fm} + 1.30 \tag{6.11}$$

where again V_{max} and V_{fm} have units of [m/s] and ψ is dimensionless. The resulting distribution of equivalent fetch x/R' obtained from (6.8) and (6.11) is shown in Figure 6.14.

6.5.3 Hurricane Wave Spectra

The vast majority of the studies discussed in Section (6.5.2) considered integral properties of the hurricane wave field such as significant wave height and peak frequency. Studies which have considered detailed analysis of recorded wave spectra are far more rare. Whalen and Ochi (1978) investigated the spectra recorded during the passage of U.S. Hurricanes Camille, Celia, Edith and Eloise. As with a number of previous observations, they found that as the hurricane approached the recording site, the spectra were initially bi-modal. As the hurricane moved closer, the spectra became uni-modal. They attempted to model the spectra using both the JONSWAP form, (5.51) (Hasselmann et al, 1973) and the 6 parameter form proposed by Ochi and Hubble (1976). The Ochi and Hubble (1976) spectral form can represent bi-modal spectra and consequently, it provided a better fit to the data than JONSWAP for such spectra. The JONSWAP spectrum, however, provided a good approximation to the data for the uni-modal spectra. They found that the data yielded mean JONSWAP parameters of $\gamma = 2.2$ and $\alpha = 0.023$. They also attempted to represent the parameters of the Ochi and Hubble (1976) spectrum in terms of the significant wave height, H_s. Although there appeared to be some dependence on H_s, there was significant scatter in the results. In view of the results of Shemdin (1980), Young (1988b) and Young and Burchell (1996), which clearly demonstrate that H_s is a function of the relative position of the hurricane, the velocity of forward movement of the hurricane, V_{fm} and the maximum wind speed within the hurricane, V_{max}, it seems unlikely that a bi-modal spectral form could be parameterized in terms of H_s, alone. The apparent dependence in the Whalen and Ochi (1978) data on H_s is most likely due to the very small number of hurricanes considered. Although there were four hurricanes in the total data base, the vast majority of the spectra were obtained from Hurricane Eloise.

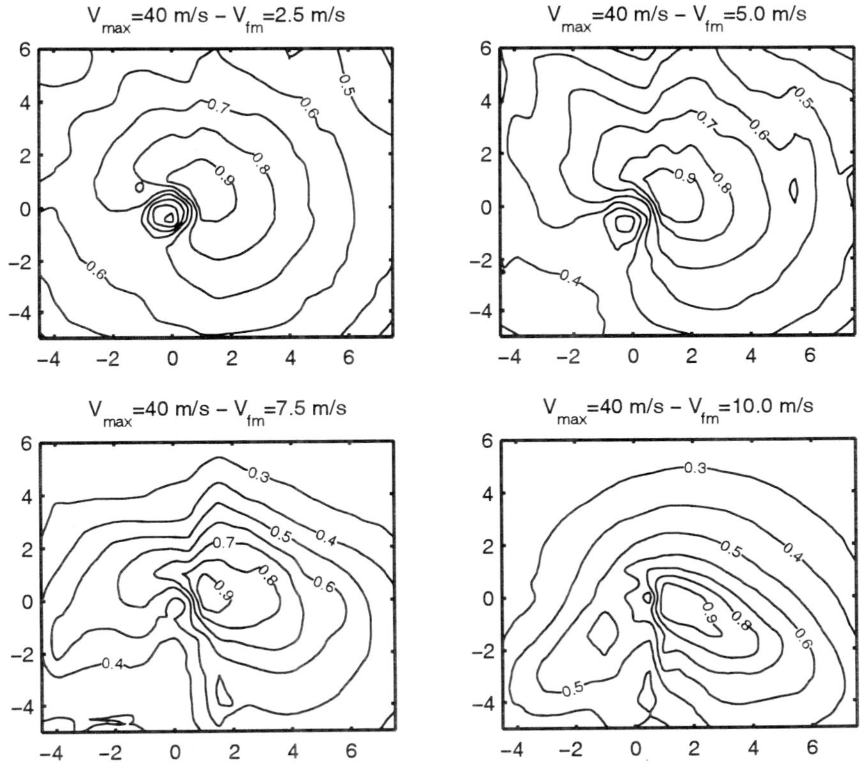

Figure 6.13: The wave height distribution as predicted by the numerical model results of Young (1988b). The spatial scale is in terms of r/R' and the contours are of normalized significant wave height. In each case the wave field has been normalized such that the maximum value is 1. In all cases the maximum wind velocity in the storm, $V_{max} = 40$ m/s. The velocity of forward movement gradually increases in each panel: (a) $V_{fm} = 2.5$ m/s, (b) $V_{fm} = 5.0$ m/s, (c) $V_{fm} = 7.5$ m/s, (d) $V_{fm} = 10.0$ m/s.

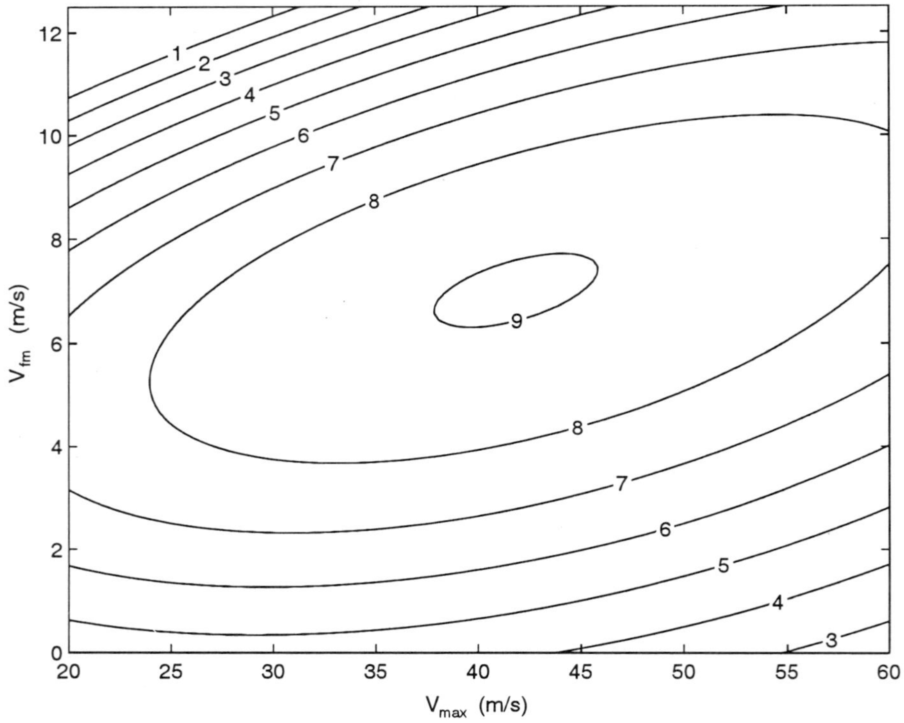

Figure 6.14: Contours of the equivalent fetch within a hurricane as proposed by Young and Burchell (1996). The contours are of normalized equivalent fetch, x/R'. The figure was generated from (6.8) and (6.11). The surface takes the form of a ridge tilted towards the top right of the figure. Therefore, as storms become more intense (increase in V_{max}) they must move increasing more rapidly (larger values of V_{fm}) to generate the maximum possible waves for that intensity storm. Note that the relationships developed by Young and Burchell (1996) are not applicable outside the parameter range shown in this figure.

Ochi and Chiu (1982) considered spectra recorded for U.S. Hurricane David. Again they confirmed that the JONSWAP form could model uni-modal spectra well and that the bi-modal spectra generated when the hurricane was distant from the measurement site were well represented by the Ochi and Hubble (1976) form.

A detailed study of the properties of the one-dimensional spectrum, $F(f)$, under hurricane conditions was presented by Young (1997a). Young (1997a) considered data recorded over a 16 year period off the north-west coast of Australia. Spectra obtained from 16 hurricanes (tropical cyclones) were considered. The hurricanes had central pressures ranging from 905 HPa to 985 HPa. Hence, the data set was quite comprehensive. Young (1997a) found that the spectra were bi-modal when the storm centres were more than 8 times the radius to maximum winds, R, from the measurement site. Spectra recorded within $8R$ of the centre of the storm, however, appeared similar to fetch limited spectra. This occured despite the fact that the wind fields were changing rapidly in direction and that the wave field consisted of a mix of locally generated waves and distant swell. Young (1997a) concluded that the shape stabilizing effects of the nonlinear source term [see Sections (4.4.2.2) and (6.4)] were sufficient to continually force the spectra back to the uni-modal form. As the swell and wind sea components became more separated in frequency and direction space (i.e. greater than $8R$ from the storm centre), the nonlinear coupling reduced in strength, and hence the spectra were bi-modal.

In order to assess whether the parametric spectral forms represented by JONSWAP (5.51) or Donelan et al (1985) (5.53) may be suitable to model hurricane spectra, Young (1997a) examined the exponent n in the expression $F(f) \propto f^n$. Figure 6.15 shows a scatter plot of n as a function of the inverse wave age, U_{10}/C_p, where C_p is the phase speed of components at the spectral peak frequency. There is significant scatter in the data and no clear evidence that either of the proposed forms f^{-4} or f^{-5} is the preferred form for the high frequency region of the spectrum. The mean value is $n = -4.56$, indicating either spectral form could be applied to the data. Such a result is consistent with the variability of the exponent n reported by Liu (1989) for fetch limited spectra [see Figure 5.25].

Also shown in Figure 6.15 is the demarcation between wind-sea and swell, $U_{10}/C_p = 0.83$ proposed by Donelan et al (1985). Based on this limit, a significant proportion of the spectra of this data set would normally be characterized as swell (ie. $U_{10}/C_p < 0.83$). Despite this, these spectra are still uni-modal.

As the mean value of n is almost mid way between -4 and -5, Young (1997a) concluded that either of the forms (5.51) or (5.53) could be used to represent the spectra. The parameters of these two formulations when applied to hurricane wave spectra are discussed in Sections (6.5.3.1) and (6.5.3.2).

6.5.3.1 JONSWAP Representation of Hurricane Spectra

As the spectral width parameter, σ, has little influence on the ultimate spectral form, and as previous fetch limited studies had found no systematic trend within

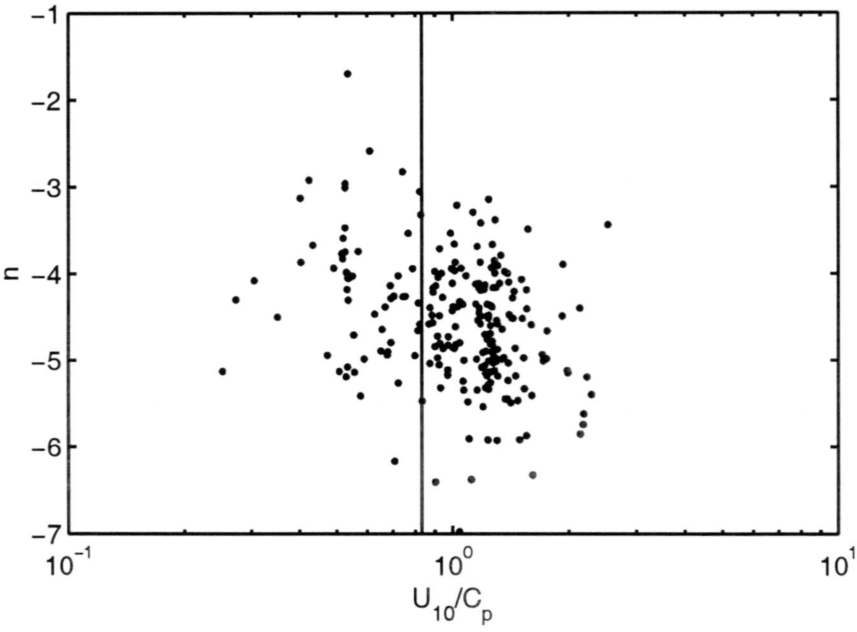

Figure 6.15: Values of the exponent n in the relationship $F(f) \propto f^n$ plotted as a function of the inverse wave age U_{10}/C_p for the hurricane data of Young (1997a). The vertical line is drawn at $U_{10}/C_p = 0.83$, the demarcation between swell and wind-sea proposed by Donelan et al (1985).

the scatter of values, Young (1997a) assumed a constant value of this parameter of $\sigma = 0.1$. Figures 6.16a and 6.16b shown the values of α and γ, respectively, for the JONSWAP form (5.51). Figure 6.16a also shows the JONSWAP (Hasselmann et al, 1973) relationship for α expressed in terms of U_{10}/C_p.

$$\alpha = 0.008\,(U_{10}/C_p)^{0.73} \tag{6.12}$$

Although there is significant scatter within the data, the values of α clearly increase with increasing U_{10}/C_p. The JONSWAP relationship (6.12), which was developed for fetch limited waves, is a remarkably good fit to the data. This is a clear demonstration of the shape stabilization provided by non-linear interactions. Even though the wind fields from which these spectra were obtained are spatially and temporarily variable, the spectra are remarkably similar to those obtained from idealized fetch limited situations.

The same dependence between α and U_{10}/C_p continues for $U_{10}/C_p < 0.83$. This suggests that either the nonlinear interactions are capable of maintaining the spectral shape in the absence of active wind input or that components with $U_{10}/C_p < 0.83$ are still receiving such input. Naturally, the frequency components in the tail of the spectrum will still be propagating significantly slower than both the wind and the components at the spectral peak. Hence, they will still receive active atmospheric input even when $U_{10}/C_p < 0.83$. The fact that the components at the spectral peak can achieve such small values of inverse wave age occurs either since the local wind speed has decreased within the rapidly changing wind field vortex or that they have propagated to the measurement site from another location.

Figure 6.16b shows γ as a function of U_{10}/C_p. As with fetch limited results, there appears to be no systematic dependence on U_{10}/C_p. The mean value obtained from the data set is $\gamma = 1.9$, consistent with the value of $\gamma = 2.2$ obtained by Whalen and Ochi (1978) from their much smaller North American data set. The values of γ are consistently smaller than the mean JONSWAP value of $\gamma = 3.3$. Therefore, it appears that although much of the recorded energy has presumably propagated to the measurement stations from other locations as swell, it has been transformed locally by a combination of atmospheric input and nonlinear interactions into relatively broad wind-sea spectra.

6.5.3.2 Donelan et al (1985) representation of Hurricane Spectra

Figures 6.17a and 6.17b show the parameters β and γ_d, respectively as functions of U_{10}/C_p, obtained for the Donelan et al (1985) spectral form (5.53). The results are qualitatively similar to those shown in Figures 6.16 for the JONSWAP spectral form. Based on their fetch limited data, Donelan et al (1985) proposed the following power law relationship between β and U_{10}/C_p.

$$\alpha = 0.006\,(U_{10}/C_p)^{0.55} \tag{6.13}$$

Equation (6.13) is shown in Figure 6.17a together with the present data. The relationship represented by (6.13) is a remarkably good fit to the data, again indicating that the spectra are similar to fetch limited wind-sea spectra.

In contrast to JONSWAP (Hasselmann et al, 1973), Donelan et al (1985) found a dependence of γ_d on U_{10}/C_p. They proposed the following functional form

$$\gamma_d = \begin{cases} 1.7 & \text{for } 0.83 < U_{10}/C_p < 1 \\ 1.7 + 6.0 \log_{10}(U_{10}/C_p) & \text{for } 1 \le U_{10}/C_p < 5 \end{cases} \tag{6.14}$$

Equation (6.14) is shown in Figure 6.17b, together with the present data. There is significant scatter in the values of γ_d obtained from the hurricane spectra and no clear trend is apparent within this scatter. Equation (6.14) is, however, consistent with the data set, indicating the peak enhancement of the present data is comparable to that observed for fetch limited spectra.

The parametric representations of the data for both the JONSWAP and Donelan et al (1985) forms yield the same general conclusions. Despite the complex temporally and spatially variable wind fields of hurricanes, the spectra are remarkably similar to those measured under idealized fetch limited conditions. In addition,

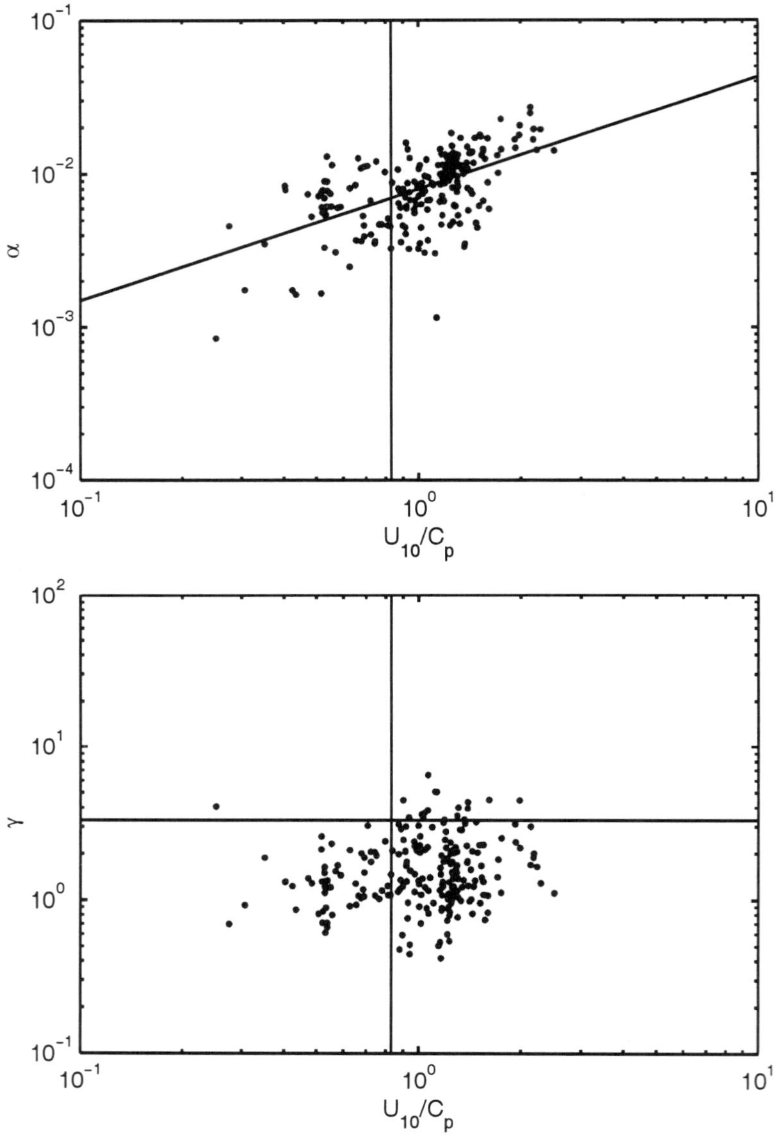

Figure 6.16: JONSWAP spectral parameters for hurricane conditions. (a) [top] Values of the parameter α as a function of the inverse wave age, U_{10}/C_p. The line through the data is the JONSWAP fetch limited relationship (6.12) and the vertical line is drawn at $U_{10}/C_p = 0.83$. (b) [bottom] Values of the parameter γ as a function of the inverse wave age, U_{10}/C_p. The horizontal line represents the mean JONSWAP value, $\gamma = 3.3$ and the vertical line is drawn at $U_{10}/C_p = 0.83$.

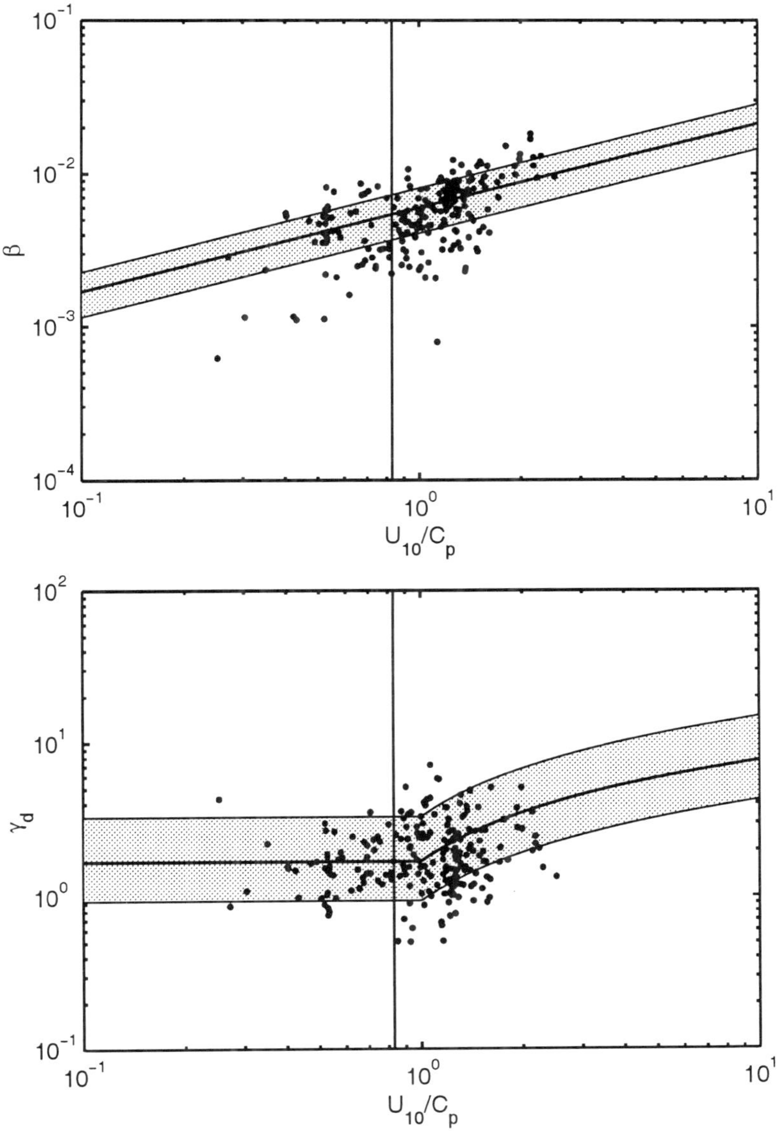

Figure 6.17: Donelan et al (1985) spectral parameters for hurricane conditions. (a) [top] Values of the parameter β as a function of the inverse wave age, U_{10}/C_p. The line through the data is the Donelan et al (1985) fetch limited relationship (6.13) and the vertical line is drawn at $U_{10}/C_p = 0.83$. The shaded region gives an estimate of the magnitude of the 95% confidence limits on the individual data points. (b) [bottom] Values of the parameter γ_d as a function of the inverse wave age, U_{10}/C_p. The line through the data is the Donelan et al (1985) fetch limited relationship (6.14) and the vertical line is drawn at $U_{10}/C_p = 0.83$. The shaded region gives an estimate of the magnitude of the 95% confidence limits on the individual data points.

the spectra would generally be characterized as mature wind-sea. The vast majority of the data are for $U_{10}/C_p < 2$. Indeed, significant data are in the region, $U_{10}/C_p < 0.83$, which would normally be classified as swell. The functional dependence of the spectral parameters for these values with $U_{10}/C_p < 0.83$ is, however, consistent with the more highly forced data. Hence, the processes which force the spectra to conform to a parametric spectral form (such as nonlinear interactions), apparently remain active in the absence of significant wind input at the spectral peak frequency.

Chapter 7

Finite Depth Effects

7.1 Introduction

In the previous chapters, attention was focused almost exclusively on deep water conditions. As shown in Section (2.2.8), waves for which $kd < \pi$ will be influenced by the bottom. Hence, additional processes become important in coastal regions. The finite depth kinematic processes of shoaling and refraction have already been considered in Section (2.3) and will not been considered further. Similarly, diffraction, in which energy spreads laterally perpendicular to the principal direction of propagation is common around islands and "man-made" coastal structures. The literature on diffraction is extensive with the classical paper by Penney and Price (1952) representing an important contribution. More recent contributions have been summarized in the comprehensive treatment of the subject provided by Mei (1989).

Neglecting the processes mentioned above, the total source term representing the physical processes active in finite depth conditions becomes [see (4.23)]

$$S = S_{in} + S_{nl} + S_{ds} + \underbrace{S_{bf} + S_{brk} + S_{tri} + \cdots}_{S_b,\ \text{Finite depth processes}} \tag{7.1}$$

where S_{in}, S_{nl} and S_{ds} are the processes of atmospheric input, quadruplet nonlinear interaction and white-cap dissipation, respectively. These processes have already been discussed in the context of deep water wave evolution in Chapter 4. Processes which occur due to the presence of finite depth are collectively represented by S_b and physically represent: S_{bf} - bottom friction, S_{brk} - depth limited wave breaking and S_{tri} - triad nonlinear interactions. In addition to these finite depth mechanisms, other processes such as: percolation within the bed material, scattering of energy from bottom objects and wave induced motion of soft bed material may also play a role (Hsiao and Shemdin, 1978; Young and Gorman, 1995). As the importance of these processes is yet to be clearly demonstrated, they will not be considered further. In certain circumstances, however, they may be of significance.

This chapter will firstly discuss the physical processes represented in (7.1) and then investigate the observational data base for finite depth waves.

7.2 Physical Processes

The first three processes represented in (7.1) [S_{in}, S_{nl}, S_{ds}] have been described as *deep water processes.* This is true in the sense that they are active in deep water. In finite depth situations their form and relative magnitudes change and these processes will be considered in the following sections. Of the finite depth processes explicitly represented in (7.1), S_{brk} and S_{tri} are generally believed to be significant only in the shoaling zone very close to shore. Hence, these two processes are generally neglected in the finite depth water of the continental shelf, seaward of the shoaling and surf zones. Bottom friction is, however, important on the shelf.

Therefore, this section firstly considers the processes of: S_{in}, S_{nl}, S_{ds} and S_{bf} followed by an examination of their relative magnitudes in finite depth situations. The two *shoaling zone* processes of S_{brk} and S_{tri} are then considered separately.

7.2.1 Atmospheric Input in Finite Depth Situations, S_{in}

All measurements of the atmospheric input term, S_{in} have been confined to deep water situations. Hence, current knowledge of this mechanism is restricted to extrapolations of the deep water theory. Deep water forms of S_{in} are generally related in some way to the ratio U_{10}/C [e.g. (4.44)]. In some proposals u_* has been used in place of U_{10} and the term has also been expressed as the square of this ratio. The phase speed, C, is reduced in finite depth water compared to deep water conditions [see (2.15)] and hence, for a given value of U_{10}, the ratio U_{10}/C and therefore S_{in} will be larger in finite depth conditions compared to deep water. Figure 7.1a shows an example of the evaluation of S_{in} using (4.44) with $\beta = 0.25$ (Snyder et al, 1981). Four cases are shown with: $k_pd = 0.50,\ 0.65,\ 1.00,\ \infty$. In each case the evaluation has been performed assuming a mean JONSWAP spectrum (5.51) with parameters: $\alpha = 0.01$, $f_p = 0.3$ Hz, $\sigma_a = \sigma_b = 0.08$ and $\gamma = 3.3$. The wind speed for this example was $U_{10} = 10$ m/s.

At the peak of the spectrum the atmospheric input increases by a factor of almost 4 as the water depth decreases from $k_pd = \infty$ (deep water) to $k_pd = 0.5$. In contrast, at higher frequencies, where the spectral components are largely unaffected by the water depth, there is little change in magnitude of the atmospheric input. In addition, atmospheric input is limited to spectral components for which $U_{10}/C > 1$. As C is reduced in magnitude in finite depth conditions, energy can be transferred to lower frequency components than in deep water.

Although it is reasonable to assume that the magnitude of the atmospheric input is related to the relative differences in speed of the wind and the waves, there is

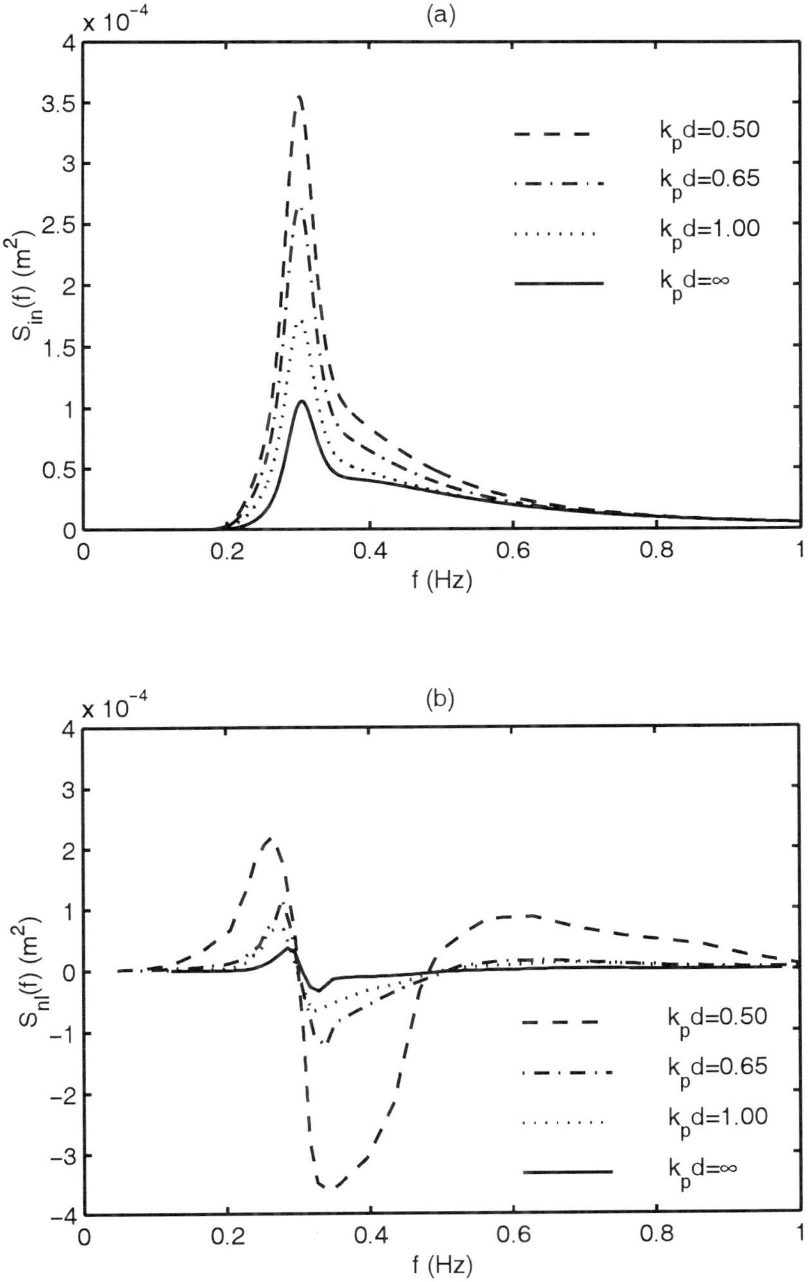
(a)
x 10^-4
S_in(f) (m^2)
f (Hz)
k_p d=0.50
k_p d=0.65
k_p d=1.00
k_p d=∞
(b)
x 10^-4
S_nl(f) (m^2)
f (Hz)
k_p d=0.50
k_p d=0.65
k_p d=1.00
k_p d=∞

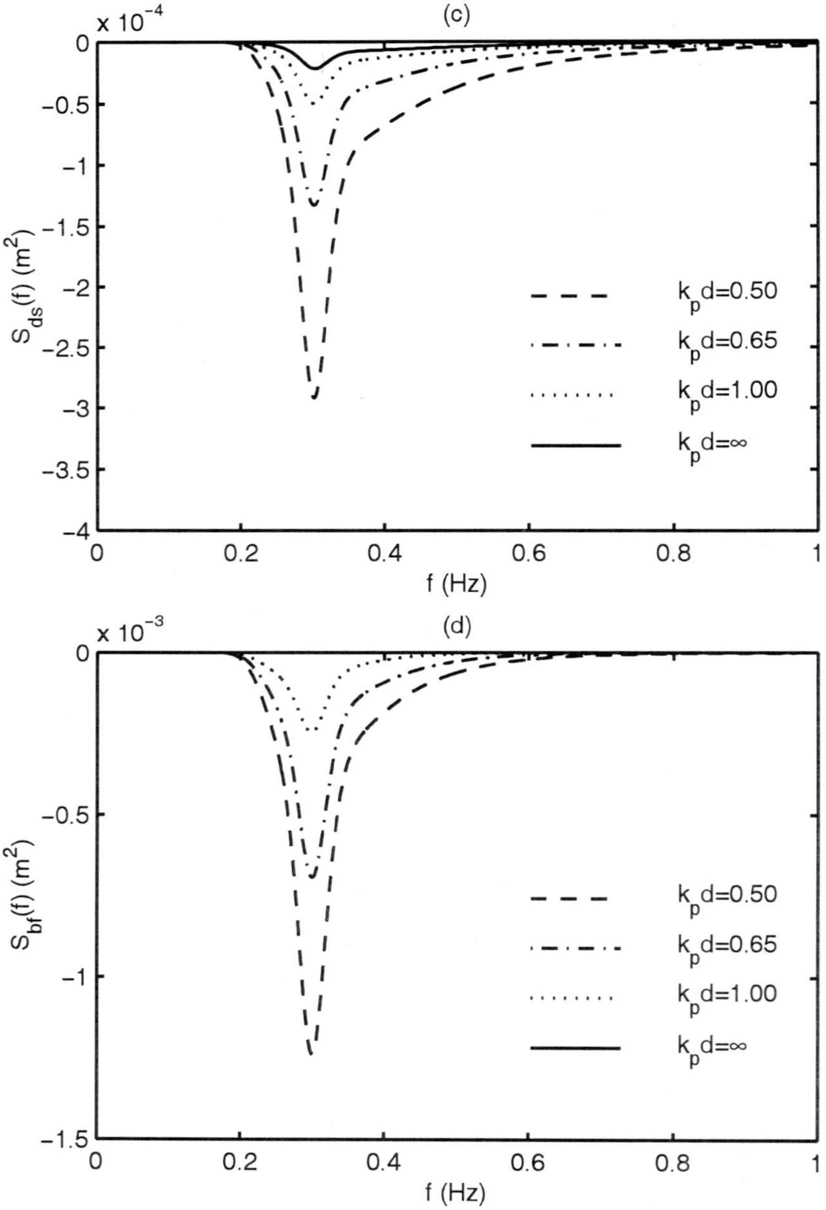

Figure 7.1: Evaluation of the magnitude of source terms as a function of water depth. Source terms are shown for $k_p d = 0.50,\ 0.65,\ 1.00,\ \infty$. In each case the evaluation has been performed assuming a mean JONSWAP spectrum (5.51) with parameters: $\alpha = 0.01$, $f_p = 0.3$ Hz, $\sigma_a = \sigma_b = 0.08$ and $\gamma = 3.3$. The source terms shown are: (a) S_{in} (4.44), (b) S_{nl} (4.55), (c) S_{ds} (7.2) and (d) S_{bf} (7.8).

no experimental evidence that the results shown in Figure 7.1a are quantitatively correct. Other processes might also influence the boundary layer flow in such conditions. For instance, finite depth waves are, on average, steeper than their deep water counterparts. This may, in turn, influence the aerodynamic roughness and hence the atmospheric input.

7.2.2 Nonlinear Quadruplet Interactions in Finite Depth Situations, $\mathbf{S_{nl}}$

Finite depth has two significant effects on the nonlinear Boltzmann integral (4.55). Firstly, the changed form of the dispersion relationship (2.13), alters the quadruplets which satisfy the resonance conditions (4.53) and (4.54). Secondly, the magnitude of the coupling coefficient, G increases with decreasing depth. Young et al (1996) investigated the changes to the resonance criteria as a function of water depth. As shown in Section (4.4), for given values of $\mathbf{k}_1$ and $\mathbf{k}_3$, the values of $\mathbf{k}_2$ and $\mathbf{k}_4$ which satisfy the resonance criteria, (4.53) and (4.54), trace out "egg-shaped" loci [see Figure 4.10]. Figure 7.2 shows the loci of the values of $\mathbf{k_2}$ as a function of water depth for the case where $f_1 = 0.3$ Hz ($\omega_1 = 2\pi f_1$) and $f_3 = 0.6$ Hz. Both f_1 and f_3 are directed along the x axis of the diagram. These values of f_1 and f_2 were chosen since $f_p \approx 0.3$Hz is the example which has been used in Figure 7.1. For this combination of f_1 and f_3, only values of $\mathbf{k_2}$ which lie on the locus for that depth will satisfy the resonance conditions and interact nonlinearly to transfer energy. The loci of $\mathbf{k_4}$ are not shown here but will be identical in shape to those of $\mathbf{k_2}$, although displaced along the x axis.

The loci of Figure 7.2 gradually increase in size as the depth decreases. Therefore, wave numbers more widely separated in both magnitude and direction have the potential to interact in shallow water compared to deep. Although only one combination of $\mathbf{k_1}$, $\mathbf{k_3}$ is presented here, the same general trend holds for other combinations. As the water depth decreases, wave number components separated by progressively larger angles can satisfy the resonance conditions and potentially interact nonlinearly.

In addition to determining the components which satisfy the resonance conditions and thus potentially interact to exchange energy, the strength of the interaction is also important. This can be measured by the coupling coefficient G in (4.55). Figure 7.3 shows the magnitude of the coupling coefficient, G as a function of $|\mathbf{k_2}|$ for each of the loci in Figure 6.1. The magnitude of the coupling generally increases with increasing $|\mathbf{k_2}|$. In addition, however, G also increases as the water depth decreases. Hence, the strength of the nonlinear coupling is larger in shallow water than in deep water.

The results shown in Figures 7.2 and 7.3 suggest that the magnitude of S_{nl} should increase with decreasing water depth. This is confirmed in Figure 7.1b which shows S_{nl} evaluated for the mean JONSWAP spectrum and a variety of water depths. The magnitude of S_{nl} clearly increases significantly as k_pd decreases.

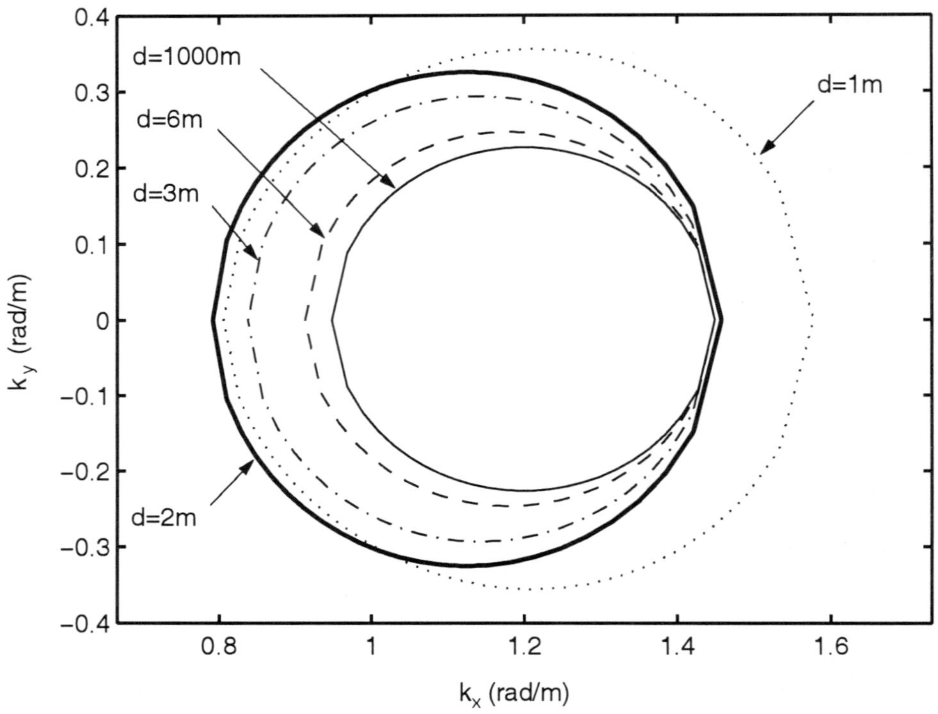

Figure 7.2: Loci of the values of $\mathbf{k_2}$ which satisfy the resonance conditions (4.53) and (4.54) for $f_1 = 0.3$ Hz and $f_3 = 0.6$ Hz. Both f_1 and f_3 are directed parallel to the x axis. Loci are shown for a number of water depths, as marked. [after Young et al (1996)].

The general "plus-minus" structure is maintained as k_pd decreases. In deep water, the transfer is very narrow, confined to the region immediately surrounding the spectral peak. As the water depth decreases, the width of both the positive and negative lobes of the transfer increases. This is consistent with the fact that the interaction locus increases in size as k_pd decreases, as shown in Figure 7.2. Therefore, components more widely separated in wave number and direction can interact nonlinearly in finite depth water than in deep water. One might expect this change in the nonlinear coupling to result in an altered shape to the spectrum in finite depth conditions [see Section (7.4)].

Hasselmann and Hasselmann (1981) neglected the changes to the shape of S_{nl} with decreasing water depth and proposed that S_{nl} could be represented in finite depth situations by a simple scaling factor applied to the deep water nonlinear transfer rate. This scaling factor was determined by comparing the magnitudes of the peaks of the positive and negative lobes of S_{nl} in deep and finite depth water, respectively. Although this approach does not capture the shape changes which occur in S_{nl} as water depth decreases, it provides a convenient, first order, repre-

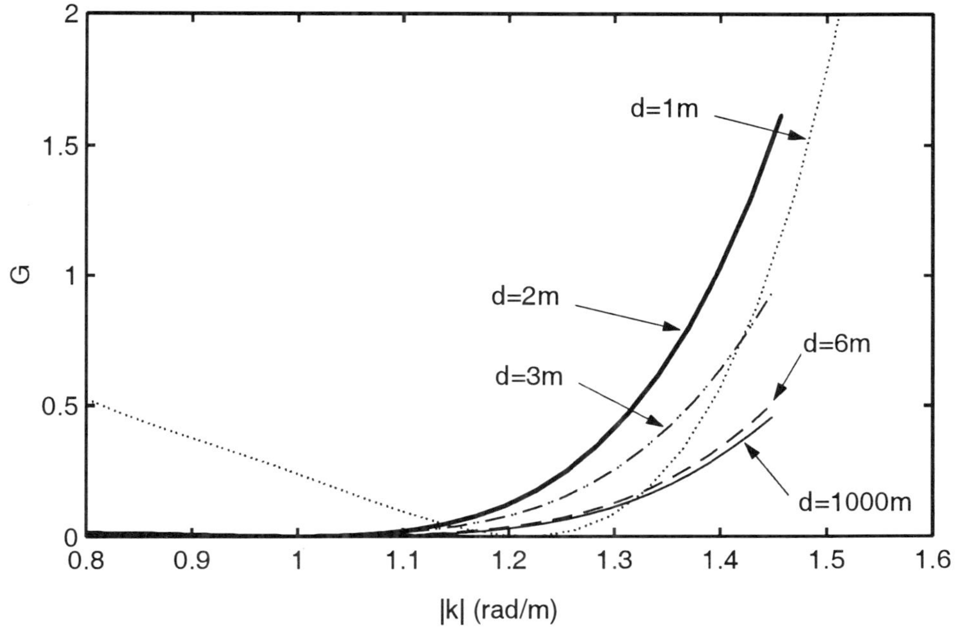

Figure 7.3: Values of the coupling coefficient, G (4.55) as a function of $|\mathbf{k_2}|$ as shown in Figure 7.2. Values are shown for the same water depths depicted in Figure 7.2. [after Young et al (1996)].

sentation of the change in magnitude of the nonlinear transfer. The Hasselmann and Hasselmann (1981) scale factor is shown as a function of $k_p d$ in Figure 7.4.

Hasselmann and Hasselmann (1981) claim that the directional distribution of S_{nl} is essentially unaltered by water depth. The results shown in Figure 7.2 would, however, suggest that a broader directional distribution would exist in finite depth conditions. Detailed calculations to determine the magnitude of such effects are, however, yet to be presented.

The enormous increase in the magnitude of S_{nl} shown in Figure 7.4 for $k_p d < 0.5$ raises questions about the applicability of the nonlinear perturbation theory upon which (4.55) is based. The theory presented in Section 4.4 is based on the assumption that the waves are weakly nonlinear and can be represented by a Taylor series expansion about the linear solution. For values of $k_p d < 0.5$ the magnitude of S_{nl} suggests a strongly nonlinear process. Hasselmann and Hasselmann (1981) suggest caution in the use of the theory for $k_p d < 0.5$, but a rigorous evaluation of the depth at which the theory becomes invalid is yet to be presented.

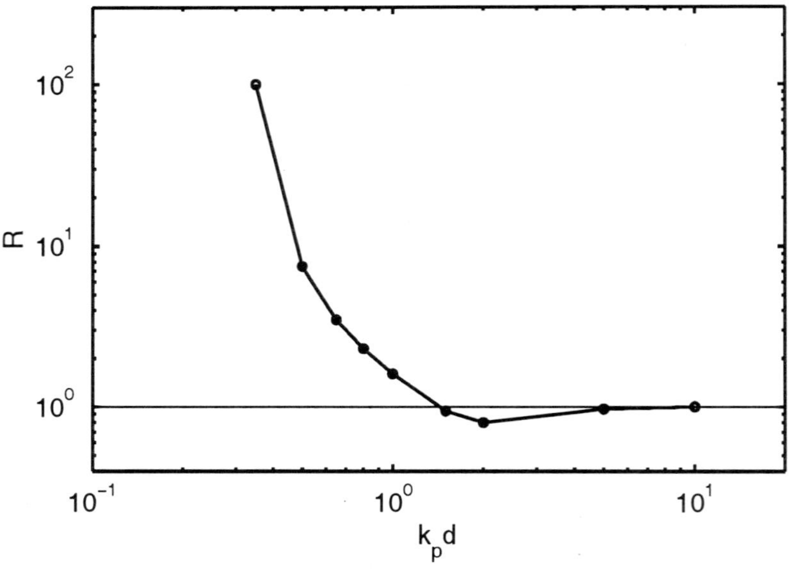

Figure 7.4: The ratio R of the nonlinear transfer for finite depth conditions to the transfer for deep water. The ratio was determine for the mean JONSWAP spectrum by Hasselmann and Hasselmann (1981) and is expressed as a function of the non-dimensional depth parameter, $k_p d$.

7.2.3 White-Cap Dissipation in Finite Depth Situations, S_{ds}

As indicated in Section (4.5), white-cap dissipation is the most poorly understood of the deep water source terms. As with atmospheric input, functional representations of S_{ds} for finite depth situations are extrapolated from the deep water formulation. Equation (4.67) defines a dissipation term which is quasi-linear in the spectrum, $F(f, \theta)$ and a function of the wave steepness. A source term predicated on white-capping being a function of wave steepness is tenuous. Numerous observations of ocean waves have shown that waves of apparently similar steepness my have quite different fates: one breaking, the other not. Despite these reservations, (4.67), represents the only plausible formulation available.

Equation (4.67) is formulated in terms of frequency and in order to ensure that the dependence on wave slope is the same in deep and finite depth conditions, it must be reformulated in terms of wave number, k (Weber, 1991). Rewriting (4.67) in this form yields (Weber, 1991).

$$S_{ds} = -1.59\overline{\omega}k/\overline{k}\left[\sigma^2\overline{k}^2\right]^2 F(f, \theta) \tag{7.2}$$

where $\overline{k}$ is the mean wave number, given by

$$\overline{k} = \left\{\frac{1}{\sigma^2}\int F(f, \theta)k^{1/2}\mathrm{d}f\mathrm{d}\theta\right\}^2 \tag{7.3}$$

As in (4.67), the total energy or variance of the spectrum is $\sigma^2 = \int F(f,\theta)\mathrm{d}f\mathrm{d}\theta$.

Section (2.3.1) indicates that as waves propagate from deep to shallow water, the wave length decreases. Hence, the wave slope and S_{ds} will increases. This is clearly seen in Figure 7.1c, where (7.2) is evaluated for a mean JONSWAP spectrum and a number of different water depths. This figure indicates that S_{ds} is approximately 10 times larger at the spectral peak frequency (f_p = 0.3 Hz) for $k_p d = 0.5$ compared to deep water. Thus, a significant increase in white-cap occurrence and severity could be expected in finite depth situations. Although such a conclusion is qualitatively consistent with observations, a quantitative validation still awaits developments in instrumentation which would allow the direct measurement of white-cap dissipation.

7.2.4 Bottom Friction Dissipation, S_{bf}

The process of the dissipation of wave energy at the seabed may be modelled by the action of a stress

$$\tau = -C_f \rho_w \mathbf{u_b} \left|\mathbf{u_b}\right| \tag{7.4}$$

on water moving at a velocity $\mathbf{u_b}$ just above the bottom boundary layer (Putnam and Johnson, 1949; Kajiura, 1968). The coefficient C_f may be expected to vary slowly with flow conditions. Its magnitude is discussed below, but a typical value of $C_f = 0.015$ has been widely used in the literature. It should be noted that (7.4) is often represented by the alternative form $\tau = -1/2 f_w \rho_w \mathbf{u_b} \left|\mathbf{u_b}\right|$, where f_w is a friction factor. Hence C_f equals $1/2 f_w$, and it is important not to confuse the two definitions (Hardy, 1993).

Hasselmann and Collins (1968) have analyzed the relationship between the bottom stress and the dissipation rate of each spectral component. They found that if the quadratic friction law (7.4) is applied, the dissipation rate due to bottom friction is

$$S_{bf}(f,\theta) = -C_f g \frac{k^2}{\omega^2} \frac{F(f,\theta)}{\cosh^2 kd} \left[\langle u \rangle + \frac{k_i k_j}{k^2} \langle \frac{u_i u_j}{u} \rangle \right] \tag{7.5}$$

assuming linear theory for the depth dependence. The angle brackets denote an ensemble mean, and the subscript b has been dropped: all fluid velocities from here on apply to the seabed and are taken to arise solely from wave action, with currents being neglected.

Collins (1972) made the approximation that the bracketed term may be replaced by the root-mean-square bed velocity $\left(\langle u^2 \rangle\right)^{1/2}$ obtained from the spectrum:

$$\langle u^2 \rangle = \langle u_1^2 \rangle + \langle u_2^2 \rangle \tag{7.6}$$

$$\langle u_i u_j \rangle = g^2 \int \int \frac{k_i k_j}{\omega^2} \frac{F(f,\theta)}{\cosh^2 kd} \mathrm{d}f \mathrm{d}\theta \tag{7.7}$$

As noted by Collins (1972) and demonstrated by Cavaleri et al (1989) the difference between this formulation and that of Hasselmann and Collins (1968) depends on the directional spreading of the wave spectrum. The formulation of Collins yields a smaller dissipation rate for the same value of C_f.

A further simplification has been used in the spectral model WAM (WAMDI, 1988), in which the source term is

$$S_{bf}(f,\theta) = -\Gamma \frac{\omega^2}{g^2} \frac{F(f,\theta)}{\sinh^2 kd} \tag{7.8}$$

with $\Gamma = 0.038$ m^{2}s^{-3}, chosen to fit the JONSWAP test cases (Hasselmann et al, 1973). In effect, this assigns the bed velocity a constant value of $\left(\langle u^2 \rangle\right)^{1/2} = 0.259$ m/s. Initially, this appears a very poor assumption, but as will be shown later, it is a reasonable approximation over a relatively wide range of conditions.

7.2.4.1 Bottom Friction Coefficient, $\mathbf{C_f}$

In order to apply any of the formulations described above, it is necessary to specify the bottom friction coefficient, C_f [or Γ in the case of (7.8)]. On the basis of the laboratory results of Jonsson (1965) and the field data of Bretschneider and Reid (1954), Hasselmann and Collins (1968) adopted a value of $C_f = 0.015$. This value has been subsequently adopted in a number of finite depth spectral wave models (Chen and Wang, 1983; Young, 1988a; Cavaleri et al, 1989). Rather than being a constant, the results of Jonsson (1965) show that C_f is a function of the near-bottom amplitude Reynolds number and the bottom roughness to semi-excursion ratio r/a. These results are presented in Figure 7.5 in the form of the Moody diagram commonly encountered in hydraulics.

The results of Jonsson (1965) have subsequently been confirmed by Kajiura (1968), Swart (1974), Kamphuis (1978), Jonsson (1980), Grant and Madsen (1982), Christoffersen and Jonsson (1985) and Weber (1991) using both laboratory data and eddy viscosity models. In the field, the bottom boundary layer is generally fully turbulent and a host of relationships between C_f and r/a have been proposed for this situation [i.e. independent of Reynolds' number as shown in Figure 7.5]. A sample of these relationships is shown in Figure 7.6.

Although these representations vary quantitatively (particularly at large r/a), they are in qualitative agreement. The bottom friction coefficient C_f is not a constant but an increasing function of r/a.

The determination of the bed roughness r is not straightforward, depending not only on the properties of the bed material but also on whether bed ripples are present. Numerous formulations for r have been proposed (Kajiura, 1968; Jonsson and Carlsen, 1976; Grant and Madsen, 1982; Swart, 1974; Nielsen, 1983; Christoffersen and Jonsson, 1985). For the purposes of illustration, the formulation proposed by Nielsen (1983) has been adopted here.

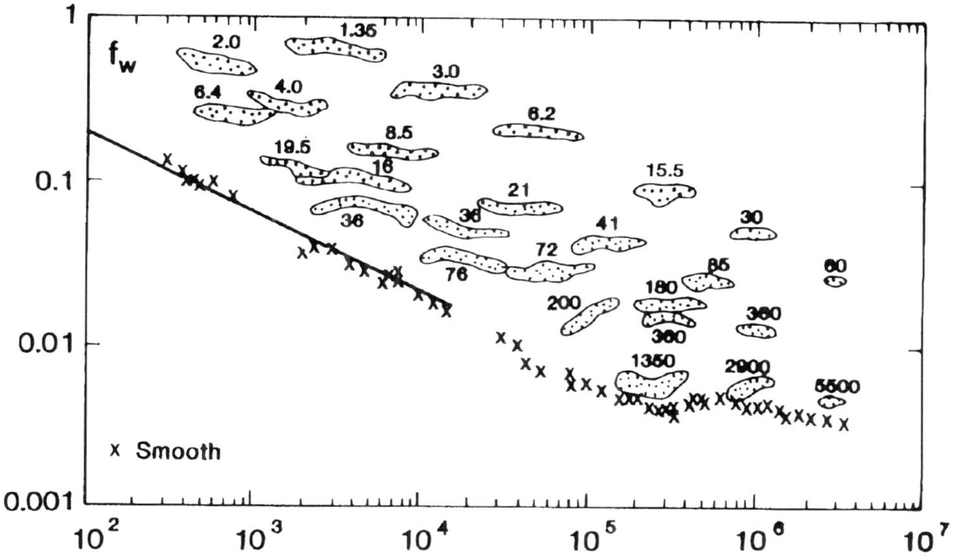

Figure 7.5: An example of the Moody Diagram for bottom friction. The vertical axis is the friction factor, $f_w = 2C_f$ and the horizontal axis is the Reynolds' number, $a^2\omega/\nu$. The numbers on the data clouds refer to the relative bed roughness, a/d_{90}. [after Kamphuis (1975)].

Nielsen (1983) represented C_f by

$$C_f = \min\left\{0.15, \frac{1}{2}\exp\left[5.213\left(\frac{r}{a}\right)^{0.194} - 5.977\right]\right\} \tag{7.9}$$

The bed roughness r depends not only on the bottom material but also on whether bed ripples are present. In an analogy to open channel hydraulics the motion of bed material can be represented in term of the Shields entrainment parameter θ_s

$$\theta_s = \frac{C_f u^2}{(S-1)gD} \tag{7.10}$$

where S is the specific gravity of the bed material, D is the particle size, and C_f can be calculated from (7.9) with $r = 2.5D$ for an initially plane bed. For values of $\theta_s < 0.05$, no sediment movement occurs and hence ripples cannot form, whereas for $\theta_s > 1$ sheet flow occurs and the ripples are "washed out."

For $0.05 < \theta_s < 1$, ripples form, their geometry being represented in terms of their height η and wavelength λ

$$\eta = 21a\left[\frac{u^2}{(S-1)gD}\right]^{-1.85} \tag{7.11}$$

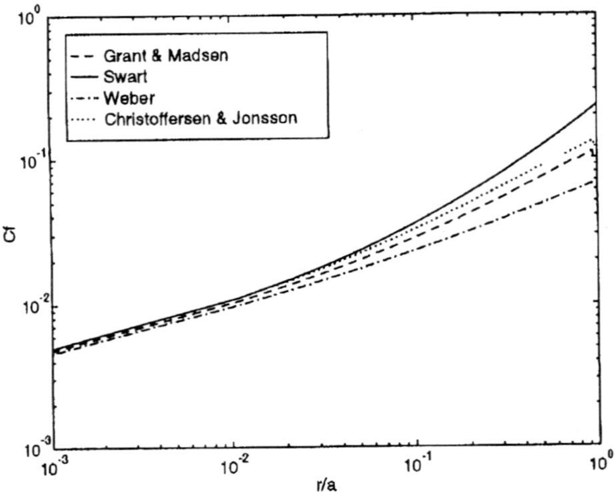

Figure 7.6: Dependence of the bottom friction coefficient C_f on the ratio of bed roughness to semi-excursion displacement r/a. The theories of Grant and Madsen (1982), Swart (1974), Christoffersen and Jonsson (1985) and Weber (1991) are shown.

$$\lambda = \frac{\eta}{(0.342 - 0.34\theta_s^{0.25})} \tag{7.12}$$

The roughness finally being defined by

$$r = \frac{8\eta^2}{\lambda} + 190D\left(\theta_s - 0.05\right)^{0.5} \tag{7.13}$$

The behaviour of such a movable bed model is illustrated in Figure 7.7 which shows the dependence of C_f on the bottom displacement a for the case where $D = 1$ mm and $S = 2.65$ (i.e., a coarse sand). Both the model of Nielsen (1983), described above, and that of Grant and Madsen (1982) are presented. At small values of bottom displacement, no ripples are present and C_f is small. The value of C_f gradually decreases as the bottom displacement increases [see also Figure 7.6]. At the critical value of the Shields parameter, ripples form, the bed roughness increases significantly, and there is a "jump" in the value of C_f. As the bed displacement continues to increase, the ripple geometry changes gradually, reducing the roughness, and C_f consequently decreases.

For given wave conditions (i.e., values of a and u), C_f also varies as a function of grain size D and specific gravity S as shown in Figure 7.8. This figure was produced assuming typical wave parameters of $u = 0.3$ m/s and $a = 0.7$ m. Moving from right to left across this figure (decreasing D) for a constant value of S, the same features highlighted in Figure 7.7 can be seen. For large D, no ripples form and C_f is low. The onset of ripples is marked by the discontinuity in the C_f surface. As

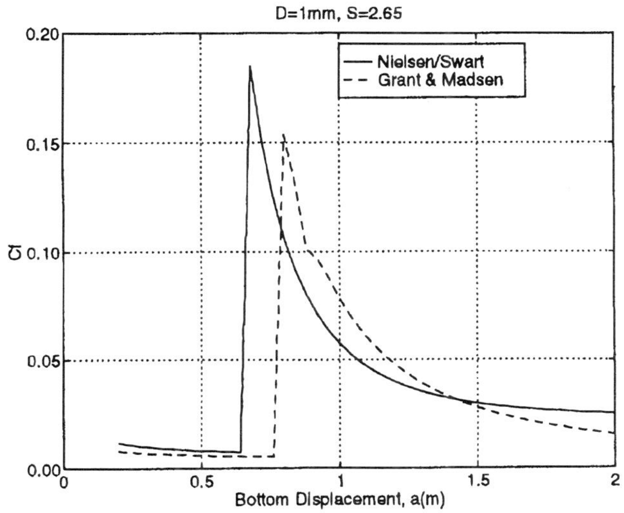

Figure 7.7: Dependence of the bottom friction coefficient C_f on the semi-excursion displacement a for a mobile bed. The figure corresponds to a coarse sand with grain diameter $D = 1$ mm and specific gravity $S = 2.65$. The theories of Nielsen (1983) and Grant and Madsen (1982) are shown.

D continues to decrease, the relative roughness falls and hence C_f decreases. For very small values of S, ripples are always present for these wave conditions.

The above discussion applies to steady state wave conditions. As the wave conditions vary over time, a form of hysteresis can be expected with ripples generated by previous events remaining as the wave bed velocity u drops below the threshold for ripple formation.

It is clear from these results that C_f is far from constant as often assumed in wave prediction models. Further, values far in excess of the commonly assumed 0.015 could often be expected to occur. A detailed field confirmation of the dependence of C_f on bed and flow conditions is, however, yet to be obtained. Hsiao and Shemdin (1978) have, however, presented field data showing C_f increasing with grain diameter and reaching values as high as 0.10. Similarly, Young and Gorman (1995) have investigated the attenuation of waves propagating across a wide continental shelf. Their data clearly indicated that the bottom friction coefficient, C_f increased in magnitude with increasing water depth [i.e., C_f increases as u or a decreases].

7.2.4.2 Typical Implementations of Bottom Friction

The results presented in Section (7.2.4.1) indicate three general approaches to the implementation of bottom friction in operational spectral wave models:

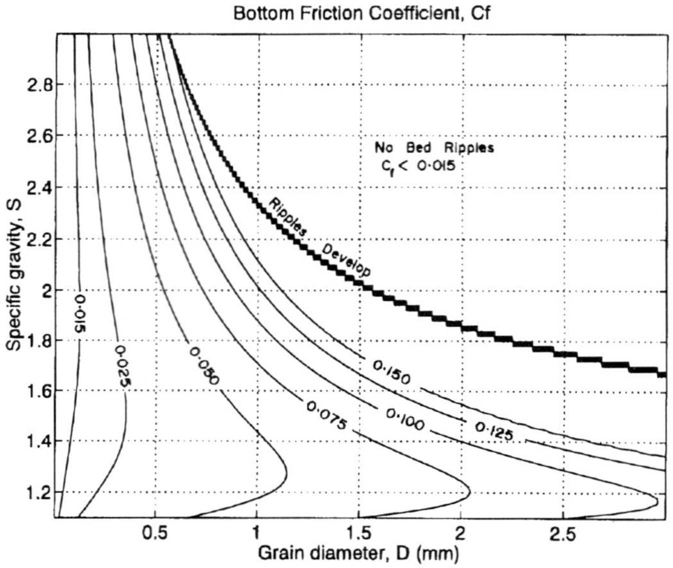

Figure 7.8: Variation in the bottom friction coefficient C_f as a function of grain diameter D and specific gravity S as predicted by Nielsen (1983). The figure corresponds to an rms bed velocity, $u = 0.3$ m/s, and a near-bottom particle semi-excursion, $a = 0.7$ m.

1. constant C_f,

2. C_f calculated from a mobile bed model, and

3. constant $\Gamma = C_f ug$.

There are examples of all three approaches in the literature.

As mentioned previously, Chen and Wang (1983), Young (1988a) and Cavaleri et al (1989) have adopted constant values of C_f. The results of Section (7.2.4.1) clearly show that C_f will not be a universal constant and indeed is likely to vary even at a particular site as wave conditions change. The fact that these models appear to yield reasonable results in their various applications is probably indicative of the relatively small amount of data available and of other possible uncertainties in the models. With data from only one or two sites available, it is likely that C_f could be tuned to yield reasonable results.

Applications of full mobile bed formulations for bottom friction are less common. An example is given by Graber and Madsen (1988). The model has a sound physical basis and performs well in idealized cases. Whether the sophistication of the approach can be justified in relation to the accuracy with which details of the bed material is known for practical problems is questionable.

Possible compromises between these two approaches are to assume that $\Gamma = C_f u g$ is a constant or that the bed roughness r is constant for a particular region. Provided the bed roughness remains relatively constant, Figure 7.6 shows that u and C_f are approximately inversely related. Hence the two approaches are very similar. Indeed, Young and Gorman (1995) show that a bottom friction formulation of the form proposed by (7.8) with Γ = constant yields good results. That is, the assumption that $\Gamma \propto C_f u$ = constant is a reasonable approximation. Tolman (1994) has compared a number of C_f formulations for idealized cases and concluded that a constant $C_f u$ is a good approximation to a full mobile bed model over a wide range of conditions. Weber (1991) similarly concluded that a constant value of r was preferable to a full mobile bed model for large scale applications. Bouws and Komen (1983) have also adopted the assumption of a constant value for Γ to model depth limited conditions in the North Sea. They found a value approximately twice that of JONSWAP was consistent with their observations.

The form of the full source term, S_{bf} as predicted by (7.8) is shown in Figure 7.1d. As expected, the magnitude of the term increases with decreasing depth (decreasing $k_p d$). The term is similar in form to S_{ds}, peaking at the frequency of the spectral peak. It is, however, more peaked than S_{ds}, decaying very rapidly at higher frequencies. It is the longer spectral components which are most influenced by bottom friction, the high frequencies components are essentially in deep water and are unaffected.

7.2.5 Spectral Balance on the Finite Depth Shelf

As with deep water, it is the sum of the individual source terms and the balance which exists between them that is responsible for spectral evolution. The results of Sections (7.2.1) to (7.2.4) indicate that the magnitudes of all source terms are enhanced in finite depth water. Figure 7.9 shows the various source terms for a mean JONSWAP spectrum with $f_p = 0.3$ Hz and $U_{10} = 10$ m/s in both deep water and finite depth water with $k_p d = 0.65$.

For the case of deep water, S_{in} is larger than the other source terms, indicating the waves are "young" and actively growing [i.e. the sum of the source terms would yield a positive total source term]. As expected, all source terms are significantly larger in the finite depth case. In particular, the bottom friction source term is much larger than the other source terms, indicating that the situation shown is not stable and that this spectrum would decay in this water depth.

The case shown in Figure 7.9 is artificial as a pre-defined spectrum has been prescribed, and the source terms evaluated for this spectrum. An alternative, and more physically sound approach is to allow the spectrum to develop using a model such as (4.62). In such an approach, the shape of the spectrum can evolve in equilibrium with the imposed source terms and the finite water depth. The end result is that the shape of the spectrum may be different to the JONSWAP spectrum specified in Figure 7.9. As S_{nl}, S_{ds} and S_{bf} are all in some way non-linear this can have a significant effect on the resulting magnitude of the source

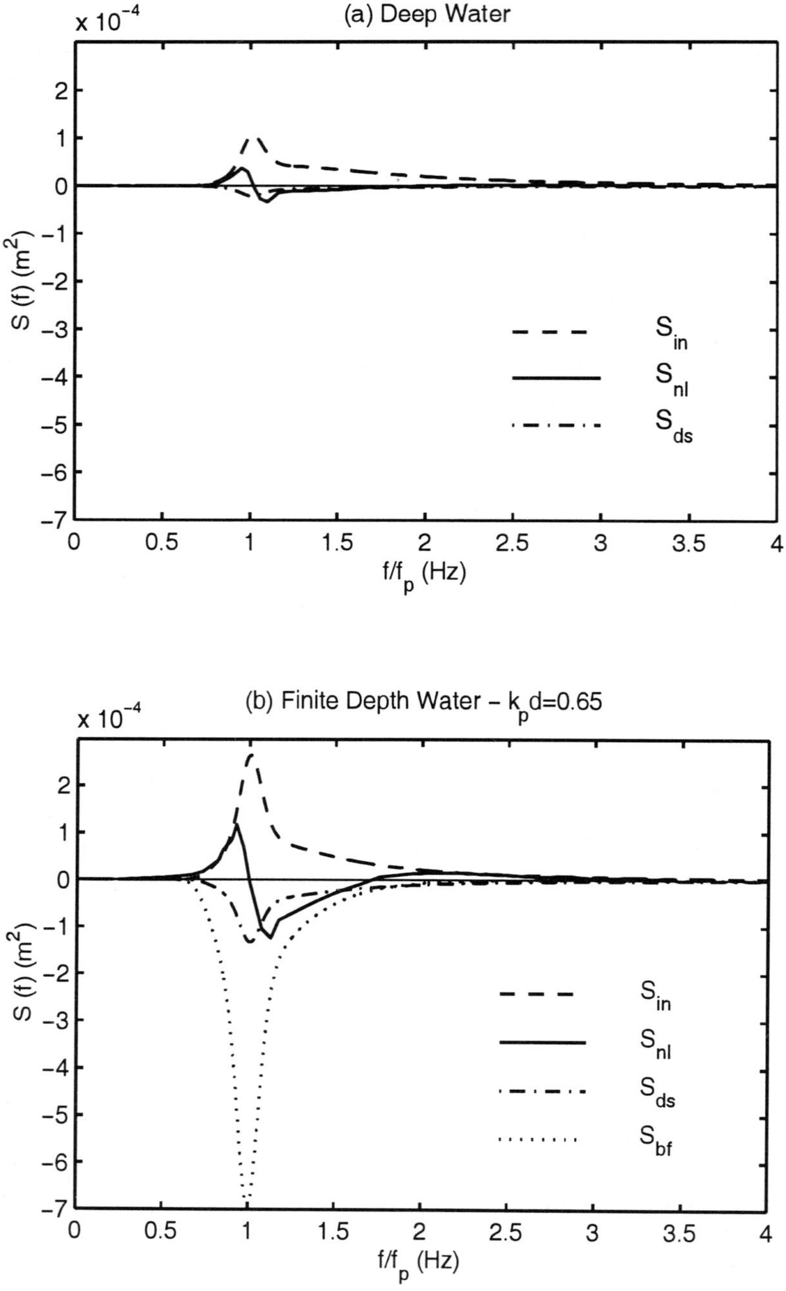

Figure 7.9: The source term balance for (a) Deep water and (b) Finite depth with $k_p d = 0.65$. The source terms have been evaluated for a mean JONSWAP spectrum with $f_p = 0.3$ Hz and $U_{10} = 10$ m/s.

terms. Khatri and Young (1997) have presented the results of such a model. They conclude that the source terms are smaller than would be expected from an analysis such as that in Figure 7.9. A detailed understanding of this balance requires both an enhanced understanding of the finite depth processes and additional numerical modelling.

7.2.6 Depth Limited Breaking, S_{brk}

As waves enter the shallow surf zone, enhanced breaking not predicted by (7.2) occurs. The process of depth limited breaking is poorly understood and our predictive capability is largely based on empirical relationships between wave height and water depth.

7.2.6.1 Limiting Total Energy

Miche (1944), Divoky et al (1970) and Chen and Wang (1983) have proposed limiting forms for the maximum wave height, H_{max} in finite depth water of the form

$$H_{max} = \frac{\gamma_1}{k} \tanh\left[\frac{\gamma_2 kd}{\gamma_1}\right] \tag{7.14}$$

where k is the wave number, $\gamma_1 \approx 0.7 - 0.9$ defines the limiting wave slope in deep water and $\gamma_2 = H_{max}/d$ defines the depth limited wave height in shallow water. A value of $\gamma_2 = 0.78$ is commonly used, although this parameter decreases with decreasing bed slope. Values as low as 0.55 have been reported by Nelson (1994). Equation (7.14) is written in terms of individual waves, rather than spectral properties. Young (1988a) reformulated (7.14) in terms of integral properties of the spectrum, to yield

$$\sigma^2_{max} = \frac{0.0355\overline{C}^4}{g^2} \tag{7.15}$$

where σ^2_{max} is the maximum total energy of the spectrum and $\overline{C}$ is the mean phase speed of the spectrum. Equation (7.15) places a limit on the total energy of the spectrum, but provides no information on the form of the limiting spectrum itself. Young (1988a) argued that as long waves "feel" the bottom to a greater extent than shorter waves, the low frequency components would be most attenuated in the breaking process. Hence, the breaking process would act like a "high pass" filter, attenuating low frequency spectral components until the limit specified by (7.15) is reached. Although this approach has some intuitive appeal, there is no conclusive experimental evidence to support this distribution of the attenuation and it has not been widely adopted.

7.2.6.2 Dissipation Rate

Battjes and Janssen (1978) considered the statistics of individual waves. They considered the individual crest to trough height of the waves and assumed that all waves above a limiting height, H_{max} were breaking. In addition, they assumed that waves below this limit followed a Rayleigh probability density distribution as given by (3.1). Therefore, from (3.5) the probability that the wave height is greater than H_{max} is

$$Q_b = P(H > H_{max}) = e^{-\frac{H_{max}^2}{8\sigma^2}} \tag{7.16}$$

where Q_b is the fraction of waves which are breaking and σ^2 is the variance or total energy of the wave record. It follows from (3.3) that the rms wave height is given by

$$\begin{aligned} H_{rms}^2 &= \int_0^{H_{max}} H^2 p(H) \mathrm{d}H + Q_b H_{max}^2 \\ &= 8\sigma^2 \left(1 - Q_b\right) \end{aligned} \tag{7.17}$$

Eliminating σ^2 between (7.16) and (7.17) yields

$$\frac{1 - Q_b}{\ln\left(Q_b\right)} = -\left[\frac{H_{rms}}{H_{max}}\right]^2 \tag{7.18}$$

Equation (7.18) has been applied to a spectral representation by Booij et al (1996) by determining H_{max} from (7.14) with $k = k_p$, the wave number of the spectral peak. The rms wave height can also be determined from the spectrum using (3.3) and hence Q_b can be determined from (7.18). Figure 7.10 shows Q_b as a function of the ratio H_{rms}/H_{max}. For $H_{rms}/H_{max} < 0.4$ almost none of the waves are breaking whereas as $H_{rms}/H_{max} \to 1$, $Q_b \to 1$.

Battjes and Janssen (1978) modelled the energy dissipation in the breaking wave as a bore of height H_{max} using the theory of Stoker (1957). The final result for the rate of change of the total energy is

$$\frac{\partial \sigma^2}{\partial t} = -\frac{1}{4} \alpha_{bj} f_p Q_b H_{max}^2 \tag{7.19}$$

where α_{bj} is a constant of order 1.

Equation (7.19) determines the rate of decay of the total energy, σ^2 as a result of depth limited breaking, it does not, however, specify how this decay is apportioned with frequency across the spectrum. Beji and Battjes (1993) have considered the evolution of spectra breaking across a laboratory bar and determined that the energy loss predicted by (7.19) is distributed in proportion to the energy of the individual spectral ordinates. Hence, most energy is lost at the spectral peak and the shape of the spectrum is preserved in the breaking process. Such a distribution of the breaking decay has been implemented in the spectral wave model SWAN

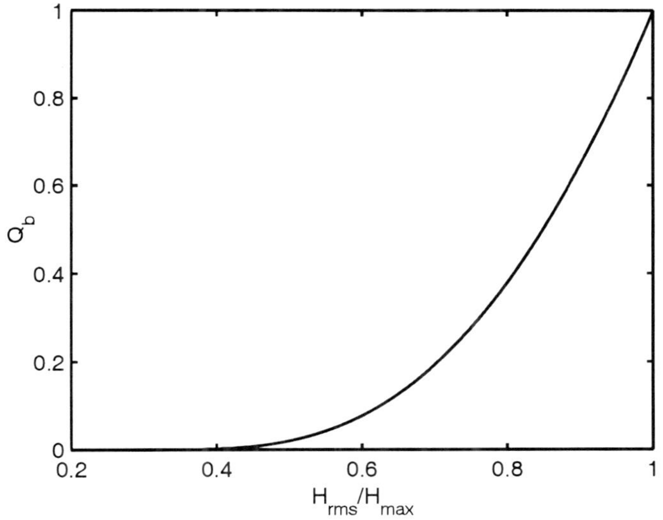

Figure 7.10: The percentage of the waves which are breaking due to the limited depth, Q_b as a function of the ratio H_{rms}/H_{max}. Evaluated from the theory of Battjes and Janssen (1978) and (7.18).

(Booij et al, 1996) and found to yield reliable results in a relatively large number of validation tests.

The approach outlined above based on the decay rate proposed by Battjes and Janssen (1978) [a similar approach has also been proposed by Dally (1990)] appears to yield reasonable predictions of the spectrum in coastal regions. A full understanding of depth limited breaking and a comprehensive source term, $S_{brk}(f, \theta)$ awaits a physical understanding of the highly nonlinear processes involved. This is a daunting task and a solution is unlikely in the short term.

7.2.7 Triad Nonlinear Interactions, S_{tri}

Nonlinear interactions can occur at all orders. The importance of quadruplet (4 wave) nonlinear interactions has been clearly demonstrated in deep [see Section (4.4)] and intermediate [see Section (7.2.2)] water depths. In contrast, triad (three wave) interactions are insignificant in deep water (Phillips, 1960) compared to quadruplet interactions. As the water depth decreases, however, triad interactions become significant.

Nonlinear triad interactions can occur among waves with frequencies and wave numbers which satisfy the relationships (Armstrong et al, 1962; Bretherton, 1964; Elgar et al, 1993)

$$f_1 \pm f_2 = f_3 \tag{7.20}$$

$$\mathbf{k_1} \pm \mathbf{k_2} = \mathbf{k_3} \tag{7.21}$$

where f_i is the scalar frequency and $\mathbf{k_i}$ is the vector wave number, of the ith wave component. The components 1 and 2 each obey the linear dispersion relationship (2.13).

Thus components which satisfy (7.20) and (7.21) can interact nonlinearly and exchange energy and momentum. Whilst the components 1 and 2 in (7.20) and (7.21) satisfy the dispersion relationship (2.13), component 3 may not necessarily satisfy (2.13). Interactions where component 3 satisfies the dispersion relationship $[\mathbf{k_3} = \mathbf{k}(f_3)]$ are termed *resonant interactions* (Armstrong et al, 1962). Resonant interactions result in components at f_3 whose amplitudes can increase to be of the same order as the primary waves at f_1 and f_2. Note that for gravity surface waves this is only possible in very shallow water where the waves are nondispersive. For all other water depth, the resonance conditions cannot be satisfied.

Interactions where component 3 does not satisfy the dispersion relationship (2.13) are termed *non-resonant or bound* [in intermediate water depths where waves are weakly dispersive – the so-called bound interaction of dispersive waves (Hasselmann, 1962)]. Non-resonant interactions result only in bound harmonics whose amplitudes remain small compared with those of the primary waves. They are called *bound* because they do not propagate at their free-wave phase speed but at the phase speed of one of the components with which they interact non-linearly. Hence, they are bound to this wave and propagate at its phase speed.

For spectra with a clearly defined major peak, one special case of the interaction conditions (7.20) and (7.21) which has been observed to be significant (Hasselmann et al, 1963; Masuda and Kuo, 1981; Elgar and Guza, 1985; Doering and Bowen, 1987; Freilich et al, 1990; Elgar et al, 1990, 1993, 1995) is that where $f_1 = f_2 = f_p$, the frequency of the spectral peak. Such self interactions result in the generation of an harmonic of the spectral peak at $2f_p$. This form of interaction represents a convenient example to examine the implications of (7.20) and (7.21) on the form of the triad interaction. In shallow water, waves become non-dispersive and the dispersion relationship, (2.13) becomes $\omega = k\sqrt{gd}$. In such cases resonant interactions occur only for co-linear waves, as this is the only geometry of wave numbers which can satisfy the resonance conditions. Hence, resonant triad interactions cannot transfer energy across directions in a manner similar to quadruplet interactions. In transitional water depths, where the waves are dispersive, only non-resonant or "bound" triad interactions can occur.

Armstrong et al (1962), Freilich and Guza (1984) and Elgar et al (1993) have shown that significant energy transfers can also occur due to near-resonant interactions, in which the sum (or difference) component of the triad nearly satisfies the dispersion relationship. A measure of the departure from resonance can be obtained by defining $|\mathbf{k}_\delta|$ as the difference between the free wave component $|\mathbf{k}(f_3)|$, which satisfies the dispersion relationship, and the sum (or difference) $(|\mathbf{k_1} \pm \mathbf{k_2}|)$

wave number magnitudes

$$|\mathbf{k}_\delta| = |\mathbf{k_1} \pm \mathbf{k_2}| - |\mathbf{k}(f_3)| \tag{7.22}$$

The wave number mis-match can be normalized as $\delta_k = |\mathbf{k}_\delta|/|\mathbf{k}(f_3)|$. The magnitude of δ_k determines the intensity of the energy exchange between the interacting waves. Zero mis-match represents the limiting case of the interaction process, in which the interacting waves remain intact and in phase (resonant interaction) during evolution. Thus, the magnitude of the energy transfer is maximum. When the mis-match is such that $\delta_k << 1$ (i.e. in the shoaling region), phase relations between the interacting waves vary slightly over a wavelength. Consequently the magnitudes and the sign of energy transfers between the interacting waves vary slowly over a wavelength, allowing significant net energy transfer over several wavelengths. Whereas resonant interactions are limited to nondispersive co-linear waves, near resonant interactions can occur between components separated by small angles (Elgar et al, 1993; Nwogu, 1994). Thus near resonant interactions can redistribute energy across direction as well as frequency.

The magnitude of triad interactions and the consequent energy transfer across the spectrum depend essentially on two factors. These are the order of magnitude of the nonlinearity determined by the relative wave amplitude, a/d and the shallowness of the water determined by the relative depth, $k_p d$, where a is the wave amplitude, d is the water depth and k_p is the wave number of the spectral peak. Increasing nonlinearity and decreasing relative depth intensifies the nonlinear rate of energy transfer across the spectrum.

Measurements to determine the magnitude of triad coupling in the wave field have concentrated on the determination of bispectra (Hasselmann et al, 1963; Lii et al, 1976; Kim and Powers, 1979).

The bispectrum is formally defined as the Fourier transform [see Section (9.3.1)] of the third-order correlation function of the time series (Hasselmann et al, 1963). For a discretely sampled time series, this becomes

$$B(f_1, f_2) = E\left[X(f_1)X(f_2)X^*(f_1 + f_2)\right] \tag{7.23}$$

where $E[\ ]$ denotes the expected value or mean, $X(f)$ is the complex Fourier coefficient and the asterisk represents the complex conjugate.

The bispectrum can be recast into its normalized magnitude (Kim and Powers, 1979)

$$b^2(f_1, f_2) = \frac{|B(f_1, f_2)|^2}{E\left[|X(f_1)X(f_2)|^2\right] E\left[|X(f_1 + f_2)|^2\right]} \tag{7.24}$$

where $b^2(f_1, f_2)$ is termed the bicoherence. The bicoherence is in the range $0 \leq b^2 \leq 1$, and is a measure of the phase coherence between spectral components at f_1, f_2 and $f_3 = f_1 + f_2$. A value of $b^2 = 1$ indicates complete coherence whilst $b^2 = 0$ indicates no coherence.

The bicoherence determines whether there is phase coherence between spectral components. The details of the phase coherence can be obtained by examination of the relative magnitudes of the real and imaginary parts of the bispectrum. One representation is in terms of the biphase, $\beta_p(f_1, f_2)$

$$\beta_p(f_1, f_2) = \arctan \left\{ \frac{Im\left[B(f_1, f_2)\right]}{Re\left[B(f_1, f_2)\right]} \right\} \tag{7.25}$$

Masuda and Kuo (1981) have shown that a spectral component and its harmonic with zero biphase is associated with the Stokes wave form with peaked crests and flat troughs. In contrast, as β_p approaches $-\pi/2$ the waves become increasingly pitched forward (a saw-tooth wave has $\beta_p = -\pi/2$).

The nature of the phase coupling can also be determined using integrated properties of the bispectrum such as skewness, S_k and asymmetry, A_s.

$$S_k = \frac{\left\{\int\int Re\left[B(f_1, f_2)\right] \mathrm{d}f_1 \mathrm{d}f_2\right\}}{\left[\overline{\eta^2}\right]^{3/2}} \tag{7.26}$$

$$A_s = \frac{\left\{\int\int Im\left[B(f_1, f_2)\right] \mathrm{d}f_1 \mathrm{d}f_2\right\}}{\left[\overline{\eta^2}\right]^{3/2}} \tag{7.27}$$

where $\eta = \eta(t)$ is the water surface elevation and the overbar indicates the mean with respect to time. Positive values of skewness and negligible asymmetry correspond to Stokes type wave forms. Negative values of asymmetry indicate waves which are pitched forward.

Hasselmann et al (1963) considered the bispectra of waves measured in 11m water depth. They found clear triad coupling with the spectral peak. The observed bispectra agreed well with the theoretical bispectra for a Stokes type expansion. Hence, it was concluded that non-resonant or bound triad interactions were responsible for the observed coupling.

Various studies have investigated shoaling waves with stations located at a number of water depths. Elgar and Guza (1985, 1986) investigated the bispectra of waves as they propagated from 9m water depth to 1m. Bicoherence values increased as the water depth decreased, indicating an increase in nonlinear coupling. In addition, as the water depth decreased the waves progressively became more "pitched forward". Examples of the spectra and bi-spectra recorded by Elgar and Guza are shown in Figure 7.11. The two cases shown have water depth of 2.7 m and 2.0 m, respectively and are separated by less than 50 m. It is clear that there are 1st, 2nd, 3rd and 4th harmonics of the spectral peak. These harmonics grow in magnitude as the water depth decreases. The magnitude of the bi-coherence also increases as the water depth decreases, indicating phase coherence between these peaks. In addition to the amplification of the harmonics, the entire tail of the spectrum has been elevated as the water depth decreases. The results clearly show significant triad interactions occuring within the spectrum. Similar results have also been reported by Hardy (1993) and Hardy and Young (1996) who investigated waves propagating across a shallow offshore reef as shown in Figure 7.12. Although the spectral peak was severely attenuated by breaking at the seaward edge of the reef, the tail of the spectrum was elevated, presumably by triad nonlinear interactions. The net result was a very flat spectrum.

Freilich et al (1990) compared high resolution directional spectra measured at 10m and 4m. They found that linear shoaling theory was incapable of predicting the increase in energy observed at $2f_p$, which they attributed to near-resonant triad interactions. Elgar et al (1993) also found significant near-resonant triad coupling for laboratory shoaling waves. Herbers et al (1992) investigated near bottom wave orbital velocity and pressure measurements in a water depth of 7m. Assuming that the waves had a relatively narrow directional spread due to the effects of refraction, they showed that forced waves at $2f_p$ were well modelled by the second order theory of Hasselmann (1962).

In a similar fashion to Hardy and Young (1996), Beji and Battjes (1993) investigated waves propagating over a bar in the laboratory. They found strong phase coupling between spectral components at the peak, f_p and $2f_p$.

The observations reported above have all been obtained within the relatively shallow shoaling zone. Young and Eldeberky (1998) investigated the bi-spectra of fetch limited finite depth wind waves developing over a horizontal bottom. These measurements were in deeper water than previously reported ($k_pd = 1.39 - 2.35$). As very long time series were recorded it was possible to obtain bi-spectra with small confidence limits. Hence, weaker triad coupling could be measured. Figure 7.13 shows an example of the spectrum and bi-coherence for a case with $k_pd = 1.39$. The spectrum is typical of wind-sea spectra with a clear dominant peak and a high frequency region decaying proportional to f^{-n}. A small discontinuity is, however, apparent at $2f_p$. The bi-coherence clearly shows that this discontinuity and most of the spectral tail are phase-coherent with the spectral peak. Young and Eldeberky (1998) concluded that even in these intermediate water depths, non-resonant or bound triad interactions can be significant.

Most attempts to model wave evolution including triad interaction effects have utilized the intrinsically nonlinear Boussinesq equations. Attention here has been focused on the evolution of spectra which, by definition, consist of a superposition of linear sinusoidal components. The whole concept of linear spectral analysis becomes questionable when nonlinear processes are introduced. Another way of considering the triad interaction process is that the originally linear sinusoidal wave form has simply changed shape into a nonlinear (Stokes) form. There is still only one wave present, however it is no longer sinusoidal in form. Nevertheless, if spectral models such as SWAN (Booij et al, 1996) are to be used to model such intermediate water depth processes, the growth of the second harmonic can only be interpreted as an energy transfer.

However one wishes to interpret the process of non-resonant triad interaction, the consequences are important. An interesting example is the Beji-Battjes bar

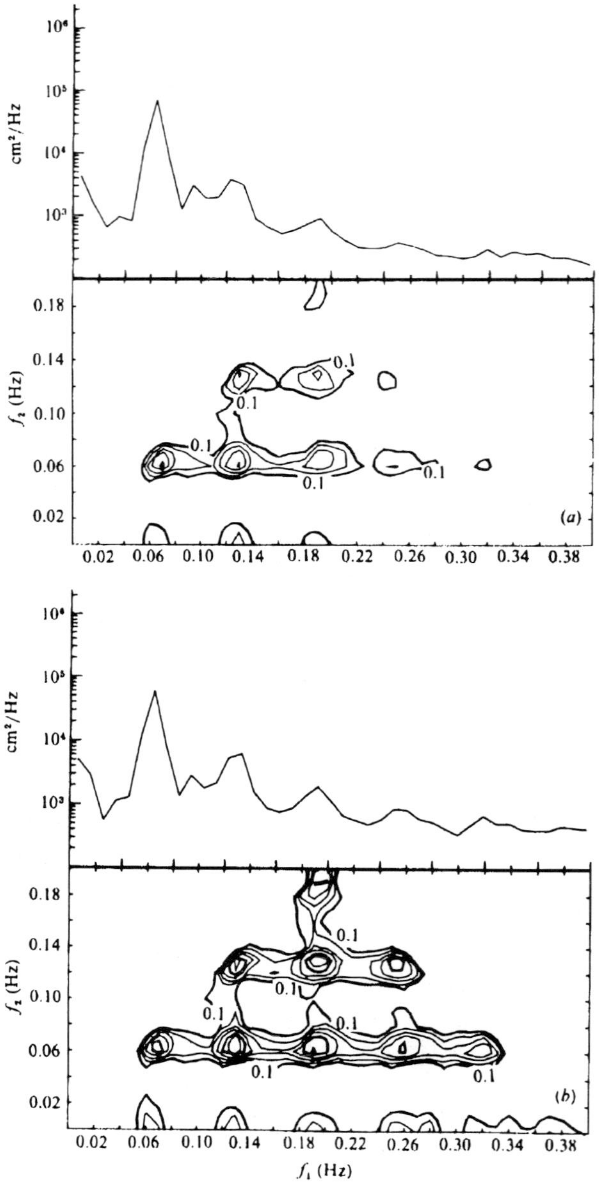

Figure 7.11: Spectra and bi-coherence obtained by Elgar and Guza (1986) for shoaling waves. (a) Water depth = 2.7 m, (b) Water depth = 2.0 m. The bi-coherence is shown as contours of $b(f_1, f_2)$. Note the development of the harmonics and the increase in the level of the bi-coherence with decreasing water depth. [after Elgar and Guza (1986)]

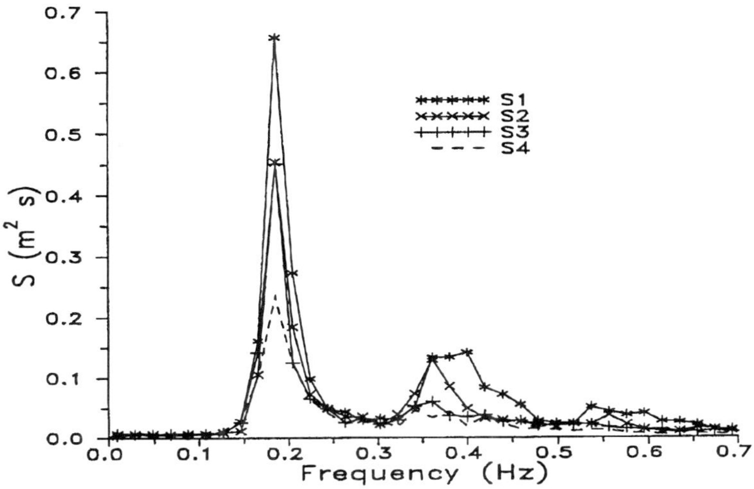

Figure 7.12: Transformation of the wave spectrum across a coral reef. Site S1 was at the seaward edge of the reef. The other stations are spaced equally along the line of propagation of the waves. The distance between each station is approximately 40 to 50m. [after Hardy (1993)]

(Beji and Battjes, 1993). Deep water linear waves were incident on a shoal. As the waves propagated across the shoal, the magnitude of the second harmonic increased and the waves were clearly observed to take a nonlinear shape. The waves were still dispersive and hence non-resonant and near-resonant interactions were responsible for the growth of the harmonic. Holthuijsen (private communication) indicates that once these waves entered deep water again, the bound components at $2f_p$ were "set free" to propagate at their respective linear group velocities. Although the appearance of the second harmonic over the shoal may be considered as a spurious artefact resulting from the linear spectral representation of a nonlinear wave, these components became real once they entered deep water. The conclusion is clear; if the linear spectral representation is to be retained in intermediate water depth, non-resonant interactions can be interpreted as an energy transfer to the bound harmonic components.

Abreu et al (1992) have abandoned the usual approach of modelling nonlinear shoaling waves in terms of the Boussinesq equations. Instead, they adopted a spectral approach. A source term, S_{tri} for the energy balance equation was developed for the case of shallow water nondispersive co-linear triad interactions. The resulting model was compared with the data of Freilich et al (1990). Model results were consistent with the data, leading Abreu et al (1992) to conclude that the nonlinear

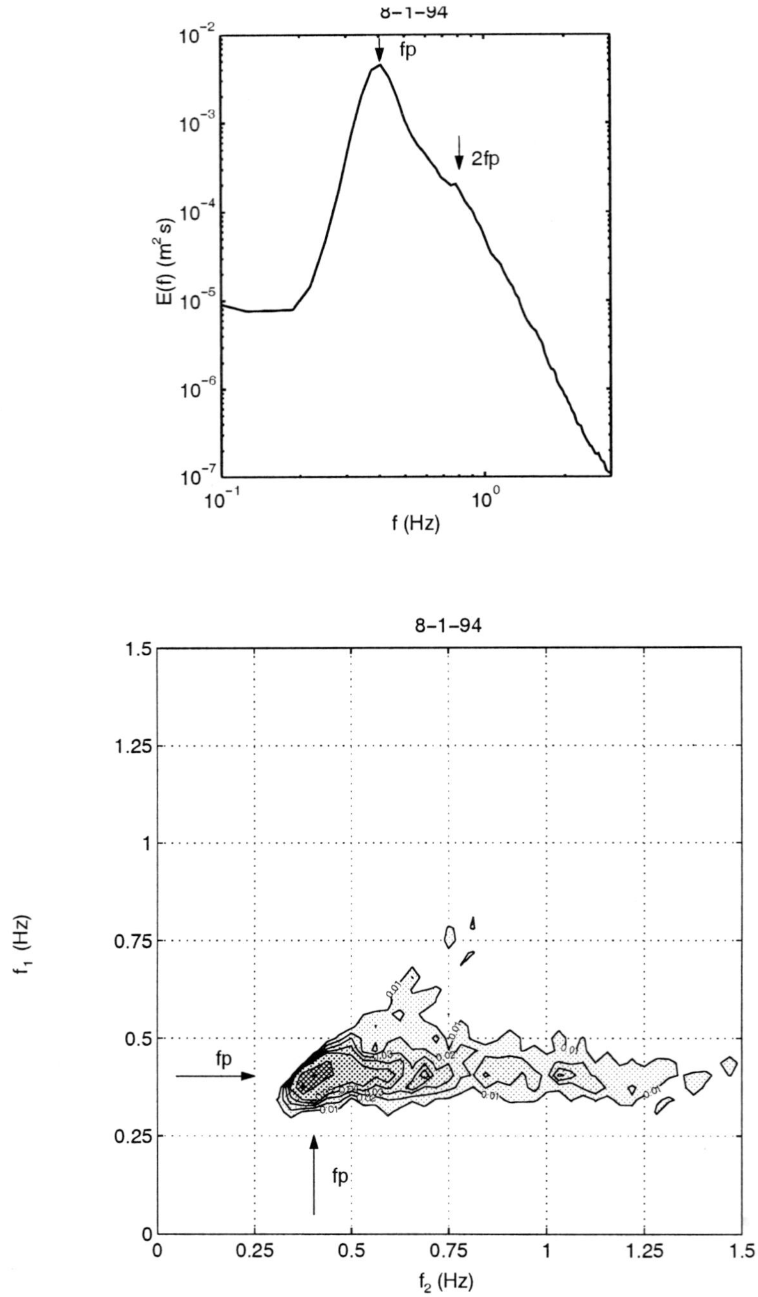

Figure 7.13: Spectrum [top] and contours of bi-coherence, $b^2(f_1, f_2)$ [bottom] for fetch limited finite depth wind waves with $k_p d = 1.39$. Note the significant phase coupling between the spectral peak and the tail. [after Young and Eldeberky (1998)].

coupling observed in these data was due to resonant co-linear interactions rather than near-resonant interactions, as concluded by Freilich et al (1990).

Recently Eldeberky and Battjes (1995) have proposed an approximate form of the triad source term, S_{tri} in which a parametric form of the bi-phase is utilized. Initial tests indicate that the source term is capable of modelling the development of the second harmonic within the spectrum in shallow coastal waters. This source term has been included within the spectral wave model SWAN (Booij et al, 1996).

Like depth limited wave breaking, the formulation of a spectral source term to represent non-linear triad interactions is still an area of active research. Although the proposals of both Abreu et al (1992) and Eldeberky and Battjes (1995) represent a promising start, considerable work is still required. One of the fundamental questions which still needs to be answered is: *Is it prudent to attempt to model intrinsically nonlinear processes using a linear spectral approach?* The advantages of the spectral approach are compelling, but the challenges and fundamental inconsistencies represented by this approach will be significant.

7.3 Finite Depth Growth Curves

As shown in Section (5.3) the composite data set for deep water fetch limited growth is extensive and there is clear evidence that the data can be represented in terms of relatively simple non-dimensional variables: non-dimensional energy, $\varepsilon = g^2\sigma^2/u^4$, non-dimensional frequency, $\nu = f_p u/g$ and non-dimensional fetch $\chi = gx/u^2$, where g is gravitational acceleration, σ^2 is the total wave energy or variance of the wave record, f_p is the frequency of the spectral peak and u is a characteristic wind velocity [see Section (5.3.1)]. Despite these relatively simple non-dimensional groupings, there appear to be inconsistencies between the various data sets and scatter within individual data sets. Attempts to explain these discrepancies (Battjes et al, 1987; Kahma and Calkoen, 1992) have been insightful, but questions about the appropriate scaling of such experimental data still exists.

In comparison to the deep water case, the available data for fetch limited growth in finite depth situations is very limited. The idealized situation to be considered is as shown in Figure 5.1 but now the water depth is of uniform depth, d.

The first study of shallow water wave growth was conducted by Thijsse (1949). This was followed by the field investigations staged in Lake Okeechobe, USA (U.S. Army, 1955; Bretschneider, 1958). This experiment was limited by available instrumentation of the time (paper tape recording) and was largely concentrated on determining the depth limited asymptotes to growth, rather than defining evolution with fetch. Using these data, Bretschneider (1958) adopted the wind velocity measured at a reference height of 10 m, U_{10} as the appropriate scaling parameter for wind speed and determined asymptotic limits to the development of non-dimensional energy and frequency. These limits can be represented as (CERC, 1977)

$$\varepsilon = 1.4 \times 10^{-3} \delta^{1.5} \tag{7.28}$$

$$\nu = 0.16 \delta^{-0.375} \tag{7.29}$$

where $\varepsilon = g^2\sigma^2/U_{10}^4$, $\nu = f_p U_{10}/g$, $\delta = gd/U_{10}^2$ and d is the water depth.

Ijima and Tang (1966) combined the results of (7.28) and (7.29) with available deep water fetch limited results and numerical modelling of the effects of bottom friction and percolation (Bretschneider and Reid, 1953) to develop a set of fetch limited finite depth growth curves for ε and ν. These results appeared in CERC (1977). They were further revised in CERC (1984) to be consistent with the JONSWAP (Hasselmann et al, 1973) deep water results, appearing as

$$\varepsilon = 5 \times 10^{-3} \left\{ \tanh A_1 \tanh \left[\frac{B_1}{\tanh A_1} \right] \right\}^2 \tag{7.30}$$

where

$$A_1 = 0.53 \delta^{0.75} \tag{7.31}$$

$$B_1 = 5.65 \times 10^{-3} \chi^{0.5} \tag{7.32}$$

and

$$\nu = 0.133 \left\{ \tanh A_2 \tanh \left[\frac{B_2}{\tanh A_2} \right] \right\}^{-1} \tag{7.33}$$

where

$$A_2 = 0.833 \delta^{0.375} \tag{7.34}$$

$$B_2 = 3.79 \times 10^{-2} \chi^{0.33} \tag{7.35}$$

and $\chi = gx/U_{10}^2$.

Equations (7.30) and (7.33) reduce to the experimentally confirmed forms (7.28) and (7.29) for the asymptotic limit of small d and large x and to the JONSWAP (Hasselmann et al, 1973) results for the limits of large d and small x. The form of the relationship between these limits, however, relies completely on the results of the numerical modelling of Ijima and Tang (1966), without any experimental corroboration.

Bouws (1986) reported the results from single point measurements in shallow (≈ 4 m) Lake Marken, The Netherlands. As there was only one experimental site, there was little variation in the fetch x. Due to the similarity of the non-dimensional scaling for $\chi = gx/u^2$ and $\delta = gd/u^2$, small δ is always associated with small χ and large δ with large χ. Hence, confirmation of results such as (7.30) and (7.33) becomes difficult. The introduction of the additional non-dimensional variable, δ strengthens the requirements for multiple fetch measurements. Nevertheless, Bouws (1986) confirmed that the growth rate for ε as a function of χ was less than that reported from deep water experiments. More surprisingly, the data indicated that the development of ν with χ was comparable to deep water. In addition, Bouws (1986) adopted a non-dimensional scaling in terms of the friction velocity, u_* rather than U_{10}, claiming this reduced scatter in the results.

The most comprehensive finite depth growth experiment to date was staged by Young and Verhagen (1996a) in Lake George, Australia. This experiment consisted of measurements of wind speed and water surface elevation made at a series of 8 stations with fetches ranging from 1.3 km to 15.3 km. The water depth in the lake was very uniform at approximately 2 m. Hence the experiment closely approximated the conditions required for finite depth fetch limited growth. In addition, data were recorded over a wide range of non-dimensional fetches and water depth and with wind measurements along the full fetch.

As the wind speed variation was measured along the fetch, Young and Verhagen (1996a) were able to confirm that the boundary layer developed in the fashion predicted by Taylor and Lee (1984) [see (5.35) and (5.36)]. Based on these measurements Young and Verhagen (1996a) adopted the wind speed averaged over the down wind fetch (Dobson et al, 1989) [see (5.37)], as the scaling wind velocity.

Based on their very large data set, Young and Verhagen (1996a) firstly investigated the asymptotic limits to growth for ε and ν which would exist at large fetch. Figure 7.14 shows a scatter plot of ε verses δ and Figure 7.15 a scatter plot of ν verses δ. In both cases approximately 65,000 data points are shown. There appear to be clear depth defined limits for both ε and ν. Hence, at long fetch, a limit will be reached where ε and ν are completely defined by δ. This represents a *depth limit to growth*. These limits were determined by Young and Verhagen (1996a) as

$$\varepsilon = 1.06 \times 10^{-3} \delta^{1.3} \tag{7.36}$$

and

$$\nu = 0.20 \delta^{-0.375} \tag{7.37}$$

Also shown in Figures 7.14 and 7.15 are the Lake Okeechobe results [(7.28) and (7.29)] (Bretschneider, 1958). Equations (7.36) and (7.37) yield consistent results. Equation (7.37), however, produces significantly larger values for ν than (7.29). This is not surprising as it is not entirely clear from Bretschneider (1958) as to how the frequency used in ν was defined.

With the asymptotic limits to growth defined, Young and Verhagen (1996a) considered only cases where the wind direction was aligned with their measurement array (i.e. fetch limited cases). This fetch limited data set was partitioned based on values of δ. For each finite range of δ, the development of ε and ν was investigated as a function of χ. Typical examples are shown in Figures 7.16 and 7.17.

Young and Verhagen (1996a) approximated these data by

$$\varepsilon = 3.64 \times 10^{-3} \left\{ \tanh A_1 \tanh \left[\frac{B_1}{\tanh A_1} \right] \right\}^{1.74} \tag{7.38}$$

where

$$A_1 = 0.493 \delta^{0.75} \tag{7.39}$$

$$B_1 = 3.13 \times 10^{-3} \chi^{0.57} \tag{7.40}$$

and

$$\nu = 0.133 \left\{ \tanh A_2 \tanh \left[\frac{B_2}{\tanh A_2} \right] \right\}^{-0.37} \tag{7.41}$$

where

$$A_2 = 0.331 \delta^{1.01} \tag{7.42}$$

$$B_2 = 5.215 \times 10^{-4} \chi^{0.73} \tag{7.43}$$

Equations (7.38) and (7.41) are shown in Figures 7.16 and 7.17 respectively. As each of the panels of these figures contains data for a finite range of δ (i.e. $\delta = 0.3-0.4$), two curves are shown, one for each of the extremes of δ for that figure. As a range of values of δ are shown, data "scatter" is expected. The data should, however, lie between these two curves. Generally, the proposed relationships [(7.38) and (7.41)] approximate the data well. As with all previous field measurements of this type there is some data scatter.

Equations (7.38) – (7.43) are presented as a series of non-dimensional design curves in Figure 7.18 and 7.19. Rather than finite depth fetch limited evolution being defined by a single power law as in deep water, a family of curves exist, one for each non-dimensional water depth, δ. At short non-dimensional fetch, the wave length is sufficiently short for waves to be in deep water and the relationships are comparable to those previously obtained for deep water. With increasing fetch, the effects of the finite depth become more pronounced. The total energy is smaller than would be expected in deep water and the peak frequency higher. With a further increase in fetch, a point is eventually reached where spectral development ceases. At this point both the non-dimensional energy and peak frequency become depth limited.

The Lake George data set is very comprehensive. Despite this, Young and Verhagen (1996a) felt compelled to force (7.38) and (7.41) to asymptote to previously recorded deep water results for $\delta \to \infty$. In a subsequent reanalysis of the data Young (1997b) abandoned this requirement and considered the differential growth between stations in the same fashion as Donelan et al (1992). An integral relationship between χ, ε and δ was developed of the form

$$\chi = 46,700 g \int \frac{C_g}{\omega_p U_{10}^3} \left(\frac{U_{10}}{C_p} \right) \left(\frac{U_{10}}{C_p} - 0.83 \right)^{-1} \tanh^{-0.45} \left(\frac{U_{10}}{C_p} - B \right) \mathrm{d}C_p \tag{7.44}$$

where C_p is the phase speed of waves at the spectral peak frequency $\omega_p = 2\pi f_p$.

Equation (7.44) has the flexibility to include variations in wind speed and water depth along the fetch. As the assumption of constant water depth along the fetch is very restrictive in practical situations, the ability to vary this quantity has significant appeal.

Although the original result of Young and Verhagen (1996a) (7.38) was a reasonable fit to the recorded data, it appeared to be anomalously high at small values of χ [see Figure 7.16]. This is surprising as these short fetch situations correspond to deep water, for which the functional form of the growth curve is reasonably

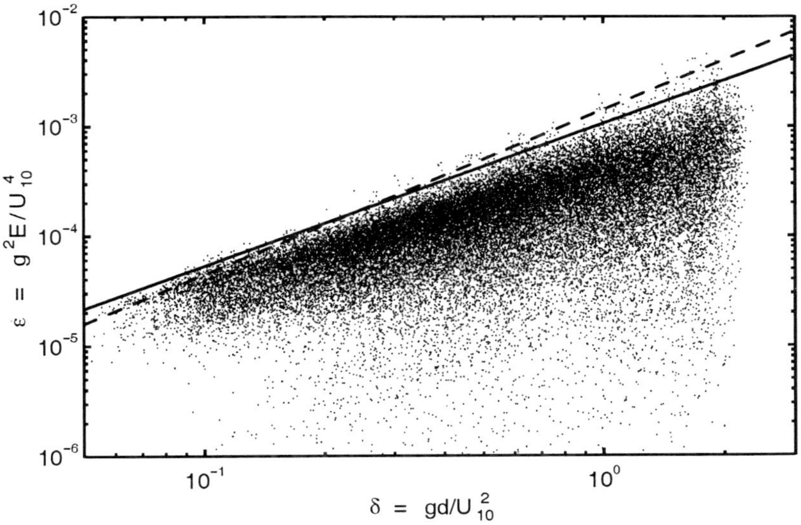

Figure 7.14: A scatter plot of non-dimensional energy, ε against non-dimensional depth, δ The envelope to the data is shown as the solid line [Equation (7.36)] and the Lake Okeechobe result (Bretschneider, 1958) [Equation (7.28)] as the dashed line. [after Young and Verhagen (1996a)].

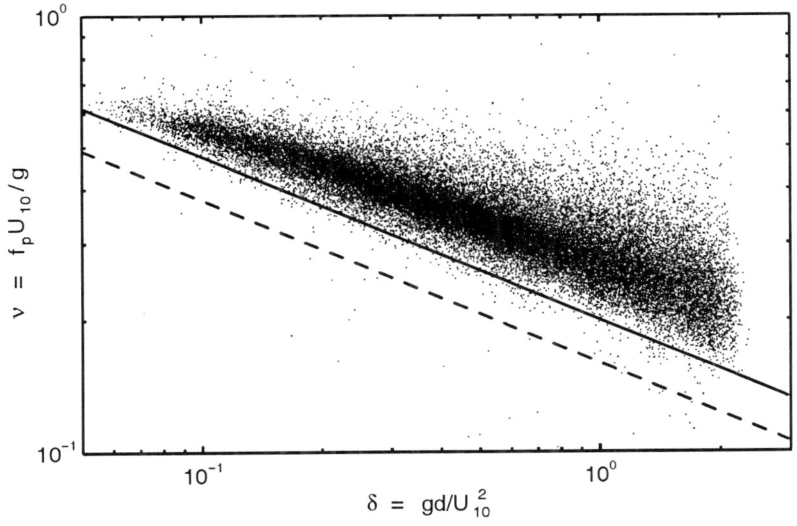

Figure 7.15: A scatter plot of non-dimensional frequency, ν against non-dimensional depth, δ The envelope to the data is shown as the solid line [Equation (7.37)] and the Lake Okeechobe result (Bretschneider, 1958) [Equation (7.29)] as the dashed line. [after Young and Verhagen (1996a)].

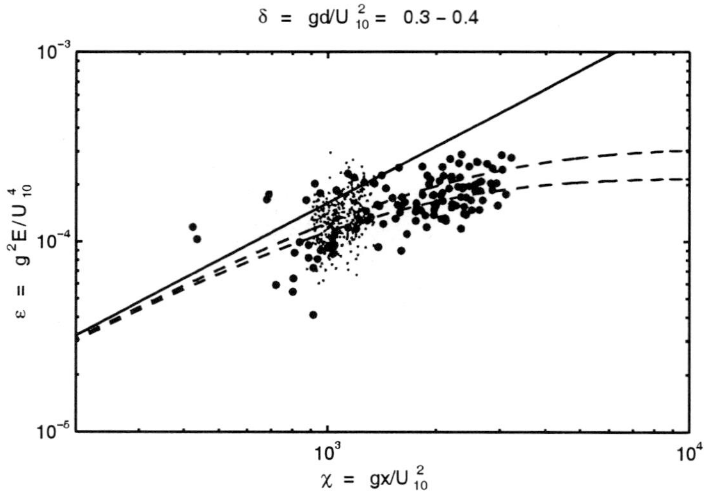

Figure 7.16: A scatter plot of non-dimensional energy, ε against non-dimensional fetch, χ. Only data with values of non-dimensional depth, δ between 0.3 and 0.4 are shown. Equation (7.38) is shown for the two extremes of δ (i.e. 0.3 and 0.4) by the two dashed lines. The deep water asymptotic form of Equation (7.38) is shown as the solid line. [after Young and Verhagen (1996a)].

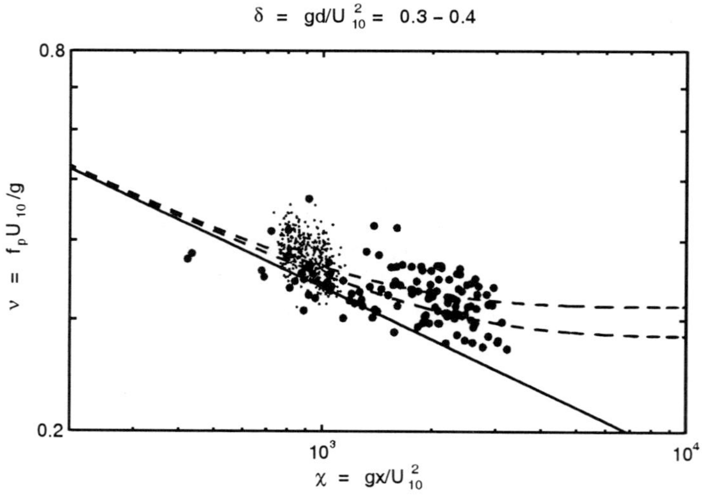

Figure 7.17: A scatter plot of non-dimensional frequency, ν against non-dimensional fetch, χ. Only data with values of non-dimensional depth, δ between 0.3 and 0.4 are shown. Equation (7.41) is shown for the two extremes of δ (i.e. 0.3 and 0.4) by the two dashed lines. The deep water asymptotic form of Equation (7.41) is shown as the solid line. [after Young and Verhagen (1996a)].

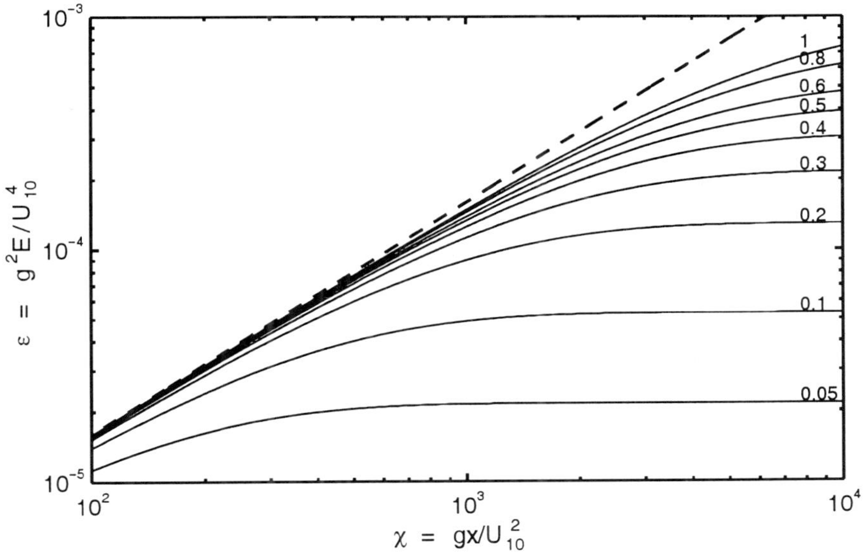

Figure 7.18: The evolution of the non-dimensional energy, ε as a function of non-dimensional fetch, χ as predicted by Equation (7.38). Each curve is for a specific value of non-dimensional depth, δ as shown at the right extremity.

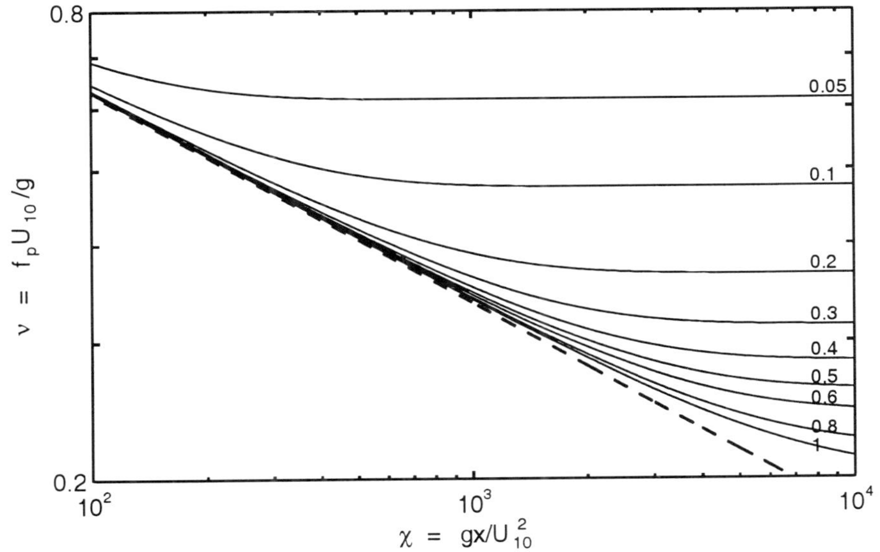

Figure 7.19: The evolution of the non-dimensional frequency, ν as a function of non-dimensional fetch, χ as predicted by Equation (7.41). Each curve is for a specific value of non-dimensional depth, δ as shown at the right extremity.

well known. Although Young and Verhagen (1996a) included the influence of the development of the boundary layer with fetch by using the average wind speed down the fetch as the scaling wind speed, Young (1997b) investigated the effects of including this type of wind speed variation in (7.44). The results are shown in Figure 7.20 for a variety of wind speeds and a constant value of $\delta = 0.25$. As shown by Donelan et al (1992) for deep water, there is no universal growth curve (even for a fixed value of δ). Rather, the development of ε with χ depends on the wind speed, particularly at short fetch.

When a realistic rate of boundary layer development was included in (7.44), the apparently low energy levels of the recorded Lake George data at short fetch could be explained. A comparison between (7.44) and the data is shown in Figure 7.21.

7.4 Finite Depth One-dimensional Spectra

In contrast to the wealth of observed deep water fetch limited spectra [see Section (5.4)], the data set in finite depth situations is very small. The well known fetch limited finite depth field experiments of Lake Okeechobe (U.S. Army, 1955; Bretschneider, 1958) and Lake Marken (Bouws, 1986) reported only on the development of the integral spectral properties of total energy and peak spectral frequency [see Section (7.3)].

A reasonably detailed study of finite depth wave spectra was performed by Bouws et al (1985, 1987). This study considered the, so called, TMA data set, comprised of measurements made at coastal sites during three separate field experiments (TEXEL - Dutch North Sea; MARSEN - German Bight; ARSLOE - US east coast). In all cases, the spectra could be described as wind sea spectra, in that no significant swell was present. As no clear fetch could be delineated in many of the cases, it is difficult to determine whether they were fetch limited. Following Kitaigorodskii (1962) and Kitaigorodskii et al (1975), they assumed that a universal spectral form existed in the wave number spectrum given by

$$Q(k) = \frac{\alpha}{2} k^{-3} \Psi(k, f_p, d) \tag{7.45}$$

where $Q(k)$ is the wave number spectrum, k is the wave number and Ψ is a non-dimensional shape function which approaches one for $k >> k_p$ where k_p is the wave number of the spectral peak. When converted to a frequency spectrum using linear wave theory, (7.45) produces a form with a variable high frequency exponent. In deep water it will be –5 and at the shallow water limit –3.

Assuming a shape function of the JONSWAP form (Hasselmann et al, 1973), (7.45) produces a frequency spectrum of the form

$$F(f) = \alpha g^2 (2\pi)^{-4} f^{-5} \exp\left[\frac{-5}{4}\left(\frac{f}{f_p}\right)^{-4}\right] \cdot \gamma^{\exp\left[\frac{-(f-f_p)^2}{2\sigma^2 f_p^2}\right]} \cdot \Phi \tag{7.46}$$

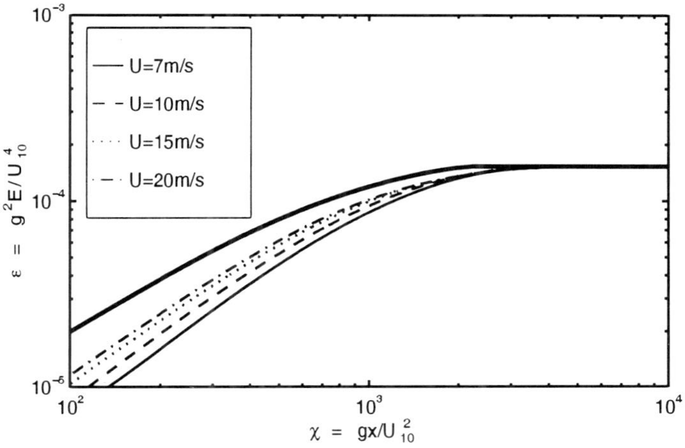

Figure 7.20: Growth curves representing the development of the non-dimensional energy, ε as a function of the non-dimensional fetch, χ for a non-dimensional water depth, $\delta = 0.25$. The wind speed, U_{10} along the fetch has been assumed to vary according to the relationship of Taylor and Lee (1984). Results are presented for four different wind speeds at full development (see legend). The result for a constant wind speed along the fetch is shown by the thick solid line. [after Young (1997b)].

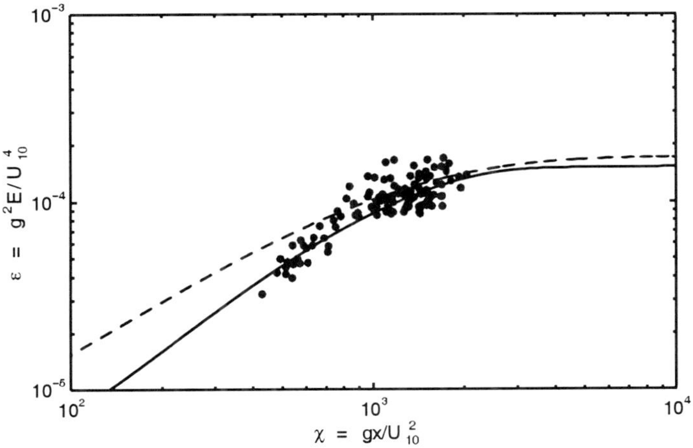

Figure 7.21: The Lake George data of Young and Verhagen (1996a) showing ε as a function of χ. Data in the interval $\delta = 0.2-0.3$ are shown. The growth curve (7.38) for $\delta = 0.25$ is shown by the dashed line. The result obtained by the integration of (7.44) with $\delta = 0.25$ and U_{10} varying along the fetch according to Taylor and Lee (1984), with a fully developed value of $U_{10} = 7\text{m/s}$, is shown by the solid line. [after Young (1997b)].

and

$$\Phi = \left\{ \frac{[k(f,d)]^{-3} \frac{\partial k(f,d)}{\partial f}}{[k(f,\infty)]^{-3} \frac{\partial k(f,\infty)}{\partial f}} \right\} \tag{7.47}$$

Equation (7.46) was termed the TMA spectral form by Bouws et al (1985).

Neglecting the shape function in (7.45), Bouws et al (1985) determined an approximation to the total spectral energy

$$\sigma^2 = \int_0^\infty Q(k)\mathrm{d}k \approx \frac{\alpha}{2} \int_{k_p}^\infty k^{-3}\mathrm{d}k \tag{7.48}$$

Introducing non-dimensional variables, (7.48) can be integrated to yield

$$\varepsilon = \frac{\alpha}{4}\kappa^{-2} \tag{7.49}$$

where $\kappa = U_{10}^2 k_p/g$ is the non-dimensional peak wave number, $\varepsilon = \sigma^2 g^2/U_{10}^4$ is the non-dimensional energy and k_p is the wave number of the spectral peak. Bouws et al (1985) demonstrated that (7.49) was a good approximation to the observed data, supporting the applicability of the TMA form, (7.46).

Bouws et al (1987) examined the evolution of the spectral parameters, α, γ and σ. They found that α was an increasing function of κ although there was significant scatter in their data. They approximated this relationship by a power law. In a similar fashion to JONSWAP (Hasselmann et al, 1973) the peak shape parameters γ and σ exhibited significant scatter with no clear trends.

A further application of the general theory of Kitaigorodskii (1962) was applied by Miller and Vincent (1990). Rather than adopting the k^{-3} scaling of (7.45), they proposed a form proportional to $k^{-2.5}$ which transforms to a deep water frequency spectrum proportional to f^{-4}. This form was compared with data from the Corps of Engineers Field Research Facility on the east coast of the United States. They found that this form, when transformed to frequency space (termed the FRF spectrum), modelled the data equally well as the TMA form. They also found that the spectral parameters of α and σ were relatively constant for this spectral formulation. In addition, a correlation existed between γ and the wave steepness ($\Lambda = \sqrt(\sigma^2)/L_p$, where σ^2 is the total energy and L_p is the wave length of components at the spectral peak). It should be noted, however, that Λ has only a relatively small dynamic range for typical condition and hence such a correlation should be treated with some caution.

An alternative approach has been proposed by Thornton (1977) in which the wave celerity is the governing parameter at high frequencies. Using similarity arguments a general spectral form is derived which is of the form f^{-5} in deep water and f^{-3} in shallow water. These asymptotic limits are consistent with Kitaigorodskii (1962).

The major problem with the various spectral forms proposed above is obtaining sufficient high quality data to determine which proposal best approximates the finite depth spectral form and determining relationships which describe the parameters of this spectral form. The extensive Lake George data set has provided some of these answers. Young and Verhagen (1996b) have examined the evolution of the one-dimensional finite depth spectrum with fetch. Figure 7.22 shows an example of the development of the spectrum, $F(f)$ with fetch. The case shown is for a wind speed of $U_{10} = 10.8$ m/s. The spectra of Figure 7.22 clearly show the self-similarity of the spectra previously discussed in deep water studies [see Section (5.2)]. The increase in total energy of the spectra and the migration of the spectral peak to lower frequencies are clearly evident in this figure. As observed in deep water, the high frequency portion of the spectrum appears reasonably well approximated by a relationship of the form $F(f) \propto f^n$. Close examination of the spectra, however, suggests that n may change in value with increasing fetch.

Young and Verhagen (1996b) firstly examined the high frequency portion of their recorded spectra, fitting the form $F(f) \propto f^n$ for $f > 2f_p$. Typical results are shown in Figure 7.23, where n is plotted as a function of χ.

The results shown in Figure 7.23 indicate that at short non-dimensional fetch, n is approximately equal to -5. As the fetch increases the value of the exponent changes and at the longer non-dimensional fetches n approximates -3. At short non-dimensional fetch, the waves are in deep water, the effects of finite depth increasing as the non-dimensional fetch increases and the waves develop. Hence, the gradual decreases in the magnitude of the exponent appears to be associated with the effects of finite depth. This is further supported by the fact that for the same value of χ, the deeper data (larger δ) yield values of n of larger magnitude than the shallower data (smaller δ). As δ increases, finite depth effects will only become significant at larger values of χ.

This type of variation in n with water depth is predicted by the TMA form (7.46). The applicability of the TMA form as an approximation to the shape of finite depth spectra was confirmed by Young and Verhagen (1996b) by representing $\alpha\kappa^{-2}$ as a function of ε. This result is shown in Figure 7.24 and is in remarkably good agreement with the TMA prediction (7.49), even though this form ignores any peak enhancement. A least squares fit to the data yielded

$$\varepsilon = 0.14 \left(\alpha\kappa^{-2}\right)^{0.91} \tag{7.50}$$

With confirmation that the one-dimensional finite depth spectral form is well represented by TMA (7.46), Young and Verhagen (1996b) investigated the values of the spectral parameters, α, γ and σ. They found that α could be represented in terms of κ by

$$\alpha = 0.0091\kappa^{0.24} \tag{7.51}$$

No clear trends could be found for γ or σ (as for deep water) within the significant scatter of the data. The mean values of the data set yield $\gamma_{mean} = 2.70$ and $\sigma_{mean} = 0.12$.

Young and Verhagen (1996b) argued that the lack of any clear dependence of γ and σ on κ was possibly due to the difficulty of determining these parameters

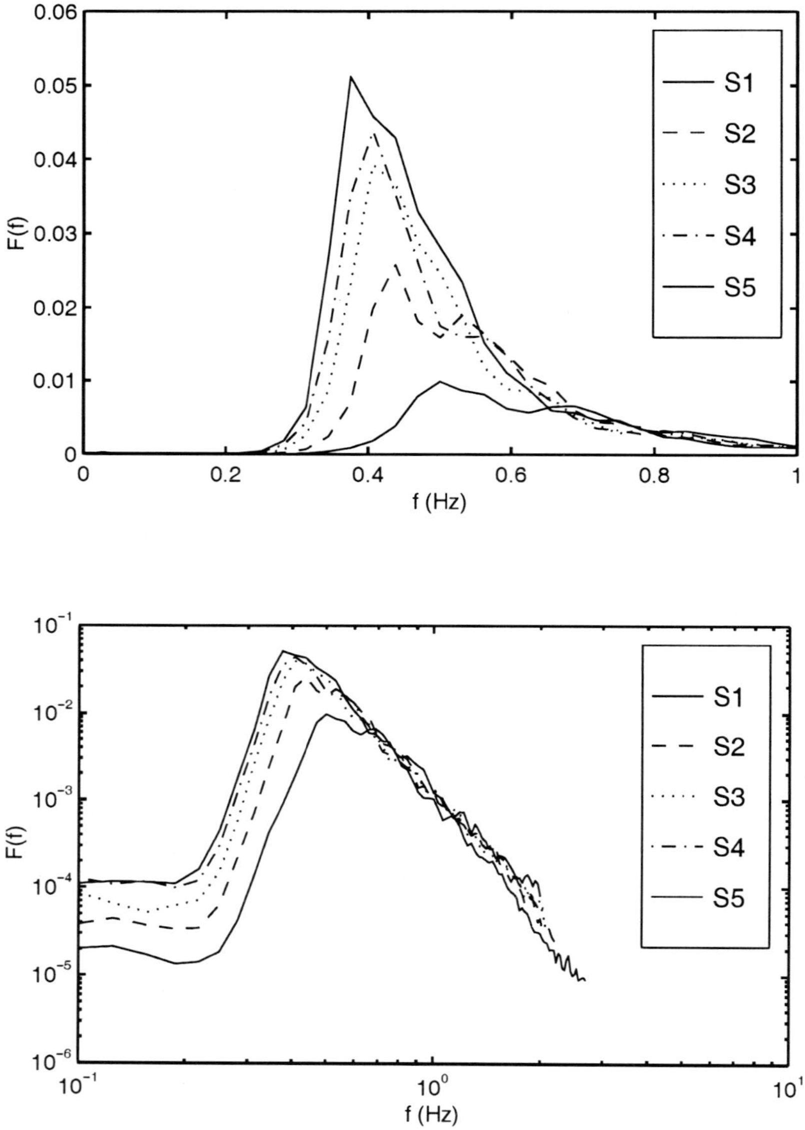

Figure 7.22: An example of the development of the finite depth wave spectrum with increasing fetch. The stations are labeled S1 to S5 (increasing fetch from $x = 1.3$ km to $x = 6.7$ km). Spectral evolution at fetches larger than S5 is very small and hence these spectra have been omitted for clarity. The case shown is for a northerly wind measured at S6 of $U_{10} = 10.8$ m/s. Top panel shows a linear scale whilst the bottom panel is a logarithmic scale. [after Young and Verhagen (1996b)].

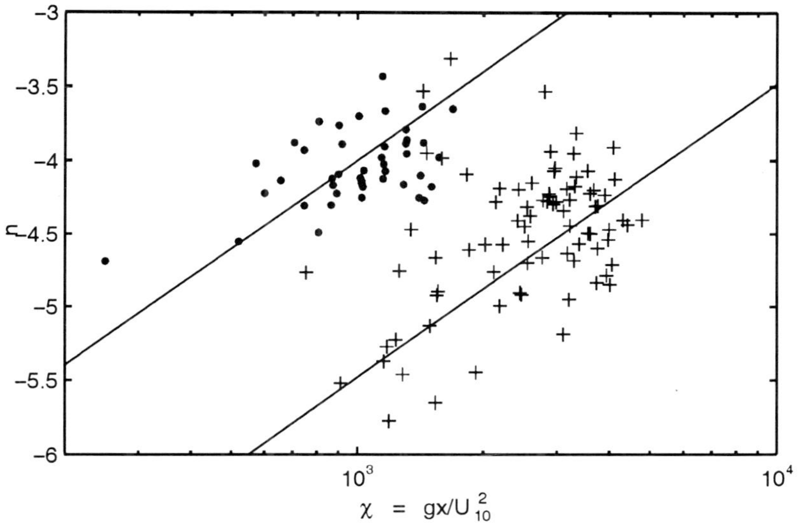

Figure 7.23: Values of the high frequency exponent, n, as a function of non-dimensional fetch, χ. Data for two ranges of non-dimensional depth, δ are shown, $0.1 < \delta < 0.2$ (dots) and $0.5 < \delta < 0.6$ (crosses). The solid lines are visual trend lines showing the decrease in the magnitude of n as finite depth effects become more significant. [after Young and Verhagen (1996b)].

accurately from spectra defined on a rather course frequency grid. In order to investigate this possibility they inverted the problem. With the total energy and peak frequency of the spectrum defined by (7.38) and (7.41), respectively and α specified by (7.51), the value of γ was determined (assuming σ was constant at 0.12). Although this approach gave a consistent trend for γ, the result was very sensitive to the exact relationship chosen for the peak frequency, ν. It was, however, clear from the results that γ was inversely related to δ (γ decreases as the water depth increases). Therefore, wind wave spectra in very shallow water could be expected to be very peaked.

7.5 Finite Depth Directional Spreading

The only detailed observations of the directional spreading of finite depth wind waves come from the Lake George data set (Young et al, 1996). The Lake George data for directional spectra are, however, far less comprehensive than for one-dimensional spectra. Young et al (1996) investigated both $\cos^{2s}$ (5.60) and sech^2 (5.65) forms and concluded that for their data set the $\cos^{2s}$ provided a slightly superior fit. Their full data set consisted of 156 directional spectra. The mean fit

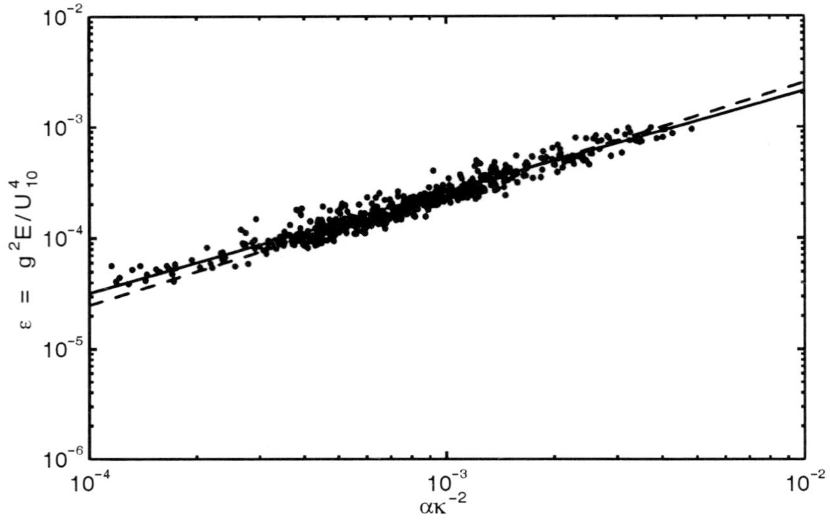

Figure 7.24: Values of the non-dimensional energy, ε as a function of the quantity, $\alpha\kappa^{-2}$. The solid line is a least squares fit to the data, (7.50) whereas the TMA form [Equation (7.49)], which ignores the contribution of peak enhancement, is shown by the dashed line. [after Young and Verhagen (1996b)].

to this data yielded the following representation for the exponent s in (5.60)

$$s = \begin{cases} 11\left(\frac{f}{f_p}\right)^{2.7} & \text{for } f < f_p \\ 11\left(\frac{f}{f_p}\right)^{-2.4} & \text{for } f \geq f_p \end{cases} \tag{7.52}$$

Equation (7.52) is constrained to yield narrowest spreading at $f/f_p = 1$. The scatter in the Young et al (1996) data was such that the actual point of narrowest spreading could not be determined with great accuracy. It was clear, however, that it was in the vicinity of the spectral peak frequency. The Young et al (1996) data set spanned inverse wave ages $1 < U_{10}/C_p < 4$. Despite this relatively wide parameter range, no dependence on U_{10}/C_p could be determined.

Young et al (1996) also investigated whether the width of the directional spreading was a function of k_pd. Figure 7.25 shows the parameter s as a function of f/f_p for a number of ranges of k_pd. Within the scatter of the data no dependence on k_pd could be determined.

Although no dependence on k_pd could be determined from the data, Young et al (1996) speculated that this may be due to the intrinsic scatter in the data and the relatively narrow range of values of k_pd. Figure 7.26 shows a comparison between (7.52) and the deep water results of Mitsuyasu et al (1975), Hasselmann et al (1980) and Donelan et al (1985).

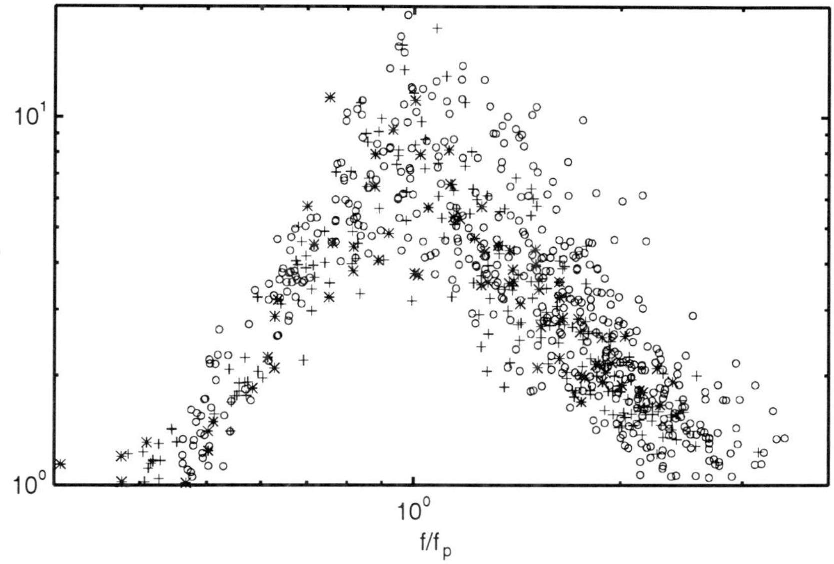

Figure 7.25: Values of the directional exponent s [Equation (5.60)] as a function of the normalized frequency, f/f_p. Values have been partitioned based on the relative depth parameter, k_pd. $1.0 < k_pd \leq 1.5$ - 'o'; $1.5 < k_pd \leq 2.0$ - '+'; $k_pd > 2.0$ - '*'. [after Young et al (1996)].

As the results of Mitsuyasu et al (1975) and Hasselmann et al (1980) are both wave age dependent, mean values typical of their respective data sets have been used to construct Figure 7.26 [Mitsuyasu et al (1975) - $U_{10}/C_p = 1.1$; Hasselmann et al (1980) - $U_{10}/C_p = 1.4$]. As Donelan et al (1985) provide no functional form expressed in terms of s, the result in Figure 7.26 was obtained from digitizing the result given in their Figure 30.

The present results are marginally narrower (higher s) than those of both Mitsuyasu et al (1975) and Hasselmann et al (1980), they are however significantly broader than the high resolution results of Donelan et al (1985). As explained in Section (5.5), the evidence suggests (Donelan et al, 1985; Young, 1994a) that the results of both Mitsuyasu et al (1975) and Hasselmann et al (1980) are excessively broad due to the instrumentation and analysis technique utilized. The inference is that finite depth wind wave spectra are broader than their deep water counterparts. Due to the relatively narrow range of k_pd spanned by the Young et al (1996) data set, a more emphatic statement cannot be made. At present it is necessary to rely on these two independent data sets [Donelan et al (1985) - deep water; Lake George - finite depth). Young et al (1996) speculated that the apparently broader

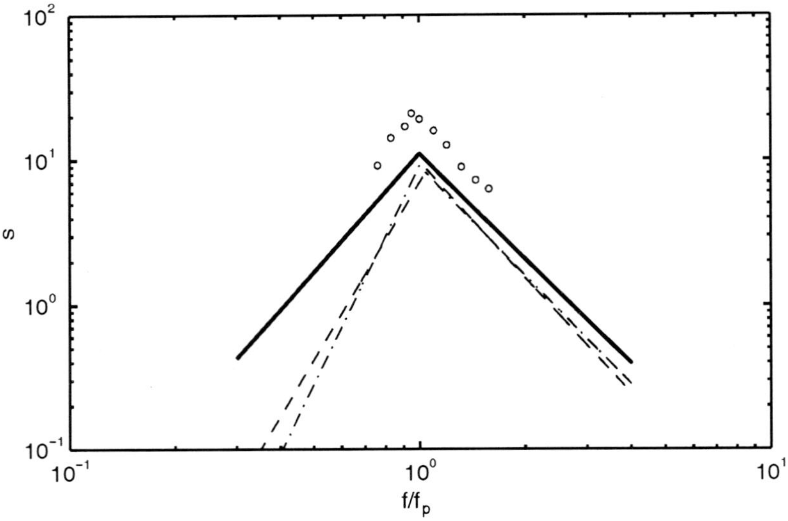

Figure 7.26: The dependence of the directional exponent s for the finite depth result of Young et al (1996) (7.52) (thick solid line). Also shown for comparative purposes are the deep water data sets: Mitsuyasu et al (1975) (5.61) - 'dash dot line'; Hasselmann et al (1980) (5.63) - 'dashed line'; Donelan et al (1985) - 'open circles'. [after Young et al (1996)].

directional spreading in finite depth situations may be due to the behaviour of S_{nl} in such situations. Figure 7.2 shows that finite depth spectral components separated in both wave number and direction are more strongly coupled than in deep water. Hence, the spectrum could be expected to be more isotropic in finite depth conditions (i.e. a flatter high frequency face and broader directional spreading). This type of coupling would be supported by the "flatter" high frequency face of the finite depth TMA spectral form (7.46), together with the broader spreading of (7.52).

Chapter 8

Numerical Modelling of Waves

8.1 Introduction

Chapters 5 and 7 presented a number of empirically based equations for the prediction of both spectra and their integral properties. Although these approaches are useful, they are limited to a very narrow range of applications (fetch and duration limited growth). These approaches were extended to some extent in Chapter 6 to include some special cases of non-stationary conditions, but the limitations are still significant. Practical wave prediction applications involve a mixture of local wind generated sea and swell, spatially and temporally varying winds and irregular bathymetry and coastlines. To predict waves in these demanding situations requires resort to a model which represents the physical processes which occur in nature (as best we understand them).

A comprehensive model which incorporates our full understanding of wind wave physics and is applicable in all situations would be prohibitively expensive. Instead, a variety of models have been proposed for applications in specific situations. In order to select the most appropriate model requires an understanding of the relative importance of the various physical processes active in each domain. Such a classification has been presented by Battjes (1994), a modified form of which appear in Table 8.1. The table divides wave prediction into four physical domains:

- *Deep Oceans* - where bottom influences can be neglected.
- *Shelf Seas* - area between the deep oceans and the shoaling zone.
- *Shoaling Zone* - the area where shoaling becomes important.
- *Harbours* - taken to represent areas where there is an interaction between the waves and any structure (e.g. breakwater, oil platform, island, reef, etc.).

Physical Process	Deep Oceans	Shelf Seas	Shoaling Zone	Harbours
Diffraction	⊗	⊗	∘	●
Depth refr./shoaling	⊗	•	●	•
Current refraction	⊗	∘	•	⊗
Quad. Interactions	●	●	∘	⊗
Triad Interactions	⊗	∘	•	∘
Atmospheric Input	●	●	∘	⊗
White-capping	●	●	∘	⊗
Depth Breaking	⊗	∘	●	⊗
Bottom Friction	⊗	●	•	⊗

Table 8.1: The relative importance of various physical mechanism in different domains: ⊗ – negligible; ∘ – minor importance; • – significant; ● – dominant. [after Battjes (1994)].

Models can be divided into two general classes: *phase resolving models*, which predict both the amplitude and phase of individual waves and *phase averaging models* which predict average quantities such as the spectrum or its integral properties (H_s, f_p, etc.). Should phase averaged properties vary rapidly (order of a few wavelengths) then it will generally be necessary to use a phase resolving model. Conversely, should the wave properties vary slowly, on a scale of many wavelengths, then phase averaging models should be adopted.

Neither of the two general classes is superior, their domains of application often do not overlap. Nevertheless, Battjes (1994) concludes that: *"Phase resolving models are computationally so much more demanding (per unit area of computational domain) that they should be used only where they are strictly needed."*. Of the processes considered in Table 8.1 only diffraction and triad nonlinear interactions require phase resolving models. Hence, the domain of application of phase resolving models is generally confined to wave-structure interactions (harbours etc.) and the nearshore zone, where triad interactions become important. The recent developments described in Section (7.2.7) to develop a spectrally based triad source term, S_{tri} may, with development, further restrict this area of application.

This chapter will consider both classes of model, although, in keeping with the spectral representation which has been adopted throughout, only a very cursory examination of phase resolving models will be included. This examination will closely follow the excellent review of Battjes (1994). The chapter will also explore the rapidly developing technique of data assimilation, in which measured data is "blended" with model results to produce enhanced accuracy.

8.2 Phase Resolving Models

Phase resolving models can be divided into a number of distinct types depending on the values of (Battjes, 1994):

- The nonlinearity as measured by ak or a/d.
- The relative depth parameter kd.
- The bottom slope α.

In the above, a is the wave amplitude, k the wave number and d the water depth. Typical values for each of the types of phase resolving models are given by Peregrine (1972), Whitham (1974) and Mei (1989).

If no assumptions are made about the magnitude of any of these quantities, then a class of so called "exact" models result. The term "exact" is used to indicate that no simplifying assumptions have been made in the model formulation. As numerical solutions are required, such models do not yield exact solutions. Models of this type are often solved using *boundary integral methods.*

If the bottom slope is small ($\alpha << kd$) and the waves are only weakly nonlinear ($ak << 1$), the velocity potential can be expanded as a Taylor series. These assumptions lead to so called *mild slope equation* models.

If the relative depth parameter and the bottom slope are both small ($kd << 1$, $\alpha << 1$) and the nonlinearity is weak ($a/d << 1$), *Boussinesq equation* models result. If only forward propagation is considered (i.e. no reflection), the Boussinesq equations reduce to the *Korteweg-de Vries equations.*

8.2.1 Boundary Integral Models

Models of this class solve the Laplace equation [first introduced in two dimensions as (2.1)]

$$\frac{\partial^2 \phi}{\partial x^2} + \frac{\partial^2 \phi}{\partial y^2} + \frac{\partial^2 \phi}{\partial z^2} = 0 \tag{8.1}$$

where ϕ is the velocity potential. As both the vertical and horizontal structure of the wave field must be computed simultaneously, such models are computationally very expensive. These models are, however, excellent tools for the investigation of detailed properties of irrotational waves (e.g. the approach to wave breaking). Numerous models have been developed to investigate such properties (Longuet-Higgins and Cokelet, 1976; Dold and Peregrine, 1986; Grilli et al, 1989; Peregrine, 1990). Although such models are powerful research tools, they are limited by the assumptions invoked in the development of the Laplace equation (8.1) [see Section (2.2.1)]. Hence, real fluid effects such as viscosity are neglected. Therefore, despite the computational expense involved, the existence of the bottom boundary layer and hence bottom friction are neglected by such models.

Battjes (1994) concludes that such models are *"... unfit for practical application to wind-wave problems"*.

8.2.2 Mild Slope Equation Models

The Mild Slope Equation was developed by Berkhoff (1972) as

$$\nabla \cdot (CC_g \nabla \phi) + k^2 CC_g \phi = 0 \tag{8.2}$$

where C is the phase velocity and C_g the group velocity. As written, (8.2)represents the effects of shoaling, refraction, diffraction and reflection of periodic gravity waves in the absence of currents. Equation (8.2) is an elliptical partial differential equation. Its solution requires specification of boundary condition around the full computational domain and solution at all interior points of the domain. As the computational grid requires a spatial resolution significantly smaller than the wavelength, solutions are computationally expensive, even for relatively small regions.

If reflection of waves propagating in the dominant direction is neglected, a computationally more efficient parabolic approximation to (8.2) can be formed (Radder, 1979)

$$\frac{\partial \phi}{\partial x} = \left[ik - \frac{1}{2kCC_g} \frac{\partial}{\partial x} (kCC_g) \right] \phi + \frac{i}{2kCC_g} \frac{\partial}{\partial y} \left(CC_g \frac{\partial \phi}{\partial y} \right) \tag{8.3}$$

where the x-axis is in the principal direction of wave propagation. Numerous extensions of this basic form have been developed to extend the range of application of the parabolic approximation to the Mild Slope Equation (Battjes, 1994). These extensions include:

- the effects of currents (Booij, 1981; Kirby, 1984);
- wider angle approximations to increase the range of directions which can be considered in the forward marching numerical schemes used to solve the parabolic form of the Mild Slope Equation (Booij, 1981; Dalrymple and Kirby, 1988; Dalrymple et al, 1989);
- curvilinear coordinates (Lozano and Liu, 1980; Isobe, 1986);
- dissipation due to wave breaking and bottom dissipation (Booij, 1981; Kirby and Dalrymple, 1986a);
- nonlinear dispersion equation (Booij, 1981; Kirby and Dalrymple, 1986b)
- wind input (Vogel et al, 1988)

Wind input and dissipation due to wave breaking and bottom interaction have been included in such models as relatively simple amplification and decay factors. The degree of sophistication with which these processes have been represented has been significantly less than shown in Chapter 4. As the spatial domains over which these models are generally applied are relatively small, these processes are generally not of significant importance.

Mild Slope Equation models have been extensively applied to predict the influence of refraction-diffraction-reflection in and around harbours, islands and submerged shoals.

8.2.3 Boussinesq Models

Boussinesq Models are applicable to shallow-water conditions with weak nonlinearity and weak frequency dispersion of the waves. The two-dimensional form of the equations were developed by Peregrine (1967)

$$\frac{\partial \eta}{\partial t} + \nabla \cdot [(d+\eta)\,\overline{u}] = 0$$

$$\frac{\partial \overline{u}}{\partial t} + \overline{u} \cdot \nabla \overline{u} + g \nabla \eta = \frac{d}{2} \nabla \left[\nabla \cdot \left(d \frac{\partial \overline{u}}{\partial t} \right) \right] - \frac{d^2}{6} \nabla \left[\nabla \cdot \frac{\partial \overline{u}}{\partial t} \right] \tag{8.4}$$

where η is the water surface elevation and $\overline{u}$ is the depth averaged horizontal velocity.

The Boussinesq equations include the effects of diffraction, refraction, reflection and wave-current interaction. Wind input and dissipation can be included in simplified forms as already indicated for Mild Slope Equation models. Triad interactions [see Section (7.2.7)] are represented by the quadratic terms. The inclusion of these terms is a particular strength of this class of model and they have been particularly successful in predicting the evolution of higher harmonics during shoaling (Beji and Battjes, 1994; Madsen et al, 1997).

8.3 Phase Averaging Models

Whereas a phase resolving models predicts the water surface elevation, thus resolving the individual waves, phase averaging models predict average or integral properties of the wave field. In there simplest form this may be quantities such as the significant wave height or peak period. More commonly, however, modern phase averaging models predict the spatial and temporal evolution of the directional spectrum, $F(f,\theta)$.

In contrast to phase resolving models, phase averaging models solve a single equation, although the sophistication of its representation may differ significantly. This basic equation is represented by (4.19) or in a simpler deep water form as

$$\frac{\partial F}{\partial t} + \mathbf{C_g} \cdot \nabla F = S_{tot} \tag{8.5}$$

where $F = F(f, \theta; x, y, t)$ is the directional spectrum and $\mathbf{C_g}$ is the group velocity vector. The source term, S_{tot} is typically represented as the summation of a number of individual processes as shown in (7.1). For simplicity, the deep water form will be considered here

$$S_{tot} = S_{in} + S_{nl} + S_{ds} \tag{8.6}$$

where S_{in} represents atmospheric input from the wind [Section (4.3)], S_{nl} represents quadruplet nonlinear interactions between spectral components [see Section (4.4)] and S_{ds} represents dissipation by white-cap breaking [see Section (4.5)].

Equations such as (8.5) are not uncommon in computational physics and are collectively classed as *advective transport equations.* Their solution requires the specification of appropriate initial and boundary conditions, the implementation of an appropriate advection scheme to represent the transport of energy within the computational domain and the specification of the forcing term, S_{tot}. The forcing or source term, S_{tot} has already been discussed at length in Chapter 4. Its practical implementation in operational wave models is, however, not always straightforward and the manner in which it has been represented has been used as a convenient parameter to class models.

8.4 Source Term Representation

The sophistication with which the source term S_{tot} has been represented has largely paralleled the increasing understanding of the physical processes responsible for wind-wave evolution and the exponential increase in available computational resources. Models are typically classed as first, second or third generation. These classifications are briefly described in Table 8.2.

Each of these classes of model are discussed in the following sections.

8.4.1 First Generation Models

The first attempt to develop a wave prediction model based on a numerical solution to (8.5) was made by Gelci et al. (1957). This model pre-dated the development of the theory of quadruplet nonlinear interactions by Hasselmann (1962) and the source term was considered to be the sum of atmospheric input and white-cap dissipation.

$$S_{tot}^{(1)} = S_{in}^{(1)} + S_{ds}^{(1)} \tag{8.7}$$

Models of this type are termed *first generation*, as indicated by the superscripts in (8.7). A number of other first generation models were later based on this same source term description (Pierson et al., 1966; Inoue, 1967; Isozaki and Uji, 1973; Cardone et al, 1975; Cavaleri and Rizzoli, 1981; Chen and Wang, 1983). The atmospheric input term, $S_{in}^{(1)}$ in these models was generally represented as the sum of a linear (Phillips) term and an exponential (Miles) term. The exponential

	S_{in}	S_{nl}	S_{ds}
First Generation	• Based on growth rate measurements • Large in magnitude		• Saturation limit
Second Generation	• Based on flux measurements • Smaller than 1st Generation	• Parametric form • Limited flexibility	• Saturation limit, as in 1st Generation
Third Generation	• Based on flux measurements • Stress coupled to sea state	• Approximate form of Boltzman integral	• Explicit form

Table 8.2: Definition of model classes based on the representation of the source terms.

growth rate terms were based on direct measurements of wave growth such as those of Snyder and Cox (1966) [see (4.44)]. As already indicated in Chapter 4, atmospheric input terms obtained in this manner implicitly included the effects of other processes such as quadruplet interactions and overestimated the magnitude of the actual contribution due to atmospheric input alone.

The dissipation term, $S_{ds}^{(1)}$ was represented as an "on-off" limiter, which prevented the spectrum exceeding a pre-defined *saturation limit* (Phillips, 1958). Thus, the high frequency regions of the spectrum (saturation range) were forced to conform to this spectral form. This first generation spectral balance is shown diagrammatically in Figure 8.1.

First generation models were applied successfully for many applications. However, models tended to produce reliable results only for the geographic regions or meteorological systems for which they were developed. When applied to other cases, they often required re-calibration. In addition, they were incapable of predicting many aspects of spectral evolution such as the existence of the "overshoot effect" observed by Barnett and Wilkerson (1967) at frequencies immediately above the spectral peak frequency. Such models also performed poorly in complex meteorological situations, such as the rapidly turning winds in hurricanes. A systematic study of these limitations and a comparison between models was undertaken as part of the SWAMP study (SWAMP Group, 1985).

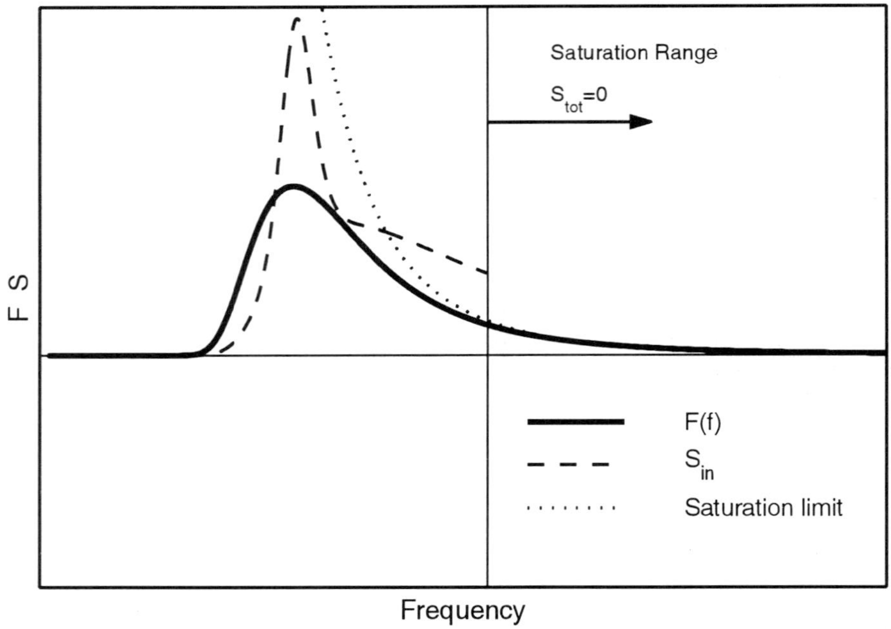

Figure 8.1: The source term balance for first generation wave models. Note that S_{in} is larger than in later generation models and that the spectral balance at high frequencies is controlled by the imposition of a saturation limit on the spectrum.

8.4.2 Second Generation Models

The JONSWAP study (Hasselmann et al, 1973) clearly showed the central role played by quadruplet nonlinear interactions in the source term balance. Hence, it became clear that this term must be included within operational models. A major impediment to the inclusion of S_{nl}, was the significant computational time required for its evaluation. In order to overcome this problem a class of wave models which utilized approximations to this term developed (Barnett, 1968; Ewing, 1971; Golding, 1983; Allender et al, 1985; Greenwood et al, 1985; Sobey and Young, 1986; Young, 1988a). This class of model, the source term balance of which is shown in (8.8) become know as *second generation models.*

$$S_{tot}^{(2)} = S_{in}^{(2)} + S_{nl}^{(2)} + S_{ds}^{(2)} \tag{8.8}$$

In contrast to first generation models, the atmospheric input term, $S_{in}^{(2)}$ was based on direct measurements of the normal stress exerted on the water surface (Snyder et al, 1981). Hence, this term was significantly smaller than in first generation models. Also, many of the second generation models neglected the linear or Phillips term as it was very small in comparison to the Miles exponential mechanism. The linear mechanism does, however, play a role in the initial growth of the spectrum from a calm sea. This initial "trigger" for growth was included in these models by defining the initial conditions for the model as some small, but finite, spectral energy.

The dissipation term, $S_{ds}^{(2)}$ was included as a saturation limit to the spectrum. As the results of JONSWAP showed that this level varied with the wave age, a number of models included variable levels for this saturation spectrum.

The significant advancement with this class of model was the inclusion of a parametric representation for the quadruplet nonlinear interaction term, $S_{nl}^{(2)}$. A number of representation were implemented. In almost all cases these approximate forms represented the model spectrum in terms of a small number (3 to 5) of parameters. The JONSWAP spectral parameters were a common choice (Barnett, 1968; Ewing, 1971; Young, 1988a). As S_{nl} could be pre-computed for spectra with a range of these standard parameters, model values of $S_{nl}^{(2)}$ could then be determined based on the model spectral parameters. Naturally, problems occurred when the model spectrum did not conform to any of those for which S_{nl} had been precomputed. A further sub-class of models, termed *hybrid models*, took this process one step further and predicted the evolution of the parameters rather than the discrete values of the two-dimensional directional spectrum. Notable examples of these hybrid models are GONO (Sanders, 1976) and HYPA (Günther et al, 1979).

A diagrammatic representation of the source terms in second generation models is shown in Figure 8.2. In contrast to first generation models [Figure 8.1], it is clear that the inclusion of the nonlinear term has resulted in a reduction in the magnitude of the atmospheric input.

The inclusion of the parametric representations for S_{nl} generally resulted in an enhanced modelling capability. Even in demanding situations such as hurricane wind fields, second generation models were able to produce reasonable results (Young, 1988a). Despite this, the relatively simple formulation for the nonlinear source term represented a significant short-coming of these models.

8.4.3 Third Generation Models

By the mid 80s a multitude of 1st and 2nd generation models were operational. Many of these were used for commercial purposes and numerous claims were made about the benefits of one model as opposed to another. In order to compare the physics of the various models and to assess their strengths and weaknesses a major international intercomparison study was initiated; the *Sea Wave Modeling Project, SWAMP*, (SWAMP Group, 1985). Rather than compare models against field data, this study utilized a series of idealized test cases, designed to highlight specific aspects of model physics. One of the significant finding of the study was: *"All present second-generation models suffer from limitations in the parameterization of the nonlinear energy transfer, S_{nl}".* The directional wave spectrum, $F(f, \theta)$ is typically represented by discrete values. In order to define the two-dimensional

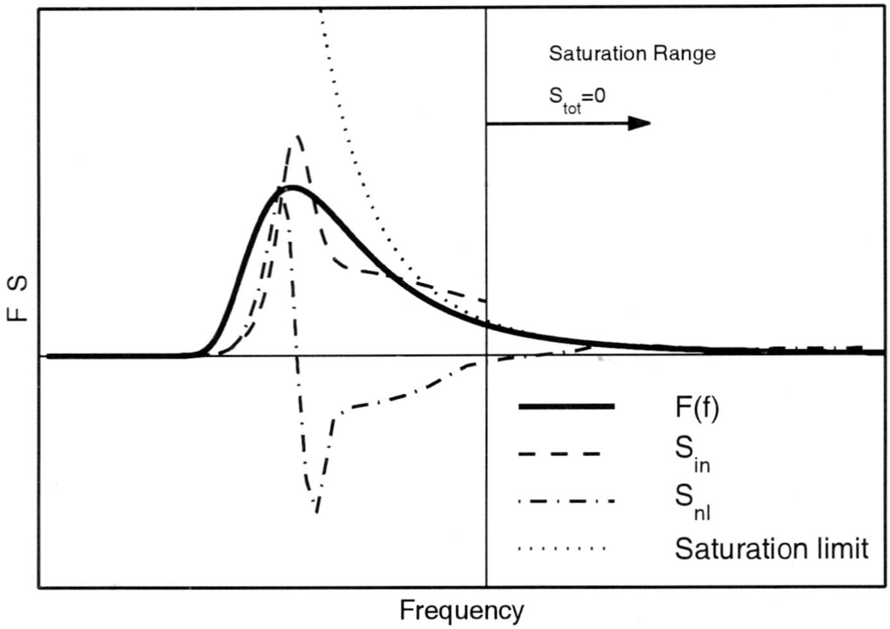

Figure 8.2: The source term balance for second generation wave models. Note that S_{in} is smaller than in first generation models [see Figure 8.1].

frequency-direction space, of order 100 points are required (typical implementation use about 300 values). The parameterizations for $S_{nl}^{(2)}$ typically use a maximum of 5 parameters. The mis-match is obvious: How can the evolution of a spectrum with hundreds of degrees of freedom be described by a source term defined by only a handful of parameters?

The solution to these short-comings was, however, a significant problem and the first steps to a solution were made in the spring of 1984 when Klaus Hasselmann invited wave researchers to Hamburg to discuss the possibility of a joint endeavour. The WAM (WAve Modelling) Group was formed. The aim of the group was to develop a *third generation* model, free from many of the assumptions which constrained earlier generation models. This significant international collaboration resulted in the WAM Model (WAMDI, 1988). Variants of this basic model were also developed by Tolman (1991) (WAVEWATCH) and more recently by Booij et al (1996) (SWAN).

The source term balance in third generation models is essentially the same as in second generation models, that is

$$S_{tot}^{(3)} = S_{in}^{(3)} + S_{nl}^{(3)} + S_{ds}^{(3)} \tag{8.9}$$

The details of the source terms are, however, significantly different to second generation models. The early versions of the WAM model utilized the Snyder et al

(1981) form of S_{in} as in second generation models. Later versions, however, incorporated the coupled air water drag model of Janssen (1991) [see Section (4.3.1.3)]. In principal, a third generation model should have a full solution to the nonlinear quadruplet interaction term [see Section (4.4)]. In practice, however, this is still not computationally feasible in a two-dimensional model. Third generation models utilize the discrete interaction approximation (DIA) to the term (Hasselmann et al, 1985a) (4.61). Although the DIA is still an approximation to S_{nl}, it is very different in its formulation to the simple parametric forms used in second generation models. The DIA retains the basic physics of the nonlinear interaction process, but considers a very small sub-set of all the possible interactions. In contrast to second generation parameterization for S_{nl}, which have only a few degrees of freedom, the DIA has as many degrees of freedom as there are values in the discretely specified directional spectrum, $F(f, \theta)$.

As both the input and nonlinear terms of second generation models had significantly fewer degrees of freedom than the spectrum, the only way that stable spectral evolution could be achieved was by imposing the saturation limit at high frequencies. This is not necessary with third generation models and explicit forms for the still poorly understood white-cap dissipation term, S_{ds}, can be utilized. The general form which has been used has been that of Komen et al (1984) (4.67).

The form of the source terms utilized in third generation models is shown in Figure 8.3. The significant difference compared to second generation models is that no artificially defined saturation limit is required to maintain a spectral balance at high frequencies. The spectrum is allowed to evolve without constraint until, at high frequencies a balance is achieved where $S_{tot}^{(3)} = 0$.

8.5 Computational Aspects

In addition to defining the source terms there are a number of computational aspects which need to be considered in the development of any numerical model. These include: the advection scheme to be used, the specification of the computational grid and the initial and boundary conditions. Many of these aspects are common to the solution of advective transport equations found in related areas of computational physics and the model designer can draw extensively from this literature. There are, however, some unique problems associated with the wave prediction problem which require specific attention. The vast bulk of the international wave modelling effort has been concentrated on the specification of the source terms. In comparison, the numerical aspects have received minimal attention.

8.5.1 Advection of Energy

The second term on the left hand side of (8.5) represents the advection of wave spectral energy at the group velocity, C_g. The advection of energy is more clearly seen when the transport equation is written in its characteristic form. Equation (4.22)

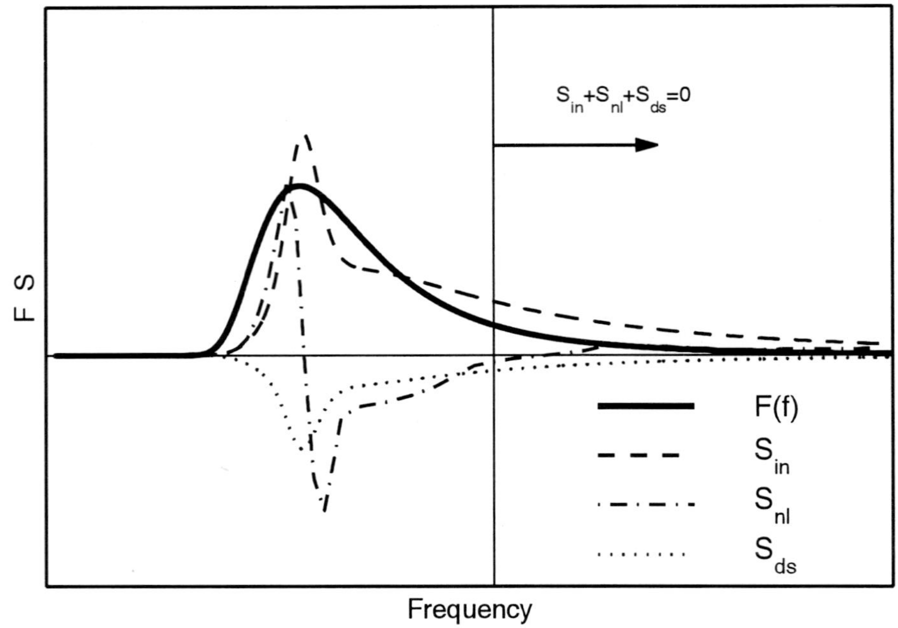

Figure 8.3: The source term balance for third generation wave models. Note that at high frequencies a balance is achieved between the source terms, such that, $S_{tot}^{(3)} = 0$.

is the finite depth form of (8.5). The set of characteristic equations, (4.21) define wave rays along which the wave action CC_gF is advected. In the absence of forcing (i.e. $S_{tot} = 0$), the wave action will be conserved along these rays. In deep water the rays become straight lines (or great circles if a spherical co-ordinate system is used) and due to the invariance of the group and phase velocities, the quantity conserved along the rays becomes the energy density, F. In finite depth situations or in the presence of spatially variable currents, the rays become bent, reflecting the effects of depth or current refraction.

The directional wave spectrum, $F(f, \theta)$ is typically represented by a series of discrete frequency and direction bins. Thus, each of these bins will have some finite size (in frequency-direction space), $\Delta f - \Delta\theta$, as shown in Figure 8.4.

A numerical scheme is required which can advect these discrete bins of energy at their respective group velocities. There are essentially four methods which have been used: finite differences, finite elements, full ray methods and piecewise ray methods. The vast majority of models have adopted finite difference techniques, WAM (WAMDI, 1988) being a good example. Finite elements have been less commonly used, although they possess some advantages in terms of fitting the grid to irregular coastlines.

Full ray methods construct rays which propagate throughout the complete grid.

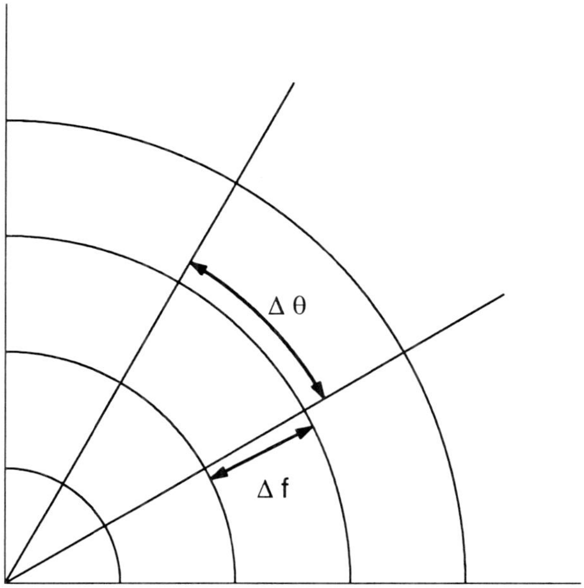

Figure 8.4: The directional wave spectrum, $F(f, \theta)$ is typically represented with wave models as a series of discrete computational bins, each of size $\Delta f - \Delta\theta$.

Equation (8.5) is then integrated numerically along these rays. This process decouples the rays, thus making it difficult to use source terms requiring coupling across spectral components (i.e. S_{nl}). Thus, full ray methods have generally not been used for second and third generation models. Piecewise ray methods trace all wave ray components from each computational point backwards for one time step. The energy at each terminating ray point is then determined by interpolation amongst surrounding computational points. Although the method has a number of advantages, including being unconditionally stable, it has not been commonly used (Young, 1988a).

Whichever technique is used, the goal is the same: to advect the energy with minimal numerical distortion. Convective transport equations are well known for the problems associated with numerical dispersion. This is generally overcome by resort to higher order schemes. In the field of wave modelling, however, some special problems exist. As indicated above, the desire is to advect the bin of energy which has finite size $\Delta f - \Delta\theta$. As shown in Figure 8.5, because of the finite size of the bin some natural dispersion will occur. As the bin represents wave components with frequencies in the range, $f_0 \pm \Delta f/2$, there will be a range of advection velocities for these components. The energy contained within the bin will naturally disperse, progressively being spread over an ever increasing spatial extent as it propagates.

As shown in Figure 8.5, spatial extent of the bin of energy increases linearly with time. A consistent numerical propagation scheme should reproduce these linear dispersion rates. Figure 8.5 also shows the spatial distribution resulting

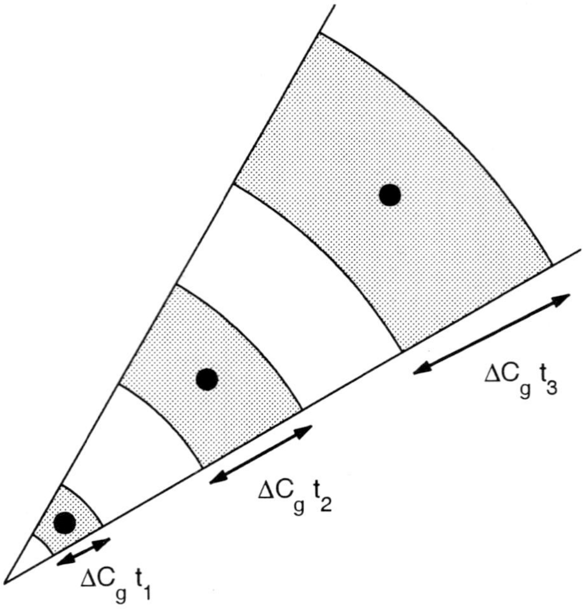

Figure 8.5: The natural dispersion of the bin of energy shown in Figure 8.4 as it is advected. Due to the finite range of group velocities and propagation directions contained within the bin, the spatial extent of the bin will increase with time. The single dots at the centre of the bin show the result of the use of a propagation scheme which has no numerical dispersion.

from a propagation scheme with no dispersion. The energy will be concentrated at a single point at the centre of the continually enlarging bin of energy. The resulting "spotty" distribution of energy has been termed the *garden sprinkler effect*. Booij and Holthuijsen (1987) have shown that the required linear spreading rates may be achieved by introducing a variable diffusion coefficient proportional to the age of the bin of energy. To date, this concept has not been tested in the context of a full model.

As a result of the garden sprinkler effect, use of higher order propagation schemes will not necessarily improve results. To date, operational models have generally used relatively low order propagation schemes in the hope that the numerically dispersion inherent in such schemes is not too dis-similar to the required natural dispersion. As an example, the WAM model uses a first order upwinding scheme. In justification for this choice Janssen et al (1994) take the pragmatic stance: *"... we have chosen the first order upwinding scheme because it is the simplest scheme to implement (requiring less computer time and memory) and because in practice it gives reasonable results"*. Higher order schemes have, however, been reported to yield improved results: Tolman (1992) – second order and Bender (1996) – third order.

It is clear from the discussion above that relatively little effort has been devoted to the propagation algorithms used in present day wave prediction models. Although it is claimed that present day models, such as WAM, predict swell propagation to a reasonable level of precision, other influences may also be active. The white-cap dissipation terms in third generation models have been "tuned" to ensure growth rates of wind sea and decay rates of swell agree with measurements. Hence, the apparently acceptable results may be a result of this artificial tuning. It is reasonable to conclude that further research into the development of appropriate propagation algorithms for use in wind-wave models is required.

8.5.2 Computational Grids

Equation (8.5) must be solved on a two-dimensional spatial grid. The precise details of this computational grid can have significant influence on the solution. The grid must be sufficiently fine to ensure that the spatial distribution of the wind field is adequately resolved. An often conflicting requirement is that the extent of the model should remain computationally tractable. Often, conflicts of scale also occur. For instance, a goal may be to model the waves generated by an extreme hurricane. Adequate resolution of the tight vortex wind field of the hurricane may require a very fine resolution grid (e.g. $\Delta x < 5$ km). At the same time, swell generated in this intense wind region can propagate large distance. Thus the spatial extent of the grid may need to be large (e.g. 2,000 km). Similarly, an extremely fine resolution grid may be required to accurately define the bathymetry in a coastal area. The meteorological system which generated swell which entered this region may, however, have occurred many thousands of kilometres away.

These conflicts of scale are often accommodated by the use of nested grids, particularly for coastal applications. A relatively course grid is initially used to model the generation of distant waves. Any number of finer nested grids can then be successively embedded within this course grid. Each of these finer grids takes their respective boundary conditions from the coarser grid in which they are nested.

Numerical wave prediction models can be divided into two general classes: global scale models and regional models. Global models operate on spherical grids. A typical example is the global WAM model which runs operationally at many centres, the most long standing implementation be that at the European Centre for Medium Range Weather Forecasting (ECMWF). In its original implementation at ECMWF, WAM operated on a 3° grid with a time step of 20 minutes, the spectral grid being represented by 25 frequencies and 12 directions. In later versions of the model, the spatial resolution was increased to 1.5°.

At the other extreme, regional models often operate on Cartesian grids with resolutions as fine as a few hundred metres. Although the resolution can be reduced further, some consideration of the rationale for such a grid is required. Phase averaging models assume that the spectrum is changing relatively slowly, compared to the wave length. Therefore, if it is expected that the wave field will change significantly over a length, of order, 10 wave lengths, than the validity of the phase averaged model is questionable.

In addition to the selection of appropriate spatial resolution, the time step also needs to be matched to the particular problem. In models which use explicit finite difference schemes, such as WAM, the upper bound on the time step Δt is specified by the requirements of numerical stability. The Courant limit requires $\Delta t < \Delta x / C_g$, where Δx is the spatial resolution. Hence, the energy is constrained to propagate, at most, one grid element during the time step. At first appearances this may seem an overly restrictive limitation and the use of implicit schemes may seen advantageous. Such schemes may remain numerically stable with larger time steps but they are not necessarily accurate. If the spatial scale has been chosen to resolve the bathymetry and wind forcing, the time step should be such that wave energy propagating through the area can appropriately interact with the local source terms. Hence, the Courant limit seems an appropriate upper bound for the time step to ensure, not only numerical stability, but also accuracy of the numerical solution.

8.5.3 Initial and Boundary Conditions

Models which do not have a Phillips mechanism require some initial spectral level to initiate growth. This is usually achieved by the introduction of a "seed" spectrum which has very low energy. The model then goes through a "spin-up" period where a wind field is imposed whilst the waves gradually reach a realistic level. The exact details of the seed spectrum are not important. One of the characteristics of spectral wave models (especially third generation models) is that they quickly modify the sea state to be in equilibrium with the local forcing wind.

All models require the specification of boundary conditions. Land boundaries are typically taken as completely absorbing with all energy being dissipated. Unless a global model is used, open boundary conditions must also be imposed. A common choice in such circumstances is to impose radiation conditions, in which energy can leave the computational domain but none can enter. If such boundaries are used, it is necessary to ensure that the computational domain is sufficiently large to ensure the boundary conditions do not influence the wave field at the points of interest.

8.6 The WAM Model

The discussion above has been largely generic, dealing with the general physical concepts underlying the various model classes. The details of the implementation of these processes in specific models has been avoided. Despite this, there have been a number of references to the WAM model (WAMDI, 1988). This model now occupies such a central role in spectral wave modelling that a short description of the model and some indication of its capability is required.

The principles underlying the model are those of third generation models: to attempt to remove all *add hoc* assumptions concerning the source term balance. This has largely been achieved through the introduction of the DIA approximation to S_{nl}. The DIA has allowed the use of a specific source function for S_{ds} and the removal of the saturation limit imposed at high frequency in earlier generation models. For an operational implementation, however, some compromises of this goal have been necessary. These operational limitations will firstly be discussed. It will then be shown that, despite these shortcomings, the WAM model represents an impressive predictive tool.

8.6.1 Prognostic Tail

At high frequencies, the wind-sea portion of the spectrum should be proportional to f^{-n}, where $n = 4 - 5$ [see Section (5.4.3)]. When this equilibrium level is reached the source terms within the model should be such that $S_{tot} \approx 0$. Slight imbalances in this equilibrium, however, continuously occur. If the energy grows to rapidly, the S_{nl} term acts to force the spectrum back to the equilibrium state (Young and Van Vledder, 1993). At progressively higher frequencies, these imbalances occur more often and S_{nl} reacts quickly to balance the instability. In other words, the higher the frequency of the waves, the more quickly they respond. Third generation models are capable of achieving this balance, even at very high frequencies. Banner and Young (1994) have achieved such a balance at frequencies as high as 1 Hz. To achieve such a balance, however, the time step within the model must be comparable to the time scale of the spectral response. For a typical ocean wave model with a high frequency tail extending to 0.2 Hz, this would imply a time step of, at most, 1 minute. Clearly this is not achievable; the global version of WAM typically uses $\Delta t = 20$ minutes.

Stability at these high frequencies is maintained within the WAM model by imposing an f^{-4} tail for frequencies which are above $\max(2.5\overline{f}, 4f_{PM})$, where $\overline{f}$ is the mean frequency of the spectrum and f_{PM} is the Pierson-Moskowitz frequency. It has been claimed that the model is insensitive to the details of this prognostic tail (WAMDI, 1988). Calculations by Banner and Young (1994), however, show that the tail does have an important impact on spectral evolution. Indeed, if the tail is removed, the equilibrium level for the spectral tail increases. As a result, the magnitude of S_{nl} decreases and spectral evolution is significantly slowed. It appears that present understanding of the source term balance at these high frequencies is not sufficient to obtain a reasonable balance without the aid of such a constraint. Therefore, the prognostic tail is required, not only for computational convenience, but in order to achieve acceptable spectral evolution.

8.6.2 Discrete Interaction Approximation

A central element of any third generation model is a computationally efficient method of calculating S_{nl}. This is achieved in WAM by the use of the Discrete Interaction Approximation (DIA) (Hasselmann et al, 1985a) (4.61). Figure 8.6 shows a comparison between S_{nl} calculated by the DIA and by a full solution. There are significant differences and this has prompted criticism over the accuracy of the DIA. It is clear from Figure 8.6 that the DIA significantly overestimates the magnitude of the negative lobe. The positive lobe of S_{nl} is, however, more important in the evolution of the spectrum and the DIA models this feature reasonably well. Also, calculation of S_{nl} for a specific spectral shape is a demanding test, as small changes in the spectral shape can result in significant changes to S_{nl}. Hasselmann et al (1985a) have shown that when the model is allowed to evolve under the source term balance, the DIA yields a slightly different spectral shape to that of the full nonlinear solution. The evolution of the spectra are however comparable.

One of the differences is that the DIA produces directionally broader spectra than the full solution. This is clear in Figure 4.11. As a result of this, the WAM model could be expected to produce spectra which are too broad.

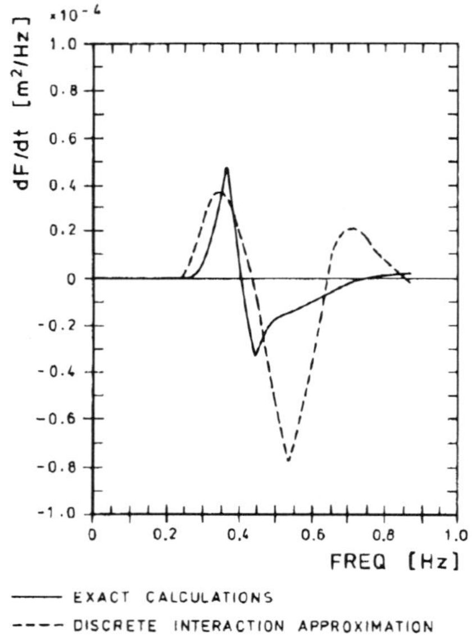

Figure 8.6: A comparison between S_{nl} calculated using the Discrete Interaction Approximation and that calculated using a full solution. The evaluations have been performed for a mean JONSWAP spectrum with $\gamma = 3.3$ [after Hasselmann et al (1985a)].

8.6.3 Evaluation of Performance

Despite the shortcomings outlined above, the WAM model has a proven record as a reliable and remarkably accurate wave prediction model. An example of model performance is shown in Figure 8.7 which shows a comparison between buoy data and model output for two locations for the one month period of December 1991. The ability of the model to accurately predict both extreme and normal conditions is clearly shown by this figure.

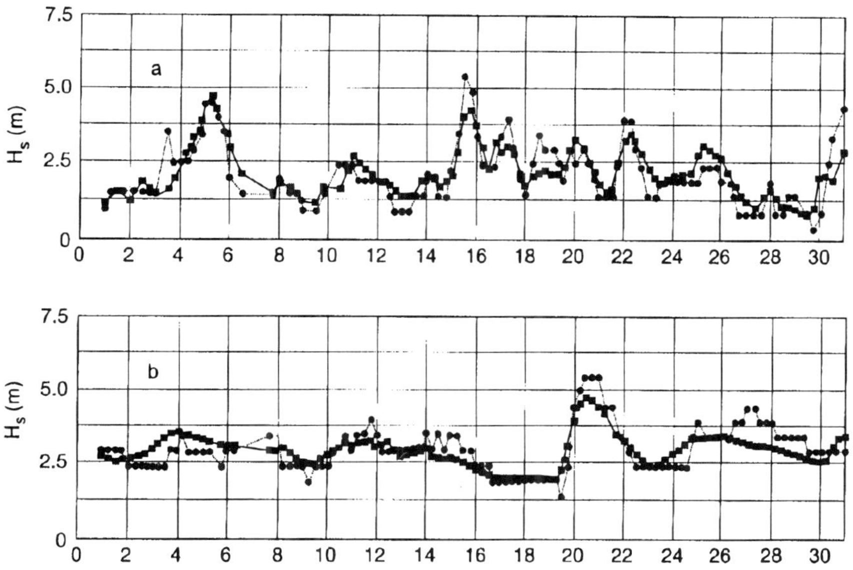

Figure 8.7: An intercomparison between WAM model results (squares) and buoy data (dots) for: (a) east coast of USA (40.5°N, 69.4°W) and (b) Hawaii (19.3°N, 160.8°W) [after Cavaleri et al (1994)].

The global version of the model has been more extensively validated than any previous model. The ECMWF have produced a series of reports comparing the model against both buoy and satellite data. The most recent of these reports is Janssen et al (1996). Figures 8.8 and 8.9 compare the model against buoy data collected in the North Atlantic and North Pacific as well as around Hawaii. Figure 8.8 shows a comparison of buoy wind speed and wave height with the model for the month of August 1995. The same information is shown for December 1995 in Figure 8.9. The agreement between buoy and model values of H_s is impressive with the model slightly underpredicting the wave height in the mean. The level of scatter of H_s values is comparable to that in the driving wind field, indicating that the model is performing as well as could be expected. More complete statistical evaluations of the model performance are contained in Janssen et al (1996).

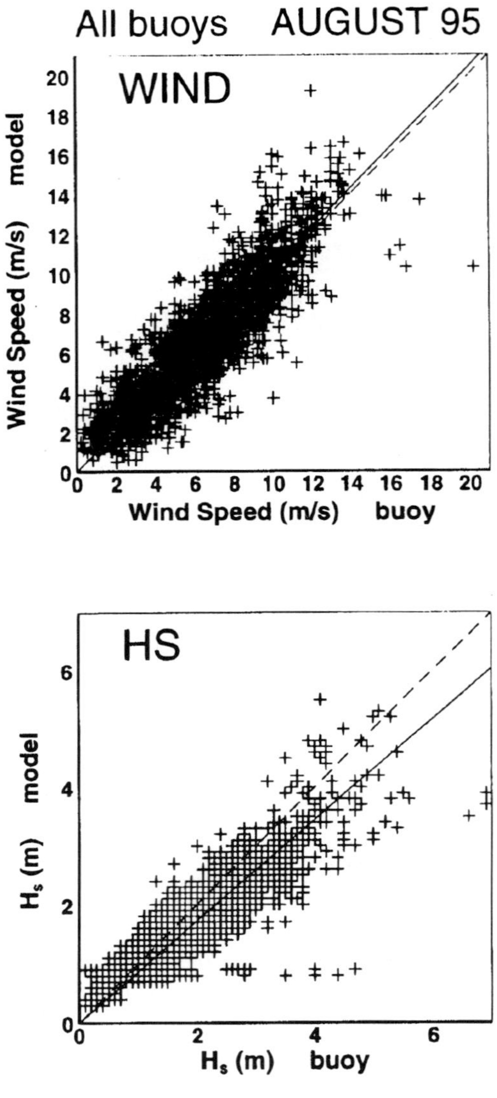

Figure 8.8: A comparison between buoy values of wind speed and wave height and WAM model results for the month of August 1995. [after Janssen et al (1996)].

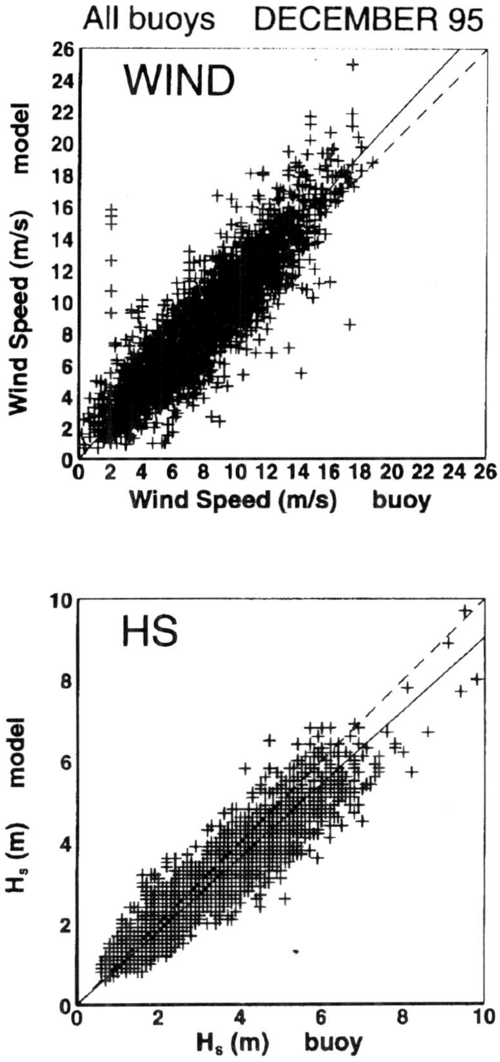

Figure 8.9: A comparison between buoy values of wind speed and wave height and WAM model results for the month of December 1995. [after Janssen et al (1996)].

8.7 Data Assimilation

The developments outlined above have shown a steady increase in the accuracy of spectral wave models, largely as a result of an improved understanding of the physical processes responsible for wind wave evolution. This continued development, however, exhibits diminished returns: the added computational expense of a more sophisticated model yielding only a marginal increase in performance. In addition, there are external parameters influencing model performance which will not be affected by refined physics. An example of this is the quality of the wind fields used to drive the wave model. Even the most sophisticated wave model will yield poor results if forced by poor quality wind fields. It is reasonable to say that the greatest single source of error within third generation models is the forcing wind field.

One possible means of improving model performance within such constraints is to assimilate measured data into the model. These data could be measured *in situ* with conventional instruments such as wave buoys or remotely sensed by orbiting satellites. The much greater spatial coverage obtained from satellites favours this data source.

One of the first published attempts at the assimilation of wave height data into a wave prediction model was made by Komen (1985). He investigated the case of swell propagating from a generation region in the Norwegian Sea to a measurement point at the entrance of Rotterdam harbour. The prediction of the swell at Rotterdam harbour using the GONO model (Janssen et al, 1984) was significantly higher than measured. Updating the model with measured buoy data at a point approximately 500 km north of the measurement site significantly improved the model swell prediction for this event.

Thomas (1988) proposed the first scheme to update the full wave spectrum using buoy measurements of significant wave height and wind speed. A JONSWAP parameterization was assumed for the wind sea part of the spectrum, which was then updated by using the buoy wind speed to rescale the energy and peak frequency (maintaining its 'wave age'). The remaining swell part of the spectrum was updated by changing only the energy scale. This technique yielded promising results, indicating that operational wave data assimilation schemes may become feasible.

Hasselmann et al (1988) and Bauer et al (1992b) made the first attempt at the global assimilation of satellite data, using SEASAT altimeter wave height data for a 30 day period in August, 1978. Their update scheme simply changed the scale of the model spectrum

$$F_{new}(f, \theta) = \gamma_s F_{model}(f, \theta) \tag{8.10}$$

where $F_{new}(f, \theta)$ is the new or modified spectrum and $F_{model}(f, \theta)$ is the initial model spectrum. The scale factor γ_s is proportional to the square of the ratio between the altimeter wave height and the model first-guess wave height. The scaling had an extreme value at the sub-satellite altimeter measurement point and relaxed back linearly to unity at the boundary of some specified 'region of influence'. For a continuous one-month assimilation experiment, good results were achieved using a region of influence of $9^\circ \times 9^\circ$. The assimilation scheme was successful in updating regions dominated by swell, yielding globally averaged correction decay times of the order of five days, but had little impact on the wind sea, which reverted rather rapidly to an equilibrium corresponding to the uncorrected local wind. It was concluded that the improvement of the wind sea predictions would require a scheme which also corrected the erroneous wind forcing.

This problem was addressed by Janssen et al (1989), who used altimeter wave height data to introduce corrections to both the wave and wind fields. Their scheme firstly partitioned the spectrum into wind sea and swell, the swell being defined as the region of the spectrum in which the atmospheric input source term was zero. The swell spectrum was modified by a change of the energy scale, maintaining its shape and peak frequency, while the wind sea part of the spectrum was modified in a manner similar to that of Thomas (1988), using JONSWAP duration limited growth relationships. Introducing the additional assumption that the energy of the wind sea and swell were changed by the same factor, a new wind sea and swell spectrum were then determined. At the same time the wind speed (or friction velocity u_*) was corrected to conform with the new wind sea. This information was then distributed to surrounding grid points, as in Hasselmann et al (1988), but using in this case a correlation region of only one wave model grid square ($3^\circ \times 3^\circ$). Although both the wind and sea state had now been adjusted, only relatively short correction decay times of the order of one day were achieved. This was presumably largely due to the much smaller region of influence used. If the scale of the region of influence is significantly smaller than the spacing between successive ascending or descending satellite orbits ($\sim 25^\circ$ longitude), only a small fraction of the ocean is corrected in one day's sweep of the satellite over the ocean. Wave energy from uncorrected regions between the satellite tracks can then propagate into the corrected regions and degrade the quality of the updated regions.

Lionello et al (1992) proposed a modification of the scheme of Janssen et al (1989) using alternative scaling relations. When swell leaves a generation region, it initially has a relatively large steepness characteristic of a wind sea. However, linear dispersion and the nonlinear processes of dissipation and energy transfer, which are both strongly dependent on the wave steepness, rapidly reduce the steepness to a level where all the source terms become negligible. If the wave height is under predicted by the model (which was normally the case for the SEASAT studies) the assimilation schemes mentioned above all increase the swell steepness. This immediately activates the dissipation and nonlinear source terms, causing the wave energy to decrease again. If the assimilation process modifies the model spectrum such that it is not in equilibrium with the source terms, the modified spectrum will rapidly be forced back to the original state. The assimilated data will be rapidly lost by the model. In the rescaling scheme of Lionello et al (1992), both the amplitude and the frequency of the swell were altered such that the swell steepness is maintained during the assimilation process. By this modified method the correction decay times were increased significantly, and were of the order of the propagation times of swell across ocean basins.

The above schemes are all relatively simple and exhibit a number of shortcomings. All require rather *ad hoc* assumptions for the separation of wind sea and swell, which are treated separately in the assimilation scheme. Similarly, the wind sea part of the spectrum is updated by applying JONSWAP shape or scaling relations. The altimeter wave height measurements are assumed to be representative over some arbitrary distance adjacent to the satellite ground track. The "laterally spread" observations are then incorporated into the model using simple "blending" algorithms. This latter shortcoming was address by Young and Glowacki (1996) who used statistical interpolation to incorporate altimeter measurements into a second generation model. The performance of the model was substantially improved.

A more fundamental approach to the problem would involve fitting the model prediction to the observations by modifying the model control variables, i.e. the wind field, rather than the model output. A general solution to this type of problem is given by the adjoint model formalism (Marchuk, 1974; Le Dimet and Talagrand, 1986). Adjoint wave models have been developed and successfully applied to assimilate wave data in simple test cases (De Valk and Calkoen, 1989; De las Heras and Janssen, 1992; Hersbach, 1998). However, operational implementation of such an approach appears to be computationally very costly and its feasibility is yet to be determined.

Formally, the change in the wind field needed to produce a given change in the wave field can be determined by inverting the wave model. However, this is normally an even more expensive computation. The basic concept of the adjoint model approach is to determine changes needed in the control variables to achieve a desired change in the model prediction without explicitly inverting the prognostic model operator.

Bauer et al (1990, 1992a) have proposed an alternative assimilation method in which the wave transport equation operator is inverted. The central idea is to estimate the impulse response (Green's) function of the system which describes the balance of the spectral energy changes influenced by perturbations in the wind fields. This direct approach becomes computationally feasible, however, only by introducing approximations in the determination of the impulse response function. Tests for ideal cases show the method has merit but its application to "real" data is yet to be assessed.

The schemes described above show the significant improvement in model performance which can result from the assimilation of satellite data into wave prediction models. The schemes are yet to be widely adopted, although that of Lionello et al (1992) is operational within the WAM model suite implemented at ECMWF. For the full potential of assimilation to be realised, methods which utilize the full spectrum rather than integral measures such as H_s will need to be further developed. Also, the quantity of satellite data will need to increase. To date only one or two suitable satellites have been operational at the same time. Greater spatial density of satellite ground tracks is required, especially for regional applications.

Chapter 9

Ocean Wave Measurement

9.1 Introduction

The measurement of ocean waves is regularly conducted for a wide range of purposes, including:

- Long term data gathering to determine statistical properties and the local wave climate [see Chapter 3].

- Calibration and validation of models for waves and other ocean processes.

- To improve the understanding of the physical processes responsible for wind wave evolution.

- For "real time" operations of ports, engineering projects and ship operations.

For any of these applications, successful outcomes require careful selection of, not only the instrumentation to be used, but also the data processing techniques to be adopted. Over the past twenty years this field has undergone enormous changes as a result of significant advances in technology. These advances have significantly improved the resolution and reliability of *in situ* instruments. More recently, the introduction of remote sensing instruments based on land, aircraft and orbiting satellites provides the capability for a quantum increase in the amount of data which can be obtained.

This chapter can only hope to focus on some of the more important instruments and processing techniques. An extensive treatment of the field can be found in Tucker (1991).

9.2 In situ Methods

Instrumentation types presented in this chapter are divided into two broad groups: in situ and remote sensing. In situ instruments are characterized by the fact that they are physically immersed in or on the ocean.

9.2.1 Surface Following Instruments

The vertical acceleration of the water surface, η_{tt} is given by the second derivative of the water surface elevation with respect to time, $\eta_{tt} = \mathrm{d}^2\eta/\mathrm{d}t^2$. Hence, if the vertical component of acceleration of the water surface elevation can be measured, this can be integrated twice to yield, η. This is generally achieved by the use of a floating buoy containing a vertically mounted accelerometer. The buoy must be designed to accurately follow the water surface. The commercial market for such instruments has been totally dominated by the Waverider Buoy, manufactured by Datawell of The Netherlands.

The Waverider Buoy has a diameter of 0.7 m or 0.9 m. The buoy is tethered with a flexible rubber cord of 15 m length. This cord is designed so as to allow the buoy to approximately follow the water surface. One of the significant design challenges of such buoys is to ensure that the accelerometer remains vertical, despite the horizontal motions and tilting of the buoy. The Waverider achieves this through the use of a fluid filled sphere mounted on gimbals. Within the sphere, a horizontal disk is suspended on wires. The accelerometer is mounted on this disk. The disk and fluid within the sphere forms a compound pendulum which has a natural period of approximately 120 s. Therefore, at the typical frequencies of ocean waves (0.04 Hz – 0.2 Hz), the accelerometer remains approximately vertical. In addition, the system has an approximately flat transfer function over these frequencies. Data from the buoy is typically transmitted ashore using a HF or VHF radio link. The range of this radio link is essentially line of sight (typically 20 km).

Surface floating buoys, such as the Waverider, are particularly suitable in deep water applications. In shallow water (less than 8 m), steep breaking waves can result in buoys being overturned. This generally results in damage to the accelerometer mounting system. Some care must be exercised in the handling of these buoys to prevent twisting of the wire suspension system of the accelerometer mount. Should this occur, it normally results in low frequency noise in the signal. As such systems are surface floating and are generally unattended, they are suseptible to theft, vandalism and damage from shipping.

A surface buoy must follow the water surface elevation accurately. Allender et al (1989) have shown that Waverider buoys underestimate the very highest waves by being dragged through the crests or, in short crested seas, riding around the crests. In addition, rather than being constrained to move vertically at a single point, the buoy tends to follow the circular orbit of the wave. Srokosz and Longuet-Higgins (1986) and Longuet-Higgins (1986) have examined the effects of the motion of the buoy, again concluding that it results in a slight underestimation of the water surface elevation. Tucker (1991) has also concluded that the motion of the buoy reduces the magnitude of higher spectral harmonics, reducing the usefulness of buoys for recording nonlinear properties of waves.

Despite these criticisms, the Waverider Buoy has been extensively adopted internationally and has proved remarkably successful. The vast bulk of the wave data gathered using in situ instruments has used the Waverider Buoy.

9.2.2 Surface Piercing Instruments

Where a platform or other structure is present, it is possible to mount instruments which pierce the water surface. A number of such instruments have been proposed. The most common are resistance and capacitance wires. *Resistance gauges* generally consist of two vertical parallel conducting wires. At their lower ends, the wires are insulated from one another. The water effectively "shorts" the wires creating a circuit. The resistance of this circuit is related to the depth of submergence of the wires and hence the water surface elevation. The resistance is, however, also dependent on the conductivity of the water. Therefore, changes in the salinity of the water will affect the calibration of the instrument. Also, fouling of the wires by sea weed is a continual problem.

A related instrument is the capacitance wire. *Capacitance wires* generally consist of a single Teflon coated wire. The conducting wire and the water form the two plates of a capacitor, separated by the insulating coating. The dielectric constant of the capacitor which is formed, varies with the length of submergence. In contrast to resistance gauges, capacitance wires are generally insensitive to salinity changes. They do, however, suffer from problems due to marine fouling and impacts which might damage the insulating coating of the wire.

Resistance and capacitance wires can measure to very high frequencies (> 10 Hz) and hence are widely used in laboratory facilities. They have also been used for research applications in the field but, as they require continual maintenance, are not suitable for routine data gathering.

Another class of surface piercing instruments are *transmission line gauges.* One form of such an instrument is the *Zwarts Pole* (Zwarts, 1974). This instrument consists of two concentric tubes forming a co-axial transmission line. Holes in the outer tube allowed water to enter and leave the annulus. The difference in dielectric constant between the air and water is used to sense the water surface elevation. An electromagnetic wave, directed towards the water surface along a transmission line, will reflect from the water surface. The time it takes for the wave to travel from the source to the water surface and back is a direct measure of the water surface elevation. Rather than transmit a pulse down the transmission line and attempt to measure the time delay, the instrument sets up an electromagnetic oscillation in the line, the length of the line determining the frequency of oscillation.

The restriction to flow represented by the holes and the physical size of the pole (of order 50 mm) obviously degrades instrument performance at high frequencies. Young et al (1997) have, however, shown that by careful design, such instruments can be used at frequencies as high as 1.2 Hz. Zwarts poles are robust although marine growth can block the holes between the inner and outer tubes, thus affecting instrument performance. Despite the fact that they are insensitive to changes in salinity and do not require regular recalibration, Zwarts poles have not been widely used.

A variation on this principle has been used in the Baylor wave gauge. This instrument consists of two vertically tensioned wire ropes. The ropes form an electrical transmission line terminated by the water surface. The impedance of the transmission line varies with the water surface elevation and can be relatively easily measured. These instruments have been used quite widely within the offshore oil industry.

9.2.3 Bottom Mounted Instruments

As shown in Section (2.2.7) the pressure induced by a linear wave is in phase with the water surface elevation and related to η by (2.28). The pressure decays with depth as given by (2.27). At the bottom, the pressure reduction factor becomes $K_p(z=-d) = 1/\cosh(kd)$. This factor is shown in Figure 9.1. From Figure 9.1 it can be seen that for the water depth greater than half the wavelength ($kd > \pi$), the factor K_p has reduced to less than 0.1.

Therefore, bottom mounted pressure transducers can be used to measure the water surface elevation, but only in relatively shallow regions. In such situations, the high frequency components of the spectrum will, however, be difficult to resolve. As the instrument is submerged it is less prone to interference than are floating or surface piercing instruments, although fishing trawlers are a significant danger.

In the above, it is assumed that linear wave theory is a reasonable representation of the pressure beneath real waves. Deviations of the pressure from linear theory as large as 10% have been measured. As a result, alternative methods have been proposed for the determination of the water surface elevation from the measured pressure record (Bishop and Donelan, 1987; Townsend and Fenton, 1995).

9.2.4 Directional Instruments

As will be shown in the following sections, analysis of data from the instruments described above yields the energy spectrum, $F(f)$. In order to understand the directional properties of the spectrum it is necessary to determine the directional spectrum, $F(f, \theta)$. An instrument such as a Waverider Buoy yields one quantity, the water surface elevation, η as a function of time. The two-dimensional directional spectrum, $F(F, \theta)$ has many degrees of freedom and hence requires the measurement of multiple water surface quantities. Invariably insufficient information is available and hence certain assumptions about the spectrum must be invoked.

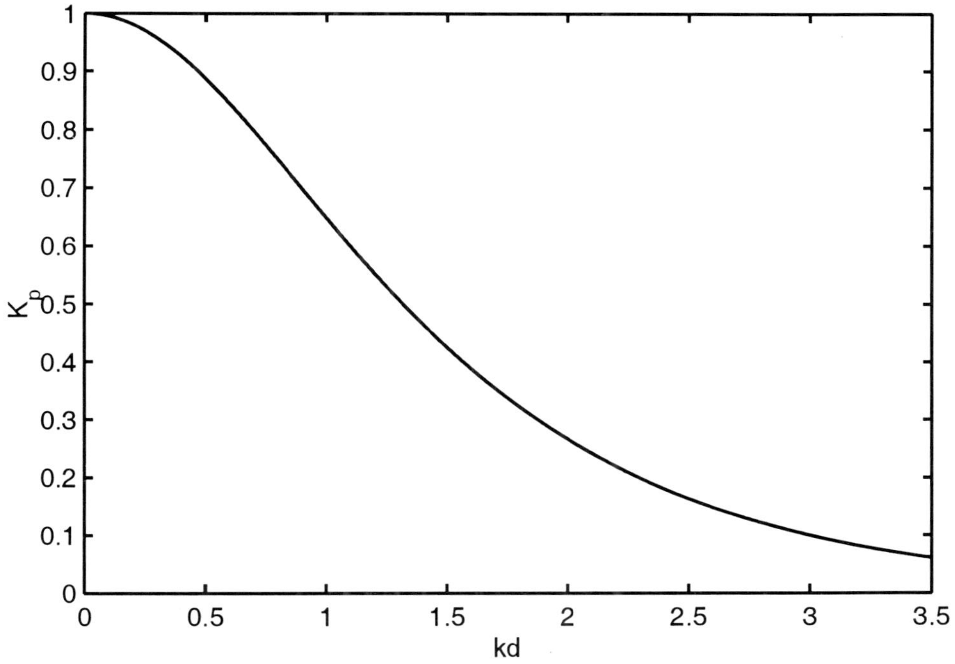

Figure 9.1: The ratio between the magnitude of the pressure measured at the bottom and that at the water surface, K_p as a function of the non-dimensional depth parameter, kd.

Directional instruments must therefore measure as many quantities related to the two-dimensional water surface as possible (a minimum of 3 is required).

9.2.4.1 Surface Following Buoys

A number of buoy systems have been designed which attempt to follow the two-dimensional water surface. The most common of these is the *heave-pitch-roll buoy*. Such buoys are of a flat toroid or discus shape with a compliant mooring system which allows the buoy to, not only follow the vertical motion of the water, but also to tilt and roll so that the flat hull remains parallel to the water surface. The sensing package typically consists of a gyroscopically stabilized platform upon which an accelerometer is mounted, together with a compass to measure the buoy heading. Hence, the buoy can measure the vertical acceleration, η_{tt} (heave) together with two orthogonal components of the water surface slope, η_x, η_y (pitch and roll). An example of such a buoy is the Wavec manufactured by Datawell.

An extension of the heave-pitch-roll buoy is the *clover-leaf buoy* (Cartwright and Smith, 1964). This buoy consists of three buoyant disks connected to a triangular frame by universal joints. Therefore, the relative motion of the apexes of the triangle can be measured. In addition to measuring the heave, pitch and roll, the clover-leaf buoy also measures three components of the water surface curvature, η_{xx}, η_{yy} and η_{xy}.

Pitch-roll-heave buoys and clover-leaf buoys both require expensive hull and mooring systems. Hence, they have generally been applied in research situations. An alternative to such systems is a *particle-following buoy* which measures the water particle motion in three orthogonal directions. Such an instrument is the *Directional Waverider* manufactured by Datawell. The hull consists of a conventional Waverider sphere, 0.9 m in diameter. This buoy is moored as for a conventional Waverider. The instrument contains a heave-pitch-roll package together with two horizontal accelerometers fixed to the hull. Analysis of these signals yields the three orthogonal components of water particle motion.

9.2.4.2 Spatial Arrays

The directional spectrum can also be obtained by measurement of the water surface elevation at a number of closely spaced points. The sensing elements can be any of the surface piercing sensors described in Section (9.2.2) or bottom mounted pressure sensors. The number of elements in the array, the geometry of the array and the spacing of the elements all affect the directional resolving power of the instrument. For applications on beaches, where there is a single predominant direction of wave approach, linear or straight line arrays have been used successfully. An example is the 17 gauge linear array at the US Army Corps of Engineers Field Research Facility at Duck, North Carolina. For more general applications two dimensional array are used (Donelan et al, 1985; Young et al, 1996). Spatial arrays have the advantage that they can be designed for the particular application. Details such as the number elements and their geometry can be optimized for the site under consideration. The size of the arrays, and the logistics of their operation, however, means that they are generally limited to research applications.

9.2.4.3 Pressure-Velocity Gauges

Bottom mounted instruments which measure the pressure together with two orthogonal components of water particle velocity (P-U-V gauges) can be used in a similar manner to heave-pitch-roll buoys to determine the directional spectrum. As with all bottom mounted instruments, they have the advantage of being concealed, but the disadvantage of relatively poor high frequency response.

9.3 Data Analysis

The instruments described in the preceeding sections all record time series, sampled at discrete times, of one or more quantities related to the water surface elevation. These time series need to be processed or analyzed to determine quantities such as the spectrum or its integral properties: significant wave height or peak period. There are two general approaches which have been used for this processing: time domain analysis and spectral analysis. *Time domain analysis* attempts to break the wave record up into individual waves, each with a measured amplitude and period. The analysis then considers the statistical distribution of these amplitudes and periods. *Spectral analysis* invokes the Fourier model that the time series can be represented as a sum of an infinite number of sinusoids, each with a defined amplitude and frequency (period). Spectral analysis is computationally more efficient and hence has largely been adopted as the standard analysis technique for ocean waves. For this reason, time domain analysis will not be considered further. Time domain analysis does, however, warrant consideration in some areas, particularly for shallow water waves. Excellent descriptions of its application can be found in Goda (1985) and Tucker (1991).

9.3.1 Spectral Analysis - F(f)

Let us assume that a time series of the water surface elevation, $\eta(t)$ has been recorded. Adopting the Fourier model, this can be represented as

$$\eta(t) = \int_{-\infty}^{\infty} X(f) e^{i\omega t} \mathrm{d}t \tag{9.1}$$

where $i = \sqrt{-1}$ and $X(f)$ is the Fourier transform of η. Noting that, $e^{i\theta} = \cos\theta + i\sin\theta$, (9.1) becomes

$$\eta(t) = \int_{-\infty}^{\infty} X(f) \left[\cos\omega t + i\sin\omega t\right] \mathrm{d}t \tag{9.2}$$

Therefore, $\eta(t)$ is represented in the same fashion as (2.45). Since, $\eta(t)$ is a real quantity, the integral in (9.2) must also be real. Hence, the Fourier transform, $X(f)$ must, in general, be a complex quantity. It can be shown that

$$X(f) = \int_{-\infty}^{\infty} \eta(t) e^{-i\omega t} \mathrm{d}t \tag{9.3}$$

or

$$X(f) = \int_{-\infty}^{\infty} \eta(t) \left[\cos\omega t - i\sin\omega t\right] \mathrm{d}t \tag{9.4}$$

As $\eta(t)$ is recorded as a discrete series (i.e. values at fixed time intervals), it is convenient to approximate the integral in (9.4) with a summation over the finite length record. (Later it will be shown that this approximation introduces a degree of uncertainty into the analysis.)

$$\underbrace{X(n/T_r)}_{=f} = \underbrace{\frac{T_r}{N}}_{=\mathrm{d}t} \sum_{j=0}^{N-1} \eta\left(jT_r/N\right) \left[\cos\left(2\pi jn/N\right) - i\sin\left(2\pi jn/N\right)\right] \tag{9.5}$$

where N is the number of points in the time series, T_r is the length of the time series and n is a counter which ranges from 0 to $N-1$. Although (9.5) represents a means for the determination of the Fourier transform, $X(f)$, it is computationally more efficient to calculate it using the Fast Fourier Transform (FFT) method (Bendat and Piersol, 1971).

With $X(f)$ defined, the spectrum follows as

$$\hat{F}(f) = \frac{2}{T_r}|X(f)|^2 \tag{9.6}$$

where $|X(f)|$ is the modulus of the complex Fourier transform and the $\hat{F}$ notation is introduced to signify that the spectrum is determined from the discretely sampled finite time series. The normalization factor $2/T_r$ in (9.6) ensures that $\int F(f)\mathrm{d}f$ is equal to the variance of the record, σ^2. As the Fourier transform is complex, it defines both the amplitude and phase. The spectrum defined in (9.6) considers only the modulus of the Fourier transform, all phase information is lost. Indeed, it is generally assumed that ocean waves have random phase.

9.3.1.1 Sampling Considerations

From (9.5) the frequency resolution of the Fourier transform, and hence the spectrum is $\Delta f = 1/T_r$. Therefore, if higher resolution is required, a longer record must be obtained. One of the implicit assumptions in spectral analysis, however, is that the record is *stationary*. That is, that the mean properties of the record do not change significantly over the duration of the record. In the context of wind waves, this implies that factors such as the wind speed, magnitude of swell and water depth should be relatively constant over the duration of the record. Naturally, as the length of the record increases, this requirement becomes progressively more difficult to maintain. Typically, ocean wave records are collected with durations of T_r=20 – 40 minutes.

It is intuitively obvious that a wave with, for instance, a period of 1 second cannot be accurately defined if it is sampled at an interval of $\Delta t = 2$ seconds. More generally, at least two sampling points are required within each wave period to define the spectral component (i.e. $\Delta t \leq T/2$). Therefore, the maximum frequency which can be represented within the frequency, called the *Nyquist frequency* is $f_N = 1/(2\Delta t)$. If there is significant energy above the Nyquist frequency it will appear as erroneous energy at frequencies below its true value. This process is called aliasing. Therefore, Δt must be selected such that aliasing is minimized and all frequencies of interest are accurately resolved. Should it be necessary, energy above the Nyquist limit can be removed by filtering prior to digitization. As the wind wave spectrum typically decays rapidly with frequency [e.g. $F(f) \propto f^{-4}$], anti-aliasing filters are often not necessary. The sampling considerations described above are shown diagrammatically in Figure 9.2.

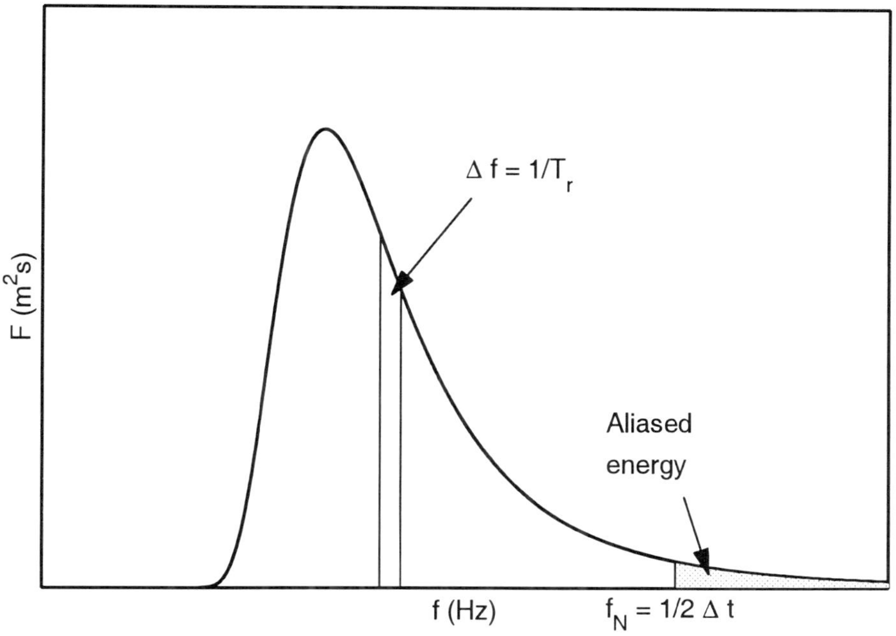

Figure 9.2: A spectrum obtained from a time series sampled at an interval of Δt. The total length of the sampled record is T_r. The shaded region at high frequencies shows energy which will be aliased as a result of the chosen sampling parameters.

9.3.1.2 Confidence Limits for Spectra

Because of the assumption that the integral (9.4) can be approximated by the finite length summation (9.5), the spectrum obtained from the discretely sampled record, $\hat{F}(f)$ is only an estimate of the true spectrum, $F(f)$. It can be shown (Bendat and Piersol, 1971) that the spectral ordinates of $\hat{F}(f)$ are stochastic variables following a chi-square probability distribution given by

$$\frac{\hat{F}(f)}{F(f)} = \frac{\chi_2^2}{2} \tag{9.7}$$

where χ_2^2 is the chi-square variable with $n = 2$ degrees of freedom. This implies that the standard deviation of each of the ordinates of $\hat{F}(f)$ is as large as the true value being estimated. A spectrum determined in this manner is shown in Figure 9.3. The spectral ordinates fluctuate wildly and clearly provide an unacceptable estimate of the true spectrum.

The variability in the spectral estimate can be improved by either ensemble averaging or frequency averaging. In *ensemble averaging* the time series is subdivided into q individual time series. Spectral estimates are individually calculated for each of these smaller time series. The q estimates of the spectral ordinate at each frequency are then averaged to produce the final spectrum. The averaged values of the spectral ordinates at each frequency now follow a chi-square distribution with $n = 2q$ degrees of freedom.

In *frequency averaging* the spectrum of the full time series is firstly calculated. Groups of l neighbouring spectral ordinates are then averaged to form a single spectral ordinate with a frequency equal to the mean of the group. Again, the resulting spectral ordinates follow a chi-square distribution with $n = 2l$ degrees of freedom.

If $q = l$, the resulting spectra are the same. Because of the nature of the FFT, it is computationally more efficient to calculate the Fourier transform of q short time series than one long time series. Hence, ensemble averaging is more computationally efficient. In both cases, the frequency resolution, Δf has been sacrificed to improve the estimate of the spectrum.

It is convenient to assign confidence limits for the resulting estimates of the spectral ordinates. The $(1-\alpha)$ confidence interval for the spectrum $F(f)$ based on the estimate $\hat{F}(f)$ is given by

$$\frac{n\hat{F}(f)}{\chi^2_{n;\alpha/2}} \leq F(f) < \frac{n\hat{F}(f)}{\chi^2_{n;1-\alpha/2}} \tag{9.8}$$

where $\chi^2_{n;\alpha}$ is the α percentage point of the chi-square distribution (tabulated in most statistical texts).

An example of the determination of the spectrum from a recorded time series is shown in Figure 9.3. Figure 9.3a shows a one hundred second portion of a time series recorded on 13 May 1989 at Macquarie Island in the Southern Ocean. It is clear from this short section of the wave record that conditions were extreme, individual crest-to-trough heights exceeding 10 m. The full wave record consisted of 1024 points recorded at $\Delta t = 1$ second ($T_r = 17.07$ minutes). Figure 9.3b shows the spectrum calculated using ensemble averaging. Two cases are shown: $q = 2$ (i.e. blocks of 512 points) and $q = 8$ (i.e. blocks of 128 points). It is clear that increasing the value of q has resulted in a much smoother spectrum. The frequency resolution is, however, significantly reduced by the averaging process.

Consider the calculation of 95% confidence limits for both these cases from (9.8). For 95% confidence limits, $(1 - \alpha) = 0.05$ is required (i.e. $\alpha = 0.95$).

For $q = 2 \rightarrow n = 4$

$$\chi^2_{n;\alpha/2} = \chi^2_{4;0.025} = 11.14$$

$$\chi^2_{n;1-\alpha/2} = \chi^2_{4;0.975} = 0.484$$

Hence, from (9.8)

$$\frac{4\hat{F}(f)}{11.14} \leq F(f) < \frac{4\hat{F}(f)}{0.484}$$

$$0.36\hat{F}(f) \leq F(f) < 8.26\hat{F}(f)$$

For $q = 8 \rightarrow n = 16$

$$\chi^2_{n;\alpha/2} = \chi^2_{16;0.025} = 28.85$$

$$\chi^2_{n;1-\alpha/2} = \chi^2_{16;0.975} = 6.91$$

Hence, from (9.8)

$$\frac{16\hat{F}(f)}{28.85} \leq F(f) < \frac{16\hat{F}(f)}{6.91}$$

$$0.55\hat{F}(f) \leq F(f) < 2.32\hat{F}(f)$$

Therefore, as a result of the averaging process, the magnitude of the confidence limits have been reduced. In the case above, for $n = 16$ one can be 95% confident that the true value of the spectral ordinates lie between 0.55 times and 2.32 times the calculated ordinates. Despite the fact that the final calculated spectrum is smooth, the necessity of using a finite length record has resulted in significant uncertainty as to the true spectrum. It is good practice to always consider the magnitude of the confidence limits when considering spectra obtained from recorded data. Also, because of the chosen sampling interval Δt=1 sec., the Nyquist frequency in the example in Figure 9.3 is f_N =0.5 Hz. Frequencies higher than this value could only be resolved if the value of Δt was reduced.

9.3.1.3 Confidence Limits for Wave Height

The significant wave height, H_s can be calculated from the spectrum ($H_s = 4\sqrt{\sigma^2} = \int F(f)\mathrm{d}f$). As the estimate of the spectrum, $\hat{F}(f)$ is a stochastic variable, quantities such as $\hat{H}_s$ which are derived from it, will also be stochastic variables.

From (9.7), $\hat{F}(f)$ follows a chi-square distribution. Integrals of the spectrum, such as those required to determine $\hat{H}_s$ will also follow a chi-square distribution (Jenkins and Watts, 1968). Young (1986) has shown that the estimate of the variance, $\hat{\sigma^2}$ determined from the spectrum follows a chi-square distribution with ν_s degrees of freedom, where

$$\nu_s = \frac{n\left[\sum\limits_{j=1}^{M}\hat{F}(f_j)\right]^2}{\sum\limits_{j=1}^{M}\left[\hat{F}(f_j)\right]^2} \tag{9.9}$$

where M is the total number of frequency bands in the spectral estimate, $\hat{F}(f)$.

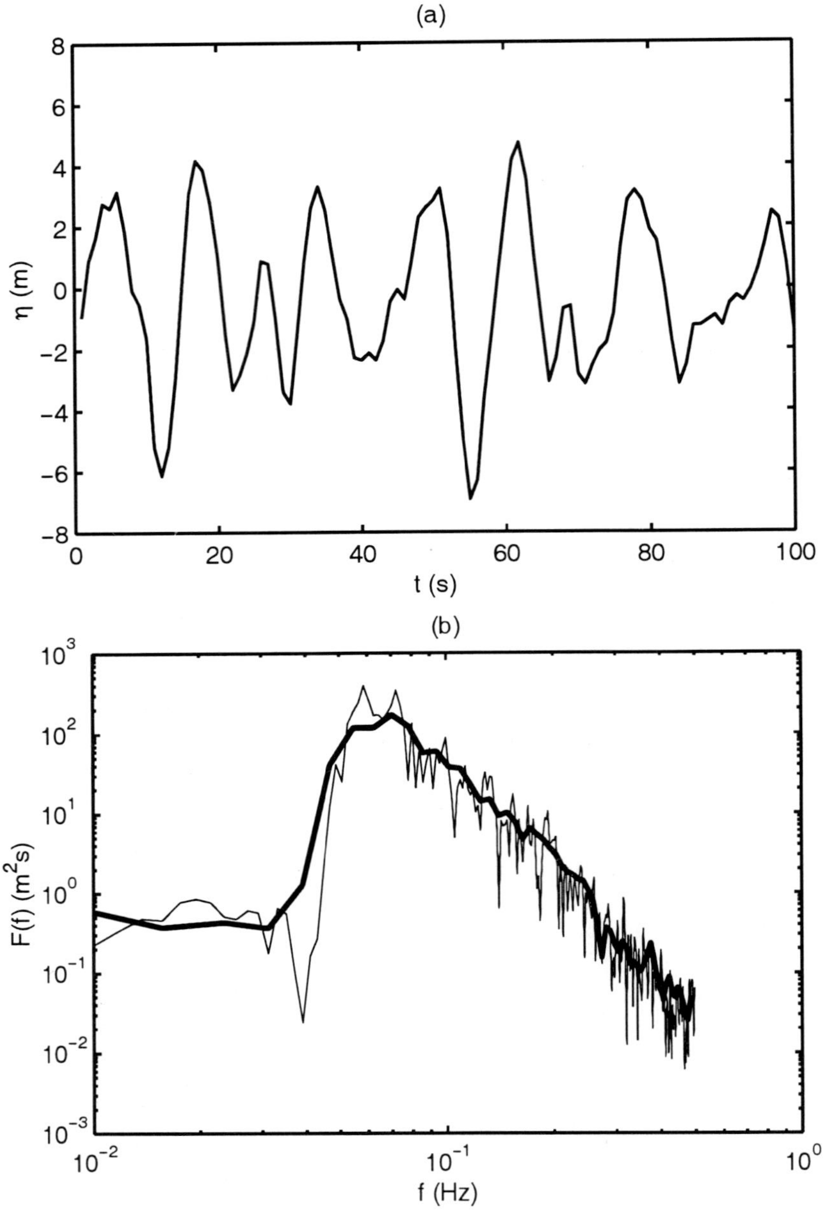

Figure 9.3: A section of a recorded time series and spectra calculated from that time series. (a) A 100 second section of a time series recorded at Macquarie Island in the Southern Ocean on 13 May 1989. (b) Spectra calculated from the time series in (a). The thin line is calculated by ensemble averaging two blocks, each of 512 points whereas the thick line is from the ensemble average of 8 blocks, each of 128 points.

It follows that the $(1-\alpha)$ confidence limits for $\hat{\sigma^2}$ are

$$\frac{\nu_s \hat{\sigma^2}}{\chi^2_{\nu_s;1-\alpha/2}} \leq \sigma^2 < \frac{\nu_s \hat{\sigma^2}}{\chi^2_{\nu_s;\alpha/2}} \tag{9.10}$$

and since $H_s = 4\sqrt{\sigma^2}$

$$\left[\frac{\nu_s}{\chi^2_{\nu_s;1-\alpha/2}}\right]^{1/2} \hat{H}_s \leq H_s < \left[\frac{\nu_s}{\chi^2_{\nu_s;\alpha/2}}\right]^{1/2} \hat{H}_s \tag{9.11}$$

Application of (9.11) to the case shown in Figure 9.3 yields $\hat{H}_s$= 10.6 m, with a 95% confidence interval from 9.5 m – 11.9 m. Hence, even when determining the significant wave height from recorded data, there is an appreciable sampling uncertainty.

9.3.1.4 Confidence Limits for Peak Frequency

The peak frequency, f_p is commonly estimated from spectra obtained from sampled data. The determination of f_p is made difficult for two reasons. Firstly, the spectrum is known only at discrete values of f and this frequency resolution may be relatively coarse as a result of the averaging processes described above. Secondly, the statistical variability of the spectrum [see for example Figure 9.3b] makes the determination of the peak difficult. As a result, a number of techniques have been proposed for the determination of the peak (Günther, 1981; Sobey and Young, 1986; Bishop and Donelan, 1988; Mansard and Funke, 1990). These techniques include: simply selecting the maximum ordinate of the spectrum, fitting curves to the peak and weighted moments. Young (1995) has critically reviewed these proposal and found that the most reliable technique involves the use of a weighted integral

$$\hat{f}_p = \frac{\int_0^\infty f F^4(f) \mathrm{d}f}{\int_0^\infty F^4(f) \mathrm{d}f} \tag{9.12}$$

Again, the $\hat{f}_p$ representation is used to signify that this is an estimate of the true peak frequency, f_p. Young (1995) found that (9.12), as with other techniques, slightly overestimates the value of f_p. The accuracy with which f_p can be determined increases with an increase in both, the number of degrees of freedom of the spectral estimate, n and the spectral resolution (measured by $\Delta f / f_p$). Further, Young (1995) showed that the 95% confidence limits for f_p determined from (9.12) are

$$a_l \hat{f}_p \leq f_p < a_u \hat{f}_p \tag{9.13}$$

where a_l and a_u are factors defining the lower and upper 95% confidence limits, respectively. These factors are shown as a function of n and $\Delta f / f_p$ in Figure 9.4.

When applied to the spectrum shown in Figure 9.3b with n = 16 degrees of freedom, (9.12) yields $\hat{f}_p$ =0.071 Hz with 95% confidence limits between 0.065 Hz – 0.078 Hz.

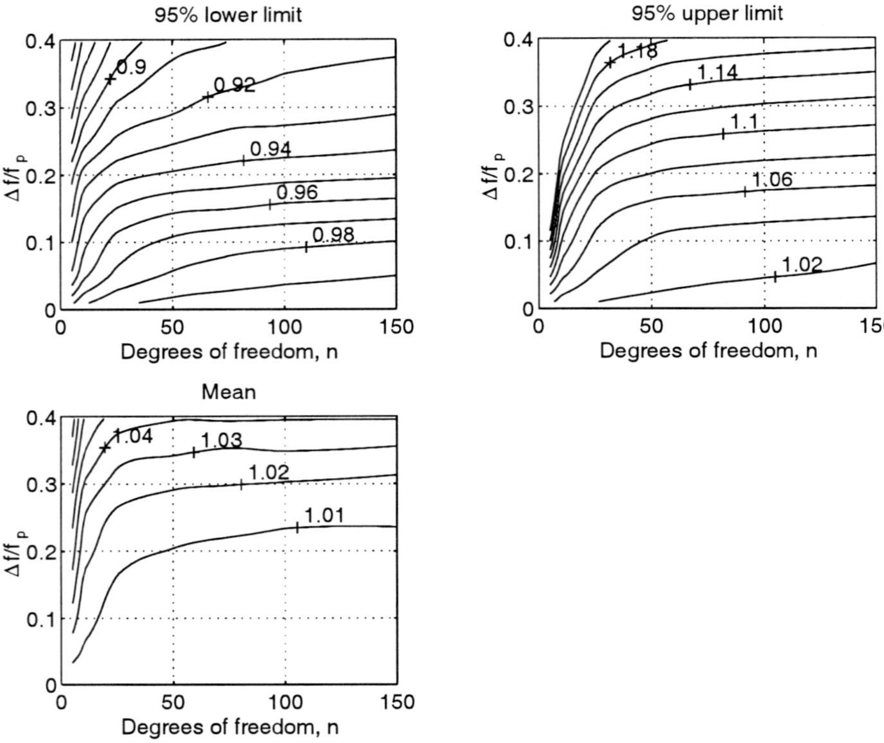

Figure 9.4: Contours of the upper (a_u) and lower (a_l) 95% confidence limit factors [see (9.13)] and the mean values for estimates of f_p from (9.12). The plots are shown as functions of the relative frequency resolution, $\Delta f/f_p$ and number of degrees of freedom of the spectral estimate, n.

9.3.2 Determination of Directional Spectra - $\mathbf{F}(\mathbf{f}, \theta)$

In Section (9.3.1), the determination of the one-dimensional or omni-directional spectrum, $F(f)$ was considered. This was based on the Fourier model for the water surface elevation (9.1). This model can be expanded to include both spatial and temporal variation of the water surface elevation.

$$\eta(\mathbf{x}, t) = \int\int X(\mathbf{k}, \omega) e^{i(\mathbf{k}\cdot\mathbf{x} - \omega t)} \mathrm{d}\mathbf{k}\mathrm{d}\omega \tag{9.14}$$

where $X(\mathbf{k}, \omega)$ is the complex Fourier amplitude of the component with wave number, $\mathbf{k}$ and frequency, $\omega = 2\pi f$.

The wave number-frequency spectrum can be represented in terms of the Fourier coefficients by

$$F(\mathbf{k}, \omega)\mathrm{d}\mathbf{k}\mathrm{d}\omega = < X(\mathbf{k}, \omega) X^*(\mathbf{k}, \omega) > \tag{9.15}$$

where the angle brackets denote an ensemble average and the star notation the

Quantity, ξ	Symbol	$H(\mathbf{k},\omega)$	$G(k,\omega)$	α	β
Water surface	η	1	1	0	0
Pressure	p	$\rho g \frac{\cosh kz}{\cosh kd}$	$\rho g \frac{\cosh kz}{\cosh kd}$	0	0
Vertical velocity	η_t	$-i\omega$	$-i\omega$	0	0
Vertical accel.	η_{tt}	$-\omega^2$	$-\omega^2$	0	0
Surface slope (x dir.)	η_x	$ik\cos\theta$	ik	1	0
Surface slope (y dir.)	η_y	$ik\sin\theta$	ik	0	1
Surface curvature (x)	η_{xx}	$-k^2\cos^2\theta$	$-k^2$	2	0
Surface curvature (y)	η_{yy}	$-k^2\sin^2\theta$	$-k^2$	0	2
Surface curvature (xy)	η_{xy}	$-k^2\cos\theta\sin\theta$	$-k^2$	1	1
Water velocity (x dir.)	u	$\omega\cos\theta\frac{\cosh kz}{\sinh kd}$	$\omega\frac{\cosh kz}{\sinh kd}$	1	0
Water velocity (y dir.)	v	$\omega\sin\theta\frac{\cosh kz}{\sinh kd}$	$\omega\frac{\cosh kz}{\sinh kd}$	0	1
Water velocity (z dir.)	w	$-i\omega\frac{\sinh kz}{\sinh kd}$	$-i\omega\frac{\sinh kz}{\sinh kd}$	0	0
Water accel. (x dir.)	u_t	$-i\omega^2\cos\theta\frac{\cosh kz}{\sinh kd}$	$-i\omega^2\cos\theta\frac{\cosh kz}{\sinh kd}$	1	0
Water accel. (y dir.)	v_t	$-i\omega^2\sin\theta\frac{\cosh kz}{\sinh kd}$	$-i\omega^2\sin\theta\frac{\cosh kz}{\sinh kd}$	0	1
Water accel. (z dir.)	w_t	$-\omega^2\frac{\sinh kz}{\sinh kd}$	$-\omega^2\frac{\sinh kz}{\sinh kd}$	0	0

Table 9.1: Values of the transfer function H for use in (9.16). [after Young (1994c)]

complex conjugate. Generally, the water surface elevation η is not measured directly, but some quantity ξ which can be related to η [e.g. pressure, p or slope, η_x, as described in Section (9.2.4)]. Following Isobe et al (1984), the generic measured quantity, ξ can be related to η through the definition of a transfer function, H

$$H(\mathbf{k},\omega) = (\cos\theta)^{\alpha}\,(\sin\theta)^{\beta}\,G(k,\omega) \tag{9.16}$$

The quantities, α, β and G depend on the quantity being measured, ξ, and are listed in Table 9.1.

Substitution of (9.16) into (9.14) yields

$$\xi(\mathbf{x},t) = \int\int H(\mathbf{k},\omega)e^{i(\mathbf{k}\cdot\mathbf{x}-\omega t)}X(\mathbf{k},\omega)\mathrm{d}\mathbf{k}\mathrm{d}\omega \tag{9.17}$$

The cross-spectrum, $\Phi_{mn}(\omega)$ between any two of the quantities measured, $\xi_m(\mathbf{x_m}, t)$ and $\xi_n(\mathbf{x_n}, t)$ is

$$\Phi_{mn}(\omega)\mathrm{d}\omega = \left\langle \int H_m(\mathbf{k},\omega)e^{i(\mathbf{k}\cdot\mathbf{x_m})}X(\mathbf{k},\omega)\mathrm{d}\mathbf{k} \times \int H_n^*(\mathbf{k}',\omega)e^{-i(\mathbf{k}'\cdot\mathbf{x_n})}X^*(\mathbf{k}',\omega)\mathrm{d}\mathbf{k}' \right\rangle \tag{9.18}$$

Since X is a random variable $\langle X(\mathbf{k},\omega)X^*(\mathbf{k}',\omega)\rangle = 0$ for $\mathbf{k} \neq \mathbf{k}'$ and (9.18) becomes

$$\Phi_{mn}(\omega) = \int H_m(\omega)H_n^*(\omega)e^{-i\mathbf{k}\cdot(\mathbf{x_n}-\mathbf{x_m})}F(\omega,\theta)\mathrm{d}\theta \tag{9.19}$$

where the dispersion relationship has been used to introduce the directional frequency spectrum $F(\omega,\theta)$ in place of the wave number-frequency spectrum, $F(\mathbf{k},\omega)$.

The determination of the directional spectrum, $F(\omega,\theta)$ can then be achieved by the inversion of (9.19). In practice, however, this is not straightforward. At any given frequency, ω, it is desired to define $F(\omega,\theta)$ at sufficient discrete values of θ to accurately define the directional spreading. The number of measured cross-spectra which can be used to determine these discrete values of the spectrum are limited by the number of independent quantities measured by the instrument. For common instruments the number of cross-spectra are far less than the desired number of discrete values of the directional spectrum. The problem is under-defined. A number of possible solutions to this problem have been proposed (Longuet-Higgins et al, 1963; Capon, 1969; Davis and Regier, 1977; Jefferys et al, 1981; Isobe et al, 1984; Brissette and Tsanis, 1992; Tsanis and Brissette, 1992), the two most commonly used being described in detail below.

9.3.2.1 Fourier Expansion Method (FEM)

Longuet-Higgins et al (1963) expanded the directional spreading function, $D(\omega,\theta)$ [see (5.58)] as a Fourier series of the form

$$D(\omega,\theta) = a_0 + \sum_{p=1}^{M} a_p(\omega)\cos p\theta + b_p(\omega)\sin p\theta \tag{9.20}$$

If there are N quantities being measured by the instrument, Φ_{mn} is an $N \times N$ matrix, but since $\Phi_{mn} = \Phi_{nm}^*$ there are effectively only $N(N+1)/2$ unique cross-spectra. Equation (9.20) has a total of $2M+1$ unknowns (the coefficients a_p, b_p). Therefore, provided $M \leq N(N+1)/4 - 1/2$ the system of equations can be directly solved for the coefficients, thus defining the directional spreading function through (9.20). As the number of measured quantities, N is decreased, the Fourier series (9.20) must be progressively truncated at lower order, M. For instance, in the case of a pitch-roll buoy, $N = 3$ and hence M must be truncated at 2. A consequence of this truncation is that negative lobes develop in the resulting directional spreading function (Longuet-Higgins et al, 1963). In order to overcome this problem, Longuet-Higgins et al (1963) proposed that D be constrained to the analytical form represented by (5.60). The two parameters in this functional form θ_m and s can be determined from the coefficients of (9.20).

$$\theta_m = \tan^{-1}\left(\frac{b_1}{a_1}\right) \tag{9.21}$$

$$s = \frac{r_1 \pi}{1 - r_1 \pi} \quad ; \qquad r_1^2 = a_1^2 + b_1^2 \tag{9.22}$$

This is the analysis technique used by both Mitsuyasu et al (1975) and Hasselmann et al (1980) to develop their directional spreading relationships considered in Section (5.5).

9.3.2.2 Solution by Maximum Likelihood Method (MLM)

Solution by the FEM as represented by (9.20) or (9.21) and (9.22) constrains the directional spreading function to a pre-determined model form. A number of model independent formulations have also been proposed. The most commonly used of these is the Maximum Likelihood Method (MLM) (Capon, 1969; Isobe et al, 1984; Young, 1994c).

As its name suggests, the MLM attempts to determine the directional spectrum which has the maximum likelihood of conforming to the limited number of cross-spectral estimates. As developed by Isobe et al (1984), the energy incident from angle θ is evaluated by minimizing the influence from all other components. This minimization is achieved through the use of Lagrange multiplier theory. The final result is

$$F(\omega, \theta) = \frac{Q(\omega)}{\sum_m \sum_n \Phi_{mn}^{-1}(\omega) H_m^*(\omega) H_n(\omega) e^{i\mathbf{k}\cdot(\mathbf{x_n} - \mathbf{x_m})}} \tag{9.23}$$

where $Q(\omega)$ is a normalization factor such that the total energy of the directional and one-dimensional spectra are equivalent [Equation (5.60)]. Details of the application of the MLM to both pitch/roll buoy systems and spatial arrays are given by Young (1994c).

9.3.2.3 Other Solution Techniques

In addition to the FEM and the MLM there are a number of alternative techniques which have been used for the extraction of directional spectra from measured data. These techniques include: the Iterative Maximum Likelihood Method (Pawka, 1983), the Normalized Maximum Likelihood Method (Brissette and Tsanis, 1992) and the Maximum Entropy Method (Lygre and Krogstad, 1986). A comparison of the advantages and disadvantages of these techniques has been reported by (Tsanis and Brissette, 1992). Based on their results, it appears that the

MLM produces reasonable results under a wide range of conditions. Although, the Normalized Maximum Likelihood Method can produce superior results, the rather subjective choices which must be made in the normalization make it unattractive. As a result, only the FEM (due to its wide usage) and the MLM are considered in the remainder of this section.

9.3.2.4 A Comparison of Solution Techniques

Young (1994c) has made a detailed comparison of the directional resolving capabilities of the FEM and MLM methods. An input directional spreading function of the form proposed by Donelan et al (1985) with $\beta_s = 3$ was defined. A small amount (1% of maximum spectral ordinate) of incoherent noise was also added to the spectrum. The techniques were then tested to see how well they could recover this known input form.

Figure 9.5 shows comparisons between the input and calculated directional spreading functions for a pitch/roll buoy system derived with both the FEM and MLM. Both analysis techniques produce directional spreading which is too broad and significantly underestimate the magnitude of the narrow directional distribution. As expected, the MLM is marginally superior but the differences are only minor. The real advantage of the MLM is that it can resolve more complex directional spreading such as bi-modal distributions. The FEM is constrained to yield uni-modal spreading.

To improve the performance of the instrument the number of measured quantities needs to be increased. This is really not practical in a floating buoy system, although clover-leaf buoys do increase the effective number of measurement elements from 3 to 6. Spatial arrays can, however, have any number of elements, although they quickly become cumbersome. Figure 9.6 shows the results for the same input spectrum as used in Figure 9.5 for spatial arrays with 4, 5, 6, 7 and 10 elements respectively. In all cases the arrays were symmetric with one gauge at the centre and the remaining gauges evenly distributed at the same radius around this central gauge. The MLM was used for the analysis. Not surprisingly, increasing the number of gauges progressively improves the results. As the number of gauges increases, the peak of the output spreading function increases and the overestimation in energy at angles away from the dominant wave direction decreases.

9.4 Remote Sensing Techniques

In situ techniques, such as those described in the previous sections, are ideal for the collection of large quantities of wave data at a specific point. Such deployments are, however, expensive and require high levels of ongoing maintenance. In addition, applications such as engineering design invariably require long term deployments. There is seldom sufficient time for a meaningful deployment once a project has been proposed. Therefore, if data are not available from previous deployments

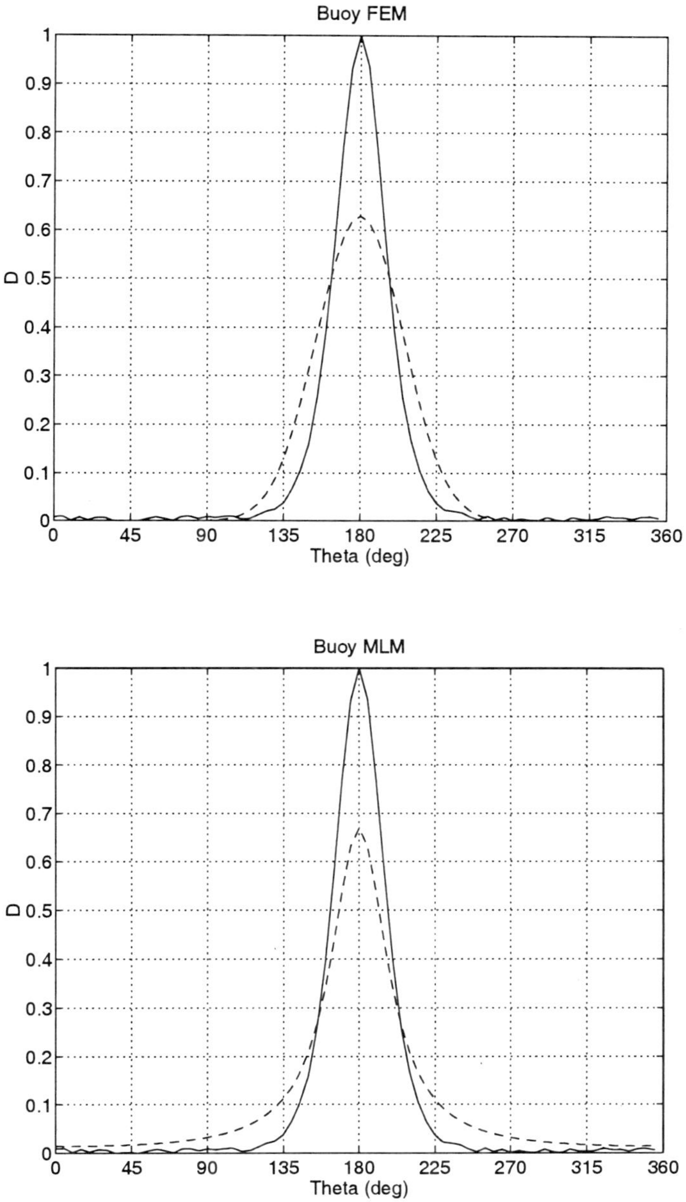

Figure 9.5: A comparison between the FEM and MLM when applied to simulated heave-pitch-roll buoy data. The solid line represents a specified input directional spreading function of the form proposed by Donelan et al (1985) with $\beta_s = 3$. The dashed lines represent the output or derived spreading function generated by the FEM (top) and MLM (bottom). [after Young (1994c)].

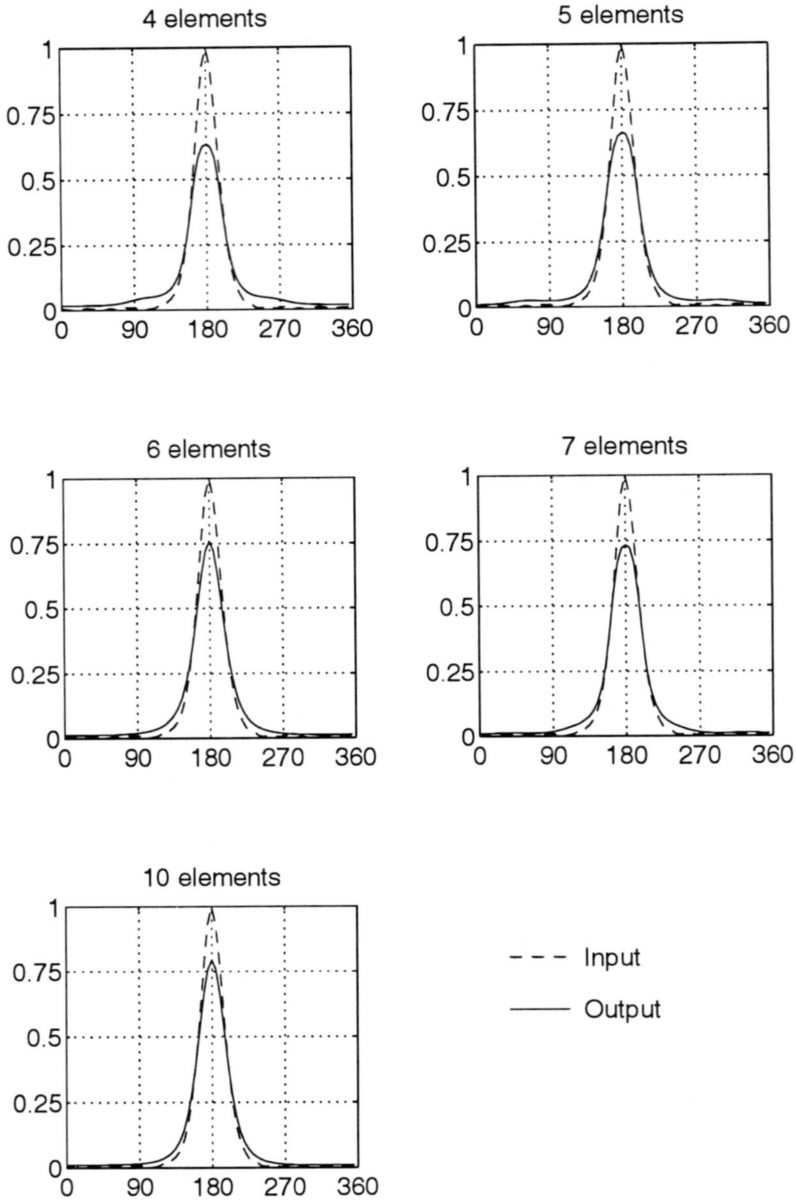

Figure 9.6: The ability of directional arrays with 4, 5, 6, 7 and 10 elements to resolve a specified directional spreading function. [after Young (1994c)].

in the area, numerical modelling is usually required to obtain design conditions. One potential solution to such problems is the advent of systems which use *remote sensing techniques* to measure the surface wave field. If such systems are mounted on orbiting satellites they offer the potential for global measurement of the Earth's wave climate. In this section the characteristics of a number of these systems will be reviewed.

9.4.1 Radar Altimeters

The first significant satellite altimeter missions were made in the 1970s. These early missions were, however, of limited success. GEOS 3 provided relatively poor coverage and SEASAT operated successfully for only a few months. Subsequent missions have, however, provided truly global measurements of the ocean wave field. These missions include

- GEOSAT: November 1986 - January 1990
- TOPEX: September 1992 - present
- ERS1: August 1991 - present
- ERS2: May 1995 - present

The coverage provided by each of these satellites varies, but indicative values can be obtained from the GEOSAT mission. In its Exact Repeat Mission (vast majority of the mission), the GEOSAT satellite was placed in an orbit at a height of approximately 800 km. The orbit was retrograde and inclined at $108°$, covering latitudes between $\pm 72°$. The satellite repeated the same ground tracks to an accuracy of approximately 1 km every 17 days (termed a 17 day repeat cycle). The resulting ground track separation at the equator was 150 km, this separation decreasing towards the poles. An example of the resulting ground track pattern for a typical 17 day repeat cycle is shown in Figure 9.7. In Figure 9.7, the ground track of the satellite is only shown where reliable ocean data were recorded.

GEOSAT transmitted 1000 pulses per second towards the Earth's surface. The returned energy from these pulses was averaged to produce data values at a frequency of 1 per second. The ground speed of the satellite was approximately 6.5 km/sec and hence one data value was obtained every 6.5 km along the ground track. The pulses radiate away from the satellite altimeter antenna as a spherical shell as shown in Figure 9.8. The pulse reflects from the water surface and the return pulse is sampled by the satellite. Rather than the return pulse being continuously sampled, it is measured in discrete intervals called "range gates". Hence it is convenient to consider the pulse during each time increment corresponding to each range gate (Dobson and Monaldo, 1996). As shown in Figure 9.8, during the first range gate, the pulse illuminates a small circular region directly below the satellite. In subsequent time increments, the surface is illuminated by a series of expanding annular rings. The area of each of these annular rings is constant.

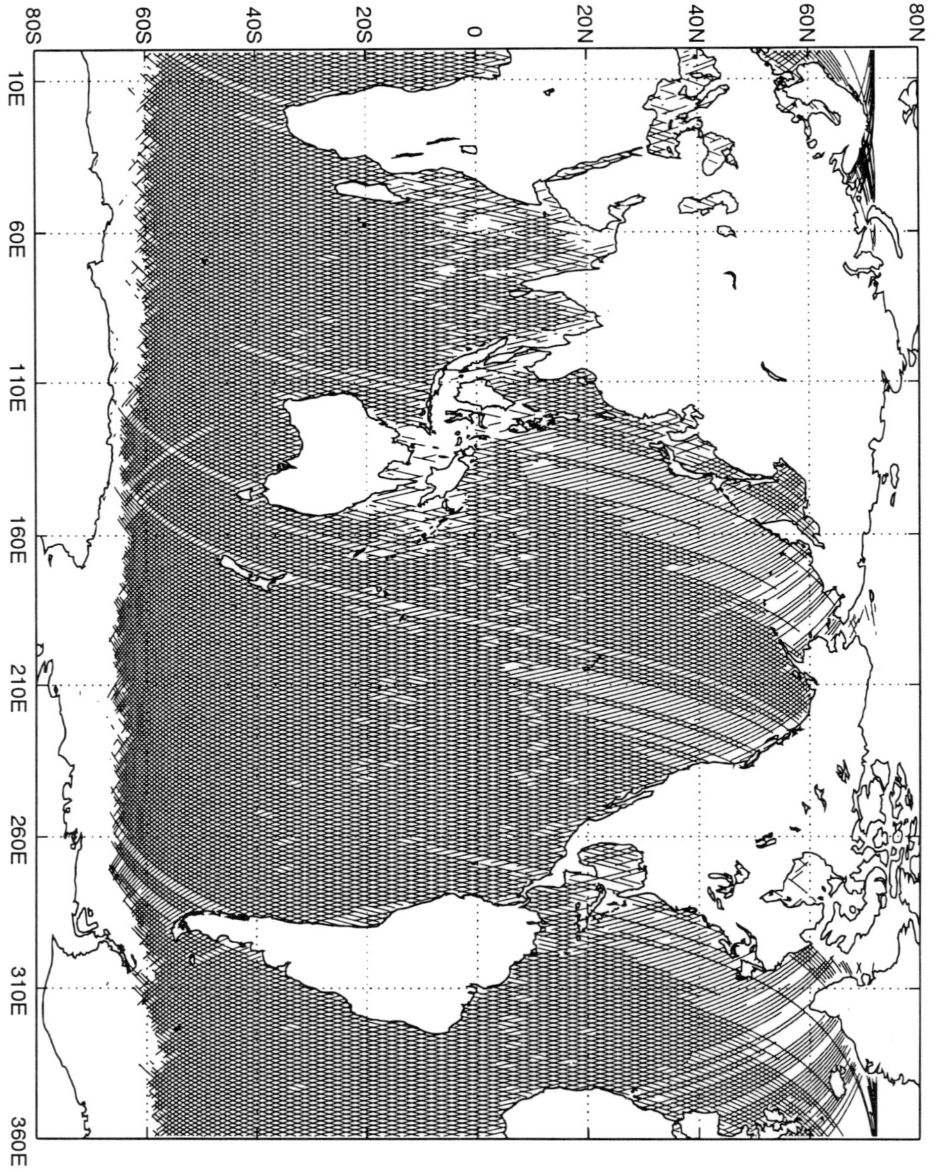

Figure 9.7: Ground track pattern for the 17 day repeat cycle of the GEOSAT satellite between 8 November, 1986 and 24 November, 1986. [after Young and Holland (1996a)].

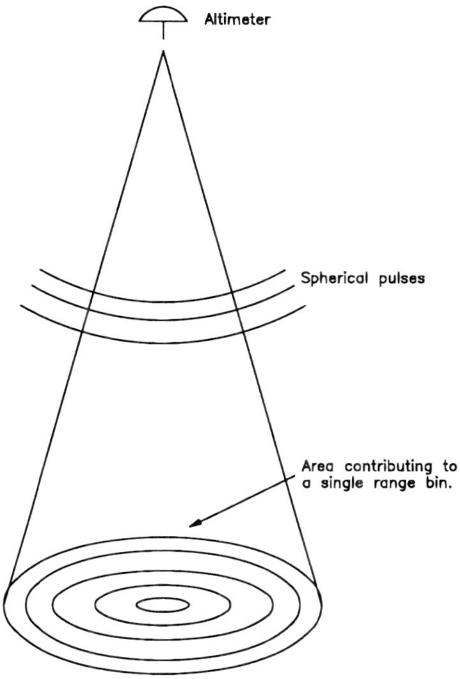

Figure 9.8: Spherical altimeter pulses radiating from the satellite antenna and areas illuminated in successive range gates. [after Young and Holland (1996a)].

If the ocean surface was perfectly flat, the return pulse would exhibit a rapid rise, corresponding to the return from the inner circular region. This would be followed by a flat section resulting from each of the constant area annular rings. If the ocean surface is roughened by the presence of waves, the leading edge of the transmitted pulse will interact with the crests of the waves a small time before interacting with the troughs. As a result, the leading edge of the return will be broadened in comparison to the flat ocean case. As the wave height increases, this broadening of the leading edge of the return pulse increases. Therefore, the slope of the leading edge of the return pulse can be used as a measure of the wave height. An example of this process is shown in Figure 9.9 (Dobson and Monaldo, 1996), which shows the one second average return pulse for the case of a relatively calm sea ($H_s = 0.36$ m) and a rough sea ($H_s = 7.47$ m). The slope of the leading edge of the pulse clearly decreases as the significant wave height increases.

A typical example of a satellite ground track and the GEOSAT altimeter values of significant wave height along this ground track are shown in Figure 9.10. The ground track in the figure is shown as the smooth line with the significant wave height as the shaded region adjacent to the ground track. The perpendicular distance from the ground track to the top of the shaded region is proportional to H_s. The case shown in Figure 9.10 occurred on 25 February, 1989 at 17.8 hours UT.

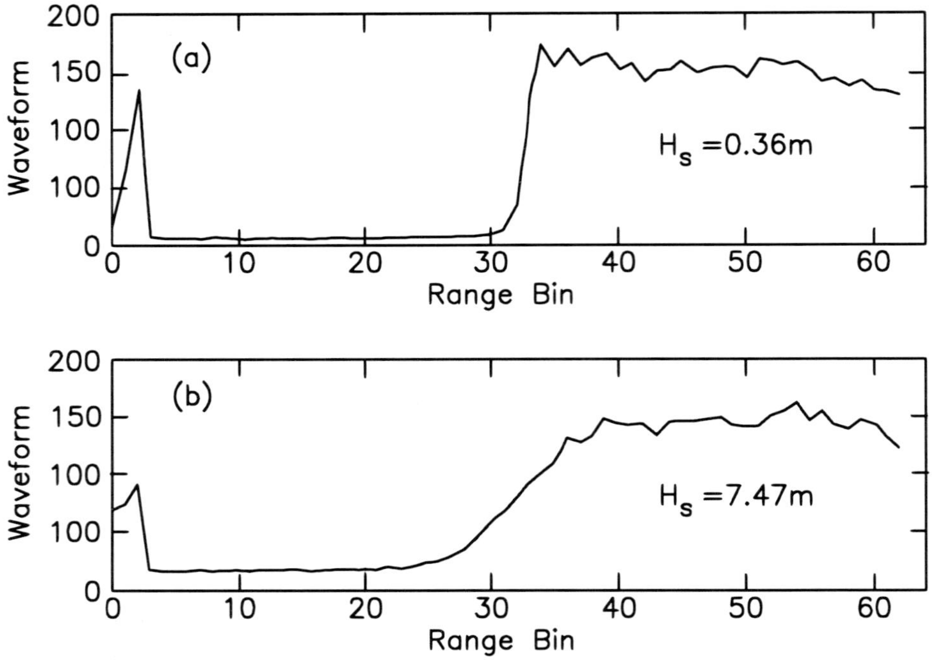

Figure 9.9: Examples of the one second average return pulse from the GEOSAT altimeter for (a) $H_s = 0.36$ m and (b) $H_s = 7.47$ m. [after Dobson and Monaldo (1996)].

The pass is south "going" over the Indian Ocean. The satellite crosses the Asian coast in the Bay of Bengal and tracks across the central Indian Ocean before crossing the Antarctic ice sheet. At approximately 20° S the satellite passes directly over Tropical Cyclone Hanitra which at the time had a central pressure of 950 HPa. The altimeter recorded a maximum significant wave height of 10.9 m and a maximum wind speed of 42 m/s (not shown) as it passed over the tropical cyclone. In addition, the rapid fall off in wave height as the ground track passed through the calm eye of the storm can be seen in the figure. Significant storm events can also be seen both to the north and south of Hanitra. In particular, the event at approximately 35° S has wind speeds of approximately 20 m/s (not shown) and significant wave heights of 5.8 m. As the satellite tracks over the Southern Ocean it reveals persistent wind speeds of 15 m/s (not shown) and resulting significant wave heights of 5 m over a large area, consistent with the year-round extreme conditions at high latitudes.

Altimeter estimates of H_s are generally regarded as having an accuracy of ± 0.5 m or 10%, whichever is greater. A detailed comparison of the altimeters on GEOSAT, TOPEX and ERS1 has been performed by Young (1998b). Although this study proposed calibration relationships to be applied to the raw satellite esti-

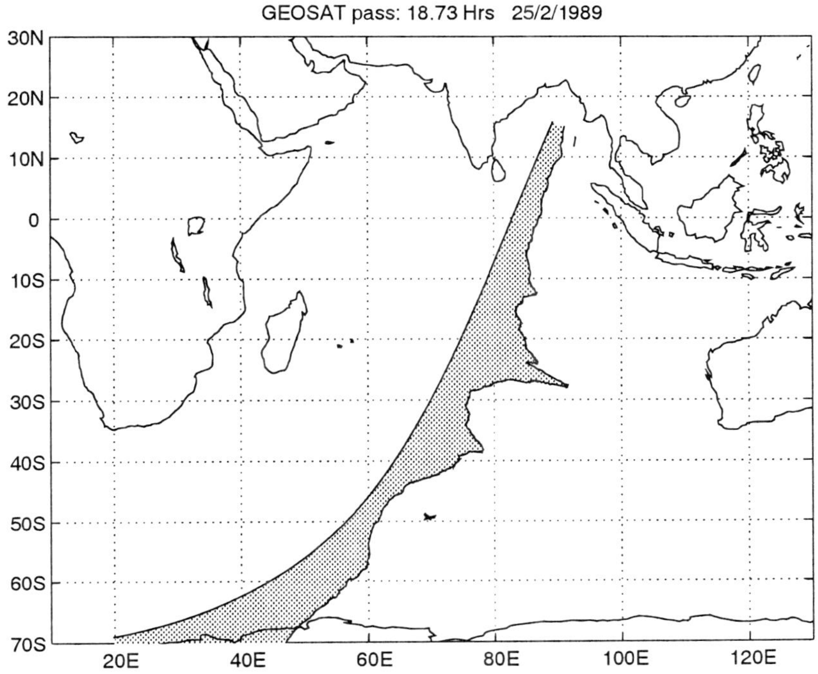

Figure 9.10: A GEOSAT ground track over the Indian Ocean recorded on 25 February, 1989 at 18.7 hours UT. Also shown is the variation in significant wave height along the track. [after Young and Holland (1996a)].

mates, the agreement with buoy data was generally good. An example of the typical scatter which can be expected is shown in Figure 9.11, which shows a comparison of TOPEX altimeter data and North American buoy data.

By combining data from the multiple satellite missions an, almost continuous, global data set of duration 10 years now exists. Hence, altimeter data is beginning to become a valuable engineering design tool. There are, however, a number of limitations associated with the data. These include:

- The altimeter footprint is approximately 10 km in diameter and hence the data is of limited applicability close to coastlines or in regions where there is significant spatial variability.

- Ground track repeat paths are relatively course.

- No information on wave period or direction is obtained.

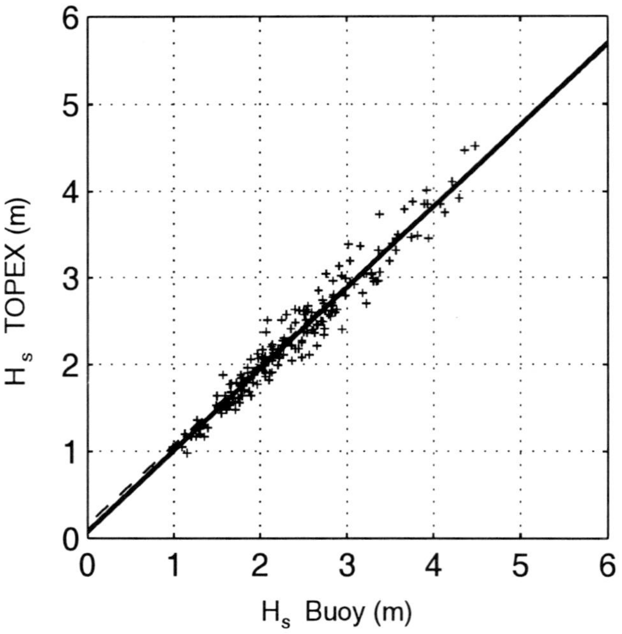

Figure 9.11: Mean monthly values of buoy H_s compared with mean monthly values of TOPEX altimeter H_s. [after Young (1998b)].

9.4.2 Synthetic Aperture Radar (SAR)

Synthetic Aperture Radars (SAR) have been flown on SEASAT and, more recently, ERS1 and ERS2. The SAR is a coherent radar which looks at right angles to the flight path, illuminating a swath typically 100 km wide and offset from the satellite track by approximately 250 km. For such a system to image waves with any precision, it must have a resolution significantly smaller than typical ocean wavelengths. In the azimuth (or along track) direction high resolution can be achieved by having a long antenna (aperture). Simply stated, such an antenna can focus the radar tightly. For a satellite flying at a height of 800 km, however, an antenna of length 4 km would be required to achieve an azimuth resolution of 25 m. Clearly, an antenna of this size is not practical. The SAR, however, takes advantage of the fact that the radar platform is moving forward. The radar transmits multiple pulses and keeps track of the phase of the received signal, thus creating a *synthetic aperture* much larger than the real aperture. In the range direction, (or across track) resolution is obtained by transmitting short pulses, thus taking advantage of the fact that the satellite observes a swath offset from the path.

For typical satellites it takes of order 0.5 s to fly 4 km and thus create the required synthetic aperture. This is sufficiently long for the sea surface to have moved significantly. This motion degrades the along track resolution of the radar. Therefore, when obtaining ocean wave spectra with the SAR, there is a high frequency azimuthal cut-off, above which it is difficult to obtain information. Range traveling waves are, however, not affected. A schematic diagram showing the geometry of the SAR imaging swath is shown in Figure 9.12.

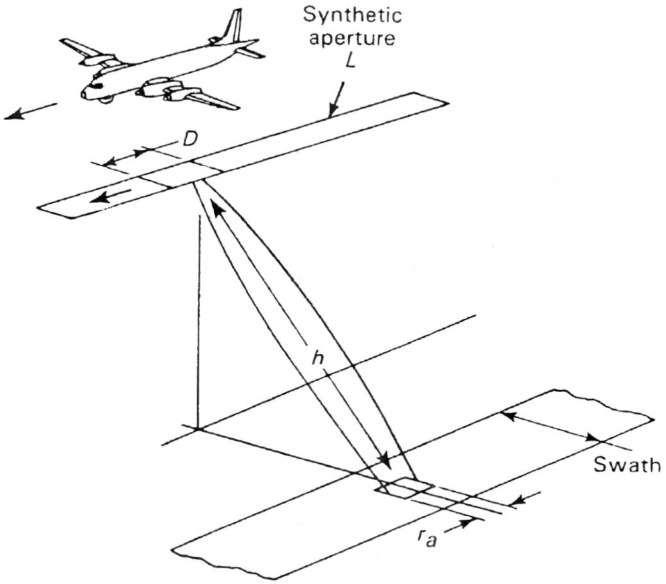

Figure 9.12: Imaging by an air borne SAR. Resolution in the range direction is achieved using short pulses, whereas in the azimuth a long Synthetic Aperture is created using Doppler mapping. The real aperture of the radar is given by D. [after Plant (1987)].

The mechanisms by which the SAR images waves are complex and have been the focus of numerous studies (Alpers and Rufenach, 1979; Alpers et al, 1981; Raney, 1981; Alpers, 1983; Hasselmann et al, 1985b; Alpers and Brüning, 1986). The primary mechanism by which oblique looking radars (such as SAR) obtain backscatter from the water surface is by Bragg scattering. Bragg scattering involves constructive interference between the radar waves and the water waves. For a radar imaging the horizontal water surface at an angle of θ to the vertical, the Bragg resonance condition is satisfied by

$$\lambda_r = 2\lambda_w \sin\theta \tag{9.24}$$

where λ_r is the radar wave length and λ_w the water wave length. For the SEASAT SAR, λ_w, the Bragg resonant wave length was 0.33 m. The longer water waves of interest are imaged by the fact that they modulate the shorter Bragg scatters.

There are three imaging or modulation mechanisms

1. *Tilt Modulation* - The water surface, upon which the Bragg scatters ride, is periodically tilted by the longer waves. This changes the strength of the radar return (backscatter coefficient). As the water surface tilts, the wave length and hence the energy of the Bragg scatters change.

2. *Hydrodynamic Modulation* - The periodic straining of the water surface causes the Bragg scatters at the long wave crests to be compressed and those in the troughs to be expanded. As a result, there is a periodic modulation of the amplitudes of the scatters.

3. *Velocity Bunching* - If the target (water surface) has a velocity component in the range (cross track) direction, it causes an apparent position offset in the azimuth (along track) direction. As the water surface velocities are maximum at the crests and troughs of the long waves, the backscatter tends to converge at these points.

The complexity of these imaging mechanisms is such that, for many years, the possibility of obtaining quantitative measurements of surface wave spectra from SAR imagery was questioned. However, the reverse problem of obtaining SAR spectra from wave spectra has been successfully achieved using Monte Carlo techniques (Brüning et al, 1990; Hasselmann and Hasselmann, 1991). More difficult, however, is the inversion of SAR image spectra to obtain wind wave spectra. The mapping process is nonlinear and hence a single SAR spectrum can map to multiple wind wave spectra. In addition, as the SAR image is "frozen in time" it has a 180° propagation ambiguity. Also, no information is available at wave numbers above the azimuthal cut-off. Hasselmann et al (1996) have, however, shown that by using a model wave spectrum as a "first-guess" wind wave spectrum, an iterative process can be used to reliably invert SAR spectra and obtain wind wave spectra.

Satellites such as ERS1 and ERS2 are providing 10 km × 5 km wave-mode imagets every 200 km along the satellite track. Hence, for the first time, the potential to obtain global directional wind wave spectra exists.

9.4.3 Other Radar Systems

Remote sensing systems can be divided in three general groups: ground based, aircraft borne and satellite borne. Satellite borne systems such as the altimeter and SAR (discussed above) have the very significant advantage of providing global coverage. Ground and aircraft based system have, however, been particularly useful in research and site specific studies. Excellent reviews of the available systems can be found in Plant (1987) and Tucker (1991).

Bibliography

Abreu, M., Larraza, A. and Thornton, E., 1992, "Nonlinear transformation of directional wave spectra in shallow water", *J. Geophys. Res.*, 97, 15,579-15,589.

Airy, G.B., 1845, "Tides and waves", *Encycl. Metrop.*, London.

Allender, J.H., Albrecht, J. and Hamilton, G., 1983, "Observations of directional relaxation of wind sea spectra", *J. Phys. Oceanogr.*, 13, 1519-1525.

Allender, J.H., Barnett, T.P. and Lybanon, M., 1985, "The DNS model: An improved spectral model for ocean wave prediction", In: 'Ocean wave modeling', SWAMP Group, *Plenum Press*, New York, 235-248.

Allender, J.H., Audunson, T., Barstow, S.F., Bjerken, S., Krogstad, H.E., Steinbakke, P., Vartdal, L., Borgman, L.E. and Graham, C, 1989, "The WADIC project: a comprehensive field evaluation of directional wave instrumentation", *Ocean Engineering*, 16, 505-536.

Alpers, W., 1983, "Monte Carlo simulations for studying the relationship between ocean wave and synthetic aperture radar image spectra", *J. Geophys. Res.*, 88, 1745-1759.

Alpers, W.R. and Rufenach, C.L., 1979, "The effect of orbital motions on synthetic aperture radar imagery of ocean waves", *IEEE Trans. Antennas Propag.*, AP-27, 685-690.

Alpers, W.R., Ross, D.B. and Rufenach, C.L., 1981, "On the detectability of ocean surface waves by real and synthetic aperture radar", *J. Geophys. Res.*, 86, 6481-6498.

Alpers, W. and Brüning, C., 1986, "On the relative importance of motion related contributions to the SAR imaging mechanism of ocean surface waves", *IEEE Trans. Geosci. Remote Sens.*, GE-24, 873-885.

Al-Zanaidi, M.A. and Hui, W.H., 1984, "Turbulent air flow over water waves – a numerical study", *J. Fluid Mech.*, 148, 225-246.

Anderson, D.M. and Prell, W.L., 1992, "The structure of the southwest monsoon winds over the Arabian Sea during the late Quaternary: Observations, simulations, and marine geologic evidence", *J. Geophys. Res.*, 97, 15,481-15,487.

Armstrong, J.A., Bloembergen, N., Ducuing, J., and Pershan, P.S., 1962, "Interactions between light waves in a nonlinear dielectric", *Phys. Rev.*, 127, 1918-1939.

Arya, S.P., 1988, "Introduction to meteorology", *Academic Press*, San Diego, 307pp.

Atkinson, G.D., and Holliday, C.R., 1977, "Tropical cyclone minimum sea level pressure-maximum sustained wind relationship for western North Pacific", *Mon. Weather Rev.*, 105, 421-427.

Banner, M.L., 1990, "Equilibrium spectra of wind waves", *J. Phys. Oceanogr.*, 20, 966-984.

Banner, M.L. and Melville, W,K., 1976, "On the separation of air flow over water waves", *J. Fluid Mech.*, 77, 825-842.

Banner, M.L., Jones, I.S.F. and Trinder, J.C., 1989, "Wavenumber spectra of short gravity waves", *J. Geophys. Res.*, 73, 513-530.

Banner, M.L. and Young, I.R., 1994, " Modeling Spectral Dissipation in the Evolution of Wind Waves - Part 1. Assessment of existing model performance", *J. Phys. Oceanogr.*, 24, 1550-1671.

Barnett, T.P., 1968, "On the generation, dissipation and prediction of ocean wind waves", *J. Geophys. Res.*, 73, 513-529.

Barnett, T.P. and Wilkerson, J.C., 1967, "On the generation of ocean wind waves as inferred by airborne radar measurements of fetch-limited spectra", *J. Marine Res.*, 25, 292-328.

Barnett, T.P. and Sutherland, A.J., 1968, "A note on an overshoot effect in wind-generated waves", *J. Geophys. Res.*, 73, 6879-6885.

Battjes, J.A., 1994, "Shallow water wave modelling", Proc. Int. Symp.: Waves – Physical and Numerical Modelling, Vancouver, 1-23.

Battjes, J.A. and Janssen, J.P.F.M., 1978, "Energy loss and set-up due to breaking of random waves", *16th Coastal Eng. Conf.*, Hamburg, 569-587.

Battjes, J.A., Zitman, T.F. and Holthuijsen, L.H., 1987, "A reanalysis of the spectra observed in JONSWAP", *J. Phys. Oceanogr.*, 17, 1288-1295.

Bauer, E., Hasselmann, S., Young, I.R. and Hasselmann, K., 1990, "Data Assimilation into a Wave Prediction Model", *25th General Assembly of the European Geophysical Society*, Copenhagen, 23-27 April.

Bauer, E., Hasselmann, K. and Young, I.R., 1992a, "Satellite Data Assimilation in the Wave Model 3G-WAM", *Proc. Central Symposium of the 'International Space Year'*, Munich, Germany, March 30 - April 4, 377-380.

Bauer, E., Hasselmann, S. and Hasselmann, K., 1992b, "Validation and assimilation of Seasat altimeter wave heights using the WAM wave model", *J. Geophys. Res.*, 97, 12,671-12,682.

Bea, R.G., 1974, "Gulf of Mexico hurricane wave heights", *6th Offshore Technology Conf.*, Houston, OTC 2110.

Beal, R.C., Gerling, T.W., Irvine, D.E., Monaldo, F.M. and Tilley, D.G., 1986, "Spatial variations of ocean wave directional spectra from the Seasat Synthetic Aperture Radar", *J. Geophys. Res.*, 91, 2433-2449.

Beji, S. and Battjes, J.A., 1993, "Experimental investigation of wave propagation over a bar", *Coastal Eng.*, 19, 151-162.

Beji, S. and Battjes, J.A., 1994, "Numerical simulation of nonlinear wave propagation over a bar", *Coastal Eng.*, 23, 1-16.

Belcher, S.E. and Hunt, J.C.R., 1993, "Turbulent shear flow over slowly moving waves", *J. Fluid Mech.*, 251, 109-148.

Belcher, S.E., Harris, J.A. and Street, R.L., 1994, "Linear dynamics of wind waves in coupled turbulent air-water flow. Part 1. Theory", *J. Fluid Mech.*, 271, 119-151.

Bendat, J.S. and Piersol, A.G., 1971, "Random data: Analysis and measurement procedures", *Wiley-Interscience*, 407pp.

Bender, L.C., 1996, "Modification of the physics and numerics in a third-generation ocean wave model", *J. Atmos. Oceanic Tech.*, 13, 726-750.

Berkhoff, J.C.W., 1972, "Computation of combined refraction-diffraction", *Proc. 13th Int. Conf. Coastal Eng.*, Lisbon, 55-69.

Bishop, C.T. and Donelan, M.A., 1987, "Measuring waves with pressure transducers", *Coastal Engineering*, 11, 309-328.

Bishop, C.T. and Donelan, M.A., 1988, "Waves and wave forecasting", In: *'Civil Engineering Practice 3 - Geotechnical/Ocean Engineering'*, Technomic Publ. Co. Inc., Lancaster, Pa., 653-695.

Black, J.L., 1979, "Hurricane Eloise directional wave energy spectra", *11th Offshore Technology Conf.*, Houston, OTC 3594.

Bliven, L.F., Huang, N.E. and Long, S.R., 1986, "Experimental study of the influence of wind on Benjamin-Feir sideband instability", *J. Fluid Mech.*, 162, 237-260.

Booij, N., 1981, "Gravity waves on water with non-uniform depth and current", *Delft Univ. of Tech.*, Rep. No. 81-1.

Booij, N. and Holthuijsen, L.H., 1987, "Propagation of ocean waves in discrete spectral wave models", *J. Comp. Phys.*, 68, 307-326.

Booij, N., Holthuijsen, L.H. and Ris, R.C., 1996, "The SWAN wave model for shallow water", *Int. Conf. Coastal Eng.*, ASCE, Orlando, 668-676.

Bouws, E., 1986, "Provisional results of a wind wave experiment in a shallow lake (Lake Marken, The Netherlands)", *KNMI Afdeling Oceanografisch Onderzoek*, OO-86-21, De Bilt, 15pp.

Bouws, E., and Komen, G.J., 1983, "On the balance between growth and dissipation in an extreme depth-limited wind-sea in the southern North Sea", *J. Phys. Oceanogr.*, 13, 1653-1658.

Bouws, E., Günther, H., Rosenthal, W. and Vincent, C.L., 1985, "Similarity of the wind wave spectrum in finite depth water, 1. Spectral form", *J. Geophys. Res.*, 90, 975-986.

Bouws, E., Günther, H., Rosenthal, W. and Vincent, C.L., 1987, "Similarity of the wind wave spectrum in finite depth water, 2. Statistical relationships between shape and growth stage parameters", *Dtsch. Hydrogh. Z.*, 40, 1-24.

Bretherton, F.P., 1964, "Resonant interactions between waves: The case of discrete oscillations", *J. Fluid Mech.*, 20, 457-480.

Bretherton, F.P. and Garrett, C.J.R., 1969, "Wavetrains in inhomogeneous media", *Proc. Royal Soc., A*, 302, 529-554.

Bretschneider, C.L., 1952a, "Revised wave forecasting relationships", *Proc. 2nd Conf. on Coastal Eng.*, ASCE, Council on Wave Research.

Bretschneider, C.L., 1952b, "The generation and decay of wind waves in deep water", *Trans. Amer. Geophys. Union*, 33, 381-389.

Bretschneider, C.L., 1958, "Revisions in wave forecasting: Deep and shallow water", *Proc. 6nd Conf. on Coastal Eng.*, ASCE, Council on Wave Research.

Bretschneider, C.L., 1959, "Hurricane design - Wave practices", *Trans. ASCE*, 124, 39-62.

Bretschneider, C.L., 1972, "A non-dimensional stationary hurricane wave model", *4th Offshore Technology Conf.*, Houston, OTC 1517.

Bretschneider, C.L. and Reid, R.O., 1953, "Change in wave height due to bottom friction, percolation and refraction", *34th Annual Meeting of American Geophysical Union.*

Bretschneider, C.L., and Reid, R.O., 1954, "Modification of wave height due to bottom friction, percolation, and refraction", *Tech. Memo.*, 28, Beach Erosion Board, Washington D.C.

Brissette, F.P. and Tsanis, I.K., 1992, "Maximum likelihood method techniques for directional analysis of heave-pitch-roll data", *3rd Int. Workshop on Wave Hindcasting and Forecasting*, Montrael, 1-11.

Brüning, C., Alpers, W. and Hasselmann, K., 1990, "Monte Carlo simulation studies of the nonlinear imaging of a two-dimensional surface wave field by a synthetic radar", *Int. J. Remote Sens.*, 11, 1695-1727.

Burgers, G. and Makin, V.K., 1993, "Boundary-layer model results for wind-sea growth", *J. Phys. Oceanogr.*, 23, 372-385.

Burling, R.W., 1959, "The spectrum of waves at short fetches", *Dtsch. Hydrogr. Z.*, 12, 96-117.

Businger, J.A., Wyngaard, J.C., Izumi, I. and Bradley, E.F., 1971, "Flux-profile relationships in the atmospheric surface layer", *J. Atmospheric Sci.*, 28, 181-189.

Capon, J., 1969, "High-resolution frequency-wavenumber spectrum analysis", *Proc. IEEE*, 57, 1408-1418.

Cardone, V.J., Pierson, W.J. and Ward, E.G., 1975, "Hindcasting the directional spectra of hurricane-generated waves", *J. Pet. Technol.*, 261, 91-127.

Cardone, V.J., Ross, D.B. and Ahrens, M.R., 1977, "An experiment in forecasting hurricane generated sea states", *Proc. 11th Tech. Conf. on Hurricanes and Tropical Met.*, Miami.

Carter, D.J.T., Foale, S. and Webb, D.J., 1991, "Variation in global wave climate throughout the year", *Int. J. Remote Sensing*, 12, 1687-1697.

Cartwright, D.E. and Smith, N.D., 1964, "Buoy techniques for obtaining directional waves from arrays at Adu Quir and Ras El-Bar, Egypt", *Oceanology*, Chapter 10, Marine Technology Society.

Cavaleri, L., 1994, "Wind variability", In *Dynamics and modelling of ocean waves*, G.J. Komen et al. (eds.), Cambridge Uni. Press, 320-331.

Cavaleri, L., Curiotto, S., Dallaporta, G. and Mazzoldi, A., 1981, "Directional wave recording in the Adriatic Sea", *Il Nuovo Cimento*, 4C(5), 519-534.

Cavaleri, L. and Rizzoli, P.M., 1981, "Wind wave prediction in shallow water: Theory and application", *J. Geophys. Res.*, 86, 10,961-10,973.

Cavaleri, L., Bertotti, L., and Lionello, P., "Shallow water application of the third-generation WAM wave model", *J. Geophys. Res.*, 94, 8111-8124.

Cavaleri, L. and Burgers, G.J.H., 1992, "Wind gustiness and wave growth", *KNMI Afdeling Oceanografisch Onderzoek memo*, OO-92-18, De Bilt, 38pp.

Cavaleri, L., Burgers, G., Günther, H., Makin, V. and Zambresky, L., 1994, "Applications to wave hindcasting and forecasting. Part 4.1: Analysis", In *Dynamics and modelling of ocean waves*, G.J. Komen et al. (eds.), Cambridge Uni. Press, 300-307.

CERC, 1977, "Shore protection manual", *U.S. Army Coastal Engineering Research Center*, 3 Volumes.

CERC, 1984, "Shore protection manual", *U.S. Army Coastal Engineering Research Center*, 2 Volumes.

Chalikov, D. and Makin, V, 1991, "Models of the wave boundary layer", *Boundary Layer Meteorol.*, 56, 83-99.

Chalikov, D. and Belevich, M.Yu., 1993, "One-dimensional theory of the wave boundary layer", *Boundary-Layer Meteorol.*, 63, 65-96.

Challenor, P.G., Foale, S. and Webb, D.J., 1990, "Seasonal changes in the global wave climate measured by the GEOSAT altimeter", *Int. J. Remote Sensing*, 11, 2205-2213.

Charnock, H., 1955, "Wind stress on the water surface", *Q. J. R. Meteorol. Soc.*, 81, 639-640.

Chen, Y., and Wang, H., "Numerical model of nonstationary shallow water wave spectral transformations", *J. Geophys. Res.*, 88, 9851-9863.

Christoffersen, J.B., and Jonsson, I.G., 1985, "Bed friction and dissipation in a combined current and wave motion", *Coastal Eng.*, 12, 387-423.

Collins, J.I., 1972, "Predictions of shallow-water spectra", *J. Geophys. Res.*, 77, 2693-2707.

Conte, S.D. and Miles, J.W., 1959, "On the integration of the Orr-Sommerfeld equation", *J. Soc. Indust. Appl. Math.*, 7, 361-369.

Darbyshire, J., 1959, "A further investigation of wind generated waves", *Dtsch. Hydrogr. Z.*, 12, 1-13.

Dally, W.R., 1990, "Random breaking waves: a closed-form solution for planar beaches", *Coastal Engineering*, 14, 233-263.

Dalrymple, R.A. and Kirby, J.T., 1988, "Models for very wide-angle water waves and wave diffraction", *J. Fluid Mech.*, 192, 33-50.

Dalrymple, R.A., Suh, K., Kirby, J.T. and Chae, J.W., 1989, "Models for very wide-angle water waves and wave diffraction: Irregular bathymetry, *J. Fluid Mech.*, 201, 299-322.

Davis, R.E., 1972, "On prediction of turbulent flow over a wavy boundary", *J. Fluid Mech.*, 52, 287-306.

Davis, R.E. and Regier, L.A., 1977, "Methods for estimating directional wave spectra from multi-element arrays", *J. Marine Res.*, 35, 453-477.

Dean, R.G. and Dalrymple, R.A., 1991, "Water wave mechanics for engineers and scientists", *World Scientific*, 353pp.

De las Heras, M.M. and Janssen, P.A.E.M., 1992, "Data assimilation with a coupled wind-wave model", *J. Geophys. Res.*, 97, 20261-20270.

De Valk, C.F. and Calkoen, C.J. 1989, "Wave data assimilation in a 3rd generation wave model for the North Sea - An optimal control approach", *Delft Hydraulics*, rep. X38, Delft.

Divoky, D., Le Mehaute, B. and Lin, A., 1970, "Breaking waves on gentle slopes", *J. Geophys. Res.*, 75, 1681-1692.

Dobson, E.B. and Monaldo, F.M. , 1996, "Radar altimeter wave height measurements", In: *'Remote sensing techniques for oceanographers'*, M. Ikeda and F. Dobson (eds.), CRC Press, New York.

Dobson, F.W., 1971, "Measurements of atmospheric pressure on wind-generated sea waves", *J. Fluid Mech.*, 48, 91-127.

Dobson, F., Perrie, W. and Toulany, B., 1989, "On the deep-water fetch laws for wind-generated surface gravity waves", *Atmosphere-Ocean*, 27, 210-236.

Doering, J.C., and Bowen, A.J., 1987, "Skewness in the nearshore zone: A comparison of estimates from Marsh-McBirney current meters and colocated pressure sensors", *J. Geophys. Res.*, 92, 13,173-13,183.

Dold, J.W. and Peregrine, D.H., 1986, "An efficient boundary-integral method for steep unsteady water waves", In: 'Numerical methods for fluid dynamics II', Morton, K.W. and Baines, M.J. (eds.), *Oxford Univ. Press*, 1-10.

Donelan, M.A., 1987, "The effect of swell on the growth of wind-waves", *Symp. on Measuring Ocean Waves from Space*, Baltimore, Johns Hopkins Univ.

Donelan, M.A., 1997, "Air-water energy flux in opposing winds", *4th Int. Waves in Shallow Environments Conf.*, San Francisco, (unpublished).

Donelan, M.A., Hamilton, J. and Hui, W.H., 1985, "Directional spectra of wind-generated waves", *Philos. Trans. R. Soc. Lond.*, A 315, 509-562.

Donelan M.A. and Pierson, W.J., 1987, "Radar scattering and equilibrium ranges in wind-generated waves with application to scatterometry", *J. Geophys. Res.*, 92, 4971-5029.

Donelan, M.A. and Hui, W.H., 1990, "Mechanics of ocean surface waves", In *Surface Waves and Fluxes*, (Geernaert, G.L. and Plant, W.J., eds.), Kluwer, 209-246.

Donelan, M., Skafel, M., Graber, H., Liu, P., Schwab, D. and Venkatesh, S., 1992, "On the growth rate of wind-generated waves", *Atmosphere-Ocean*, 30, 457-478.

Donelan, M.A., Dobson, F.W., Smith, S.D. and Anderson, R.J., 1993, "On the dependence of sea surface roughness on wave development", *J. Phys. Oceanogr.*, 23, 2143-2149.

Donelan, M.A. and Yuan, Y., 1994, "Wave dissipation by surface processes", In *Dynamics and modelling of ocean waves*, G.J. Komen et al. (eds.), Cambridge Uni. Press, 143-155.

Drennan, W.M., Donelan, M.A., Terray, E.A. and Katsaros, K.B., 1996, "Oceanic turbulence dissipation measurements in SWADE", *J. Phys. Oceanogr.*, 26, 808-815.

Duncan, J.H., 1981, "An experimental investigation of breaking waves produced by a towed hydrofoil", *Proc. Roy. Soc. London*, 337, 331-348.

Dungey, J.C. and Hui, W.H., 1979, "Nonlinear energy transfer in a narrow gravity-wave spectrum", *Proc. R. Soc. Lond.*, A 368, 239-265.

Dyer, A.J. and Hicks, B.B., 1970, "Flux-gradient relationships in the constant flux layer", *Q. J. R. Meteorol. Soc.*, 96, 715-721.

Ebuchi, N., Kawamura, H. and Toba, Y., 1992, "Growth of wind waves with fetch observed by the Geosat altimeter in the Japan Sea under winter monsoon", *J. Geophys. Res.*, 97, 809-819.

Elachi, C., Thompson, T.W., and King, D.B., 1977, "Observations of the ocean wave pattern under Hurricane Gloria with Synthetic Aperture Radar", *Science*, 198, 609-610.

Eldeberky, Y. and Battjes, J.A., 1995, "Parameterization of triad interactions in wave energy models", *Coastal Dynamics '95*, Dally, W.R. and Zeidler, R.B. (eds), ASCE, 140-148.

Elgar, S., and Guza, R.T., 1985, "Observations of bispectra of shoaling surface gravity waves", *J. Fluid Mech.*, 161, 425-448.

Elgar, S., and Guza, R.T., 1986, "Nonlinear model predictions of bispectra of shoaling surface gravity waves", *J. Fluid Mech.*, 167, 1-18.

Elgar, S., Freilich, M., and Guza, R.T., 1990, "Model-data comparisons of moments of nonbreaking shoaling surface gravity waves", *J. Geophys. Res.*, 95, 16,055-16,063.

Elgar, S., Guza, R.T., and Freilich, M.H., 1993, "Observations of nonlinear interactions in directionally spread shoaling surface gravity waves", *J. Geophys. Res.*, 98, 20,299-20,305.

Elgar, S., Herbers, T.H.C., Chandran, V. and Guza, R.T., 1995, "Higher-order spectral analysis of nonlinear ocean surface gravity waves", *J. Geophys. Res.*, 100, 4997-4983.

Elliott, J.A., 1972, "Microscale pressure fluctuations measured within the lower atmospheric boundary layer", *J. Fluid Mech.*, 53, 351-384.

Elliott, J.A., 1972, "Microscale pressure fluctuations near waves being generated by wind", *J. Fluid Mech.*, 54, 427-448.

Ewing, J.A., 1971, "A numerical wave prediction method for the North Atlantic Ocean", *Dtsch. Hydrogh. Z.*, 24, 241-261.

Fenton, J.D. and McKee, W.D., 1990, "On calculating the lengths of water waves", *Coastal Engineering*, 14, 499-513.

Forristall, G.Z., 1981, "Measurements of a saturation range in ocean wave spectra", *J. Geophys. Res.*, 86, 8075-8084.

Forristall, G.Z., Ward, E.G., Cardone, V.J., and Borgmann, L.E., 1978, "The directional spectra and kinematics of surface gravity waves in Tropical Storm Delia", *J. Phys. Oceanogr.*, 8, 888-909.

Fox, M.J.H., 1976, "On the nonlinear transfer of energy in the peak of a gravity-wave spectrum", *Proc. R. Soc. Lond.*, A 348, 467-483.

Freilich, M.H., and Guza, R.T., 1984, "Nonlinear effects on shoaling surface gravity waves", *Philos. Trans. R. Soc. London*, A, 311, 1-41.

Freilich, M.H., Guza, R.T., and Elgar, S., 1990, "Observations of nonlinear effects in directional spectra of shoaling surface gravity waves", *J. Geophys. Res.*, 95, 9645-9656.

Garratt, J.R., 1977, "Review of drag coefficients over oceans and continents", *Mon. Weather Rev.*, 105, 915-929.

Geernaert, G.L., 1990, "Bulk parameterizations for the wind stress and heat fluxes", in G.L. Geernaert and W.J. Plant (eds.), 'Surface waves and fluxes', Vol. 1, *Kluwer Academic*, Dordrecht, 336pp.

Geernaert, G.L. and Plant, W.L., 1990, "Surface waves and fluxes", 2 vols., *Kluwer Academic*, Dordrecht, 336pp.

Gelci, R., Cazalé, J. and Vassal, J., 1956, "Utilisation des diagrammes de propagation à la provision energitique de la houle", *Bull. Inf. Comitè Cen. Oceanogr. Etudes Côtes*, 8, 169-187.

Gelci, R., Cazalé, J. and Vassal, J., 1957, "Prévision de la houle. La méthode des densités spectroangulaires", *Bull. Inform. Comité Central Oceanogr. D'Etude Côtes*, 9, 416-435.

Gent, P.R. and Taylor, P.A., 1976, "A numerical model of air-flow above water waves", *J. Fluid Mech.*, 77, 105-128.

Goda, Y., 1985, "Random seas and design of marine structures", *University of Tokyo Press*, 323pp.

Golding, B.W., 1983, "A wave prediction system for real-time sea state forecasting", *Q. J. Roy. Meteorol. Soc.*, 109, 393-416.

Gonzalez, F.I., Thompson, T.E., Brown, W.E. and Weissman, D.E., 1982, "Seasat wind and wave observations of Northeast Pacific Hurricane Iva, August 13, 1978", *J. Geophys. Res.*, 87, 3431-3438.

Graber, H.C., and Madsen, O.S., "A finite-depth wind-wave model, 1, Model description", *J. Phys. Oceanogr.*, 18, 1465-1483.

Graham, H.E., and Hudson, G.N., 1960, "Surface winds near the center of hurricanes (and other cyclones)", *NHRP Rep., Govt. Print. Office*, 39, 200pp.

Grant, W.D., and Madsen, O.S., "Movable bed roughness in unsteady oscillatory flow", *J. Geophys. Res.*, 87, 469-481.

Gray, W.M., and Shea, D.J., 1973, "The hurricane's inner core region, II: Thermal stability and dynamic characteristics", *J. Atmos. Sci.*, 30, 1565-1576.

Greenwood, J.A., Cardone, V.J. and Lawson, L.M., 1985, "Intercomparison test version of the SAIL wave model", In: 'Ocean wave modeling', SWAMP Group, *Plenum Press*, New York, 221-233.

Grilli, S.T., Skourop, J. and Svendsen, I.A., 1989, "An efficient boundary element method for nonlinear water waves", *Eng. Anal. with Boundary Element Methods*, 6, 97-107.

Günther, H., 1981, "A parametric surface wave model and statistics of the prediction parameters", *Hamburger Geophys. Einzelsch.*, 55, 1-90.

Günther, H., Rosenthal, W., Weare, T,J., Worthington, B.A., Hasselmann, K. and Ewing, J.A., 1979, "A hybrid parametric wave prediction model", *J. Geophys. Res.*, 84, 5727-5738.

Günther, H., Rosenthal, W. and Dunckel, M., 1981, "Directional wave spectra observed during JONSWAP 1973", *J. Phys. Oceanogr.*, 11, 718-728.

Hamilton, R.C., and Ward, W.G., 1974, "Ocean Data Gathering Program - Quality and reduction of data", *6th Offshore Technology Conf.*, Houston, OTC 2108-A.

Hardy, T.A., 1993, "The attenuation and spectral transformation of wind waves on a coral reef", PhD thesis, James Cook Univ., 336pp.

Hardy, T.A. and Young, I.R., 1996, "Field Study of Wave Attenuation on an Offshore Coral Reef", *J. Geophys. Res.*, 101, 14,311-14,326.

Harper, B.A., Mason, L.B., and Bode, L., 1993, "Tropical Cyclone Orson - A Severe test for modelling", *11th Aust. Conf. Coastal and Ocean Eng.*, Townsville, 59-64.

Hasselmann, D.E., Dunckel, M. and Ewing, J.A., 1980, "Directional wave spectra observed during JONSWAP 1973", *J. Phys. Oceanogr.*, 10, 1264-1280.

Hasselmann, D. and Bösenberg, J., 1991, "Field measurements of wave-induced pressure over wind-sea and swell", *J. Fluid Mech.*, 230, 391-428.

Hasselmann, K., 1960, "Grundgleichugen der Seegangsvorhersage", *Schifftechnik*, 7, 191-195.

Hasselmann, K., 1962, "On the non-linear energy transfer in a gravity-wave spectrum, Part 1. General Theory", *J. Fluid Mech.*, 12, 481-500.

Hasselmann, K., 1963a, "On the non-linear energy transfer in a gravity-wave spectrum, Part 2. Conservation theorems; wave-particle analogy; irreversibility", *J. Fluid Mech.*, 15, 273-281.

Hasselmann, K., 1963b, "On the non-linear energy transfer in a gravity-wave spectrum, Part 3. Evaluation of energy flux and swell-sea interaction for a Neumann spectrum", *J. Fluid Mech.*, 15, 385-398.

Hasselmann, K., 1966, "Feyman diagrams and interaction rules of wave-wave scattering processes", *Rev. Geophys.*, 4, 1-32.

Hasselmann, K., 1969, "Weak-interaction theory of ocean waves", *Basic Developments in Fluid Mech.*, 2, 117-182.

Hasselmann, K., 1974, "On the spectral dissipation of ocean waves due to white-capping", *Boundary Layer Meteorol.*, 6, 107-127.

Hasselmann, K., Munk, W., and MacDonald, G., 1963, "Bispectra of ocean waves", in *Time Series Analysis*, edited by M. Rosenblatt, pp. 125-139, John Wiley, New York.

Hasselmann, K., and Collins, J.I., 1968, "Spectral dissipation of finite-depth gravity waves due to turbulent bottom friction", *J. Mar. Res.*, 26, 1-12.

Hasselmann, K. et al., 1973, "Measurements of wind-wave growth and swell decay during the Joint North Sea Wave Project (JONSWAP)", *Dtsch. Hydrogh. Z.*, Suppl. A, 8, 12, 95pp.

Hasselmann, K., Ross, D.B., Müller, P. and Sell, W., 1976, "A parametric wave prediction model", *J. Phys. Oceanogr.*, 6, 200-228.

Hasselmann, K., Hasselmann, S., Bauer, E., Brüning, C., Lehner, S., Graber, H.C. and Lionello, P., 1988, "Development of a satellite SAR image spectra and altimeter wave height data assimilation system for ERS-1", *Max-Planck-Institut für Meteorologie*, Rep. No. 19, 155pp.

Hasselmann, K. and Hasselmann, S., 1991, "On the nonlinear mapping of an ocean wave spectrum into a SAR image spectrum and its inversion", *J. Geophys. Res.*, 96, 10,713-10,729.

Hasselmann, S. and Hasselmann, K., 1981, "A symmetrical method of computing the nonlinear transfer in a gravity wave spectrum", *Hamburger Geophys. Einzelschriften*, A 52.

Hasselmann, S. and Hasselmann, K., 1985a, "The wave model EXACT-NL", In *Ocean wave modelling* (The SWAMP Group), pp. 249-251, New York, Plenum Press.

Hasselmann, S. and Hasselmann, K., 1985b, "Computations and parameterizations of the nonlinear energy transfer in a gravity wave spectrum. Part 1: A new method for efficient computations of the exact nonlinear transfer integral", *J. Phys. Oceanogr.*, 15, 1369-1377.

Hasselmann, S., Hasselmann, K., Allender, J.H. and Barnett, T.P., 1985a, "Computations and parameterizations of the nonlinear energy transfer in a gravity wave spectrum. Part 2: Parameterizations of the nonlinear transfer for application in wave models", *J. Phys. Oceanogr.*, 15, 1378-1391.

Hasselmann, K., Raney, R.K., Plant, W.J., Alpers, W., Schuchman, R.A., Lyzenga, D.R., Rufenach, C.L. and Tucker, M.J., 1985b, "Theory of Synthetic Aperture Radar ocean imaging – A MARSEN view", *J. Geophys. Res.*, 90, 4659-4686.

Hasselmann, S., Brüning, C. and Hasselmann, K., 1996, "An improved algorithm for the retrieval of ocean wave spectra from synthetic aperture radar image spectra", *J. Geophys. Res.*, 101, 16,615-16,629.

Hatori, M., Tokuda, M. and Toba, Y., 1981, "Experimental study on strong interactions between regular waves and wind waves", *J. Ocean. Soc. Japan*, 37, 111-119.

Herbers, T.H.C., Lowe, R.L. and Guza, R.T, 1992, "Field observations of orbital velocities and pressure in weakly nonlinear surface gravity waves", *J. Fluid Mech.*, 245, 413-435.

Hersbach, H., 1998, "Application of the adjoint of the WAM model to inverse wave modeling", *J. Geophys. Res.*, 103, 10,469-10,487.

Herterich, K. and Hasselmann, K., 1980, "A similarity relation for the nonlinear energy transfer in a finite-depth gravity-wave spectrum", *J. Fluid Mech.*, 97, 215-224.

Hidy, G.M. and Plate, E.J., 1966, "Wind action on water standing in a laboratory channel", *J. Fluid Mech.*, 25, 651-687.

Holland, G.J., 1980, "An analytical model of the wind and pressure profiles in hurricanes", *Mon. Weather Rev.*, 108, 1212-1218.

Holt, B., and Gonzalez, F.I., 1986, "SIR-B Observations of dominant ocean waves near Hurricane Josephine", *J. Geophys. Res.*, 91, 8595-8598.

Holthuijsen, L.H., 1983, "Observations of the directional distribution of ocean-wave energy in fetch-limited conditions", *J. Phys. Oceanogr.*, 13, 191-207.

Hsiao, S.V. and Shemdin, O.H., 1978, "Bottom dissipation in finite-depth water waves", *16th Int. Conf. Coastal Eng., ASCE*, Hamburg, 434-448.

Hsiao, S.V. and Shemdin, O.H., 1983, "Measurements of wind velocity and pressure with a wave follower during MARSEN", *J. Geophys. Res.*, 88, 9841-9849.

Hubbert, G.D., Holland, G.J., Leslie, L.M. and Manton, M.J., "A real-time storm-surge forecasting system for the Australian tropics", *Weather Forecasting*, 6, 86-97.

Hunt, J.N., 1979, "Direct solution of the wave dispersion equation", *J. Waterway, Port, Coastal and Ocean Div., ASCE*, WW4, 457-459.

Ijima, T. and Tang, F.L.W., 1966, "Numerical calculations of wind waves in shallow water", *10th Conf. on Coastal Eng.*, ASCE, Tokyo, 2: 38-45.

Ijima, T., Soejima, T. and Matsuo, T. , 1968, "Ocean wave distribution in typhoon area", *Proc. Coastal Eng. in Japan*, 2, 29-42.

Inoue, T., 1967, "On the growth of the spectrum of a wind generated sea according to a modified Miles-Phillips mechanism and its application to wave forecasting", *Ph.D. dissertation, New York Univ.*, New York.

Isozaki, I. and Uji, T., 1973, "Numerical prediction of ocean wind waves", *Pap. Meteorol. Geophys.*, 24, 207-231.

Isobe, M., 1986, "A parabolic refraction-diffraction equation in the ray-front coordinate system", *20th Int. Conf. Coastal Eng.*, Taipeh, 306-317.

Isobe, M., Kondo, K. and Horikawa, K. 1984, "Extension of MLM for estimating directional wave spectrum", *Symp. on Description and Modelling of Directional Seas, DHI and MMI*, Copenhagen, 1-15.

Janssen, P.A.E.M., 1989, "Wave-induced stress and the drag of air flow over sea waves", *J. Phys. Oceanogr.*, 19, 745-754.

Janssen, P.A.E.M., 1991, "Quasi-linear theory of wind-wave generation applied to wave forecasting", *J. Phys. Oceanogr.*, 21, 1631-1642.

Janssen, P.A.E.M., Komen, G.J. and de Voogt, W.J.P. 1984, "An operational coupled hybrid wave prediction model", *J. Geophys. Res.*, 89, 3635-3654.

Janssen, P.A.E.M., Komen, G.J. and De Voogt, W.J.P., 1987, "Friction velocity scaling in wind wave generation", *Boundary-Layer Meteorol.*, 38, 29-35.

Janssen, P.A.E.M., Lionello, P., Reistad, M. and Hollingsworth, A., 1989, "Hindcasts and data assimilation studies with the WAM model during the SEASAT period", *J. Geophys. Res.*, 94, 973-993.

Janssen, P.A.E.M., Hasselmann, S., Hasselmann, K. and Komen, G.J., 1994, "Numerical modelling of wave evolution. Part 4: Numerical scheme", In *Dynamics and modelling of ocean waves*, G.J. Komen et al. (eds.), Cambridge Uni. Press, 233-238.

Janssen, P.A.E.M., Hansen, B. and Bidlot, J., 1996, "Verification of the ECMWF wave forecasting system against buoy and altimeter data", *European Centre for Medium-Range Weather Forecasts*, Tech. Memor. No. 229.

Jefferys, E.R., Wareham, G.T., Ramsden, N.A. and Platts, M.J., 1981, "Measuring directional spectra with MLM", In *'Directional wave spectra applications'*, Weigel, R.L. (ed), 203-219.

Jeffreys, H., 1924, "On the formation of waves by wind", *Proc. Roy. Soc. A*, 107, 189-206.

Jeffreys, H., 1925, "On the formation of waves by wind. II", *Proc. Roy. Soc. A*, 110, 341-347.

Jenkins, G.M. and Watts, D.G., 1968, "Spectral analysis", *Holden-Day*, San Francisco.

Jonsson, I.G., 1965, "Friction factor diagrams for oscillatory boundary layers", *Prog. Rep., Tech. Univ. of Denmark*, 10, 10-21.

Jonsson, I.G., 1980, "A new approach to oscillatory rough turbulent boundary layers", *Ocean Eng.*, 7, 109-152.

Jonsson, I.G., and Carlsen, N.A., 1976, "Experimental and theoretical investigations in an oscillatory rough turbulent boundary layer", *J. Hydraul. Res.*, 14, 45-60.

Kahma, K.K., 1981, "A study of the growth of the wave spectrum with fetch", *J. Phys. Oceanogr.*, 11, 1503-1515.

Kahma, K.K. and Calkoen, C.J., 1992, "Reconciling discrepancies in the observed growth of wind-generated waves", *J. Phys. Oceanogr.*, 22, 1389-1405.

Kahma, K.K. and Pettersson, H., 1996, "Influence of the fetch geometry on wave growth", In *"The air-sea interface: Radio and acoustic sensing, turbulence and wave dynamics"*, Donelan, M.A., Hui, W.H. and Plant, W.J. (eds), Univ. of Miami, 91-96.

Kajiura, K., 1968, "A model of the bottom boundary layer in water waves", *Bull. Earthquake Res. Inst., Univ. Tokyo*, 46, 75-123.

Kamphuis, J.W., 1975, "Friction factors under oscillatory flows", *J. Waterways, Harbours, Coastal Div., ASCE*, 101, 135-144.

Kamphuis, J.W., 1978, "Attenuation of gravity waves by bottom friction", *Coastal Eng.*, 2, 111-118.

Kawai, S., Okada, K. and Toba, Y., 1977, "Field data support of three-seconds law and $gu_*\sigma^{-4}$ spectral form for growing wind waves", *J. Oceanogr. Soc. Japan*, 33, 137-150.

Kendall, J.M., 1970, "The turbulent boundary layer over a wall with progressive surface waves", *J. Fluid Mech.*, 41, 259-281.

Keulegan, G.H. and Patterson, G.W., 1940, "Mathematical theory of irrotational translation waves", *National Bureau of Standards*, Rep. No. 1272, Washington D.C., 47-101.

Khatri, S. and Young, I.R., 1997, "Modelling finite depth fetch limited growth", *4th International Waves in Shallow Environments Conference*, San Francisco, April 13-17 (unpublished).

Kim, Y.C., and Powers, E.J., "Digital bispectral analysis and its application to nonlinear wave interactions", *IEEE Trans. Plasma Sci.*, 120-131.

King, D.B., and Shemdin, O.H., 1978, "Radar observations of hurricane wave directions", *16th Int. Conf. on Coastal Eng.*, Hamburg, 209-226.

Kinsman, B., 1965, "Wind Waves - Their generation and propagation on the ocean surface", *Prentice-Hall*, 676pp.

Kirby, J.T., 1984, "A note on linear surface wave-current interaction over slowly varying topography", *J. Geophys. Res.*, 89, 745-747.

Kirby, J.T. and Dalrymple, R.A., 1986a, "Modelling waves in surfzones and around islands", *ASCE J. Waterway, Port, Coastal and Ocean Eng.*, 112, 78-93.

Kirby, J.T. and Dalrymple, R.A., 1986b, "An approximate model for nonlinear dispersion in monochromatic wave propagation models", *Coastal Engineering*, 9, 545-561.

Kitaigorodskii, S.A., 1962, "Applications of the theory of similarity to the analysis of wind-generated wave motion as a stochastic process", *Bull. Acad. Sci. USSR, Geophys. Ser.*, 1, 105-117.

Kitaigorodskii, S.A., 1970, "The physics of air-sea interaction", English edition, 1973, A. Baruch, translator, P. Greenberg, Ed., *Israel Program for Scientific Translations*, 237pp.

Kitaigorodskii, S.A., 1983, "On the theory of the equilibrium range in the spectrum of wind-generated gravity waves", *J. Phys. Oceanogr.*, 13, 816-827.

Kitaigorodskii, S.A. and Strekalov, S.S., 1962, "Contribution to an analysis of the spectra of wind-caused wave action", *Izv. Akad. Nauk SSSR Geophys.*, 9, 1221-1228.

Kitaigorodskii, S.A., Krasitskii, V.P. and Zaslavskii, M.M. , 1975, "On Phillips' theory of equilibrium range in the spectra of wind-generated gravity waves", *J. Phys. Oceanogr.*, 5, 410-420.

Kirby, J.T. and Dalrymple, R.A., 1986, "Modelling waves in surfzones and around islands", *ASCE J. Waterway, Port, Coastal and Ocean Div.*, 112, 78-93.

Komen G.J., 1985, "Activities of the WAM (Wave Modelling) group", In *Advances in underwater technology. Ocean science and offshore engineering*, Vol. 6, Oceanology, Graham and Trotman, 121-127.

Komen, G.J., Hasselmann, S. and Hasselmann, K., 1984, "On the existence of a fully developed wind-sea spectrum", *J. Phys. Oceanogr.*, 14, 1271-1285.

Komen, G.J., Cavaleri, L., Donelan, M., Hasselmann, K., Hasselmann, S. and Janssen, P.A.E.M., 1994, "Dynamics and modelling of ocean waves", *Cambridge Univ. Press*, 532pp.

Kononkova, G.E., Nikitina, E.A., Poborchaya, L.V. and Speranskaya, A.A., 1970, "On the spectra of wind driven waves at small fetches", *Fizika Atmosfery i Okeana*, VI, 7, 747-751 (Eng. Trans. Royal Aircraft Establishment, No. 1634).

Korteweg, D.J. and De Vries, G., 1895, "On the change of the form of long waves advancing in a rectangular canal, and on a new type of long stationary wave", *Philosophical Magazine*, 5th series, 422-443.

Kusaba, T. and Mitsuyasu, H., 1986, "Nonlinear instability and evolution of steep water waves under wind action", *Rep. Inst. Appl. Mech.*, 33, 33-64.

Lamb, H., 1932, "Hydrodynamics", Cambridge Univ. Press, 738pp.

Large, W.G. and Pond, S., 1982, *Sensible and latent heat flux measurements over the ocean*, J. Phys. Oceanogr., 12, 464-482.

Larson, T.R. and Wright, J.W., 1975, "Wind-generated gravity-capillary waves: Laboratory measurements of temporal growth rates using microwave backscatter", *J. Fluid Mech.*, 70, 417-436.

LeBlond, P.H. and Mysak, L.A., 1978, "Waves in the Ocean", *Elsevier*, Amsterdam, 602pp.

Le Dimet, F.-X. and Talagrand, O., 1986, "Variational algorithms for analysis and assimilation of meteorological observations: Theoretical aspects", *Tellus*, 38A, 97-110.

Lewis, A.W. and Allos, R.N., 1990, "JONSWAP's parameters: sorting out inconsistencies", *Ocean Engineering*, 17, 409-415.

Lii, K.S., Rosenblatt, M., and Van Atta, C., 1976, "Bispectral measurements in turbulence", *J. Fluid Mech.*, 77, 45-62.

Lin, C.C., 1955, "The theory of hydrodynamic stability", *Cambridge Univ. Press.*

Lin, R.Q. and Huang, N.E., 1998, "The Goddard coastal wave model. Part III: Nonlinear interactions", submitted to *J. Phys. Oceanogr.*.

Lionello, P., Günther, H. and Janssen, P.A.E.M., 1992, "Assimilation of altimeter data in a global third generation wave model", *J. Geophys. Res.*, 97, 14,253-14,474.

Liu, P.C., 1971, "Normalized and equilibrium spectra of wind waves in Lake Michigan", *J. Phys. Oceanogr.*, 1, 249-257.

Liu, P.C., 1989, "On the slope of the equilibrium range in the frequency spectrum of wind waves", *J. Geophys. Res.*, 94, 5017-5023.

Longuet-Higgins, M.S., 1952, "On the statistical distributions of sea waves", *J. Marine Res.*, 11, 3, 245-265.

Longuet-Higgins, M.S., 1976, "On the nonlinear transfer of energy in the peak of a gravity-wave spectrum: A simplified model", *Proc. R. Soc. Lond.*, A 347, 311-328.

Longuet-Higgins, M.S., Cartwright, D.E. and Smith, N.D., 1963, "Observations of the directional spectrum of sea waves using the motions of a floating buoy", In *Ocean wave spectra*, Prentice Hall, 111-136.

Longuet-Higgins, M.S. and Cokelet, E.D., 1976, "The deformation of steep surface waves on water. Part I. A numerical method of computation", *Proc. Roy. Soc. London*, A 350, 1-26.

Longuet-Higgins, M.S. and Cokelet, E.D., 1978, "The deformation of steep surface waves on water. II. Growth of normal mode instabilities", *Proc. Roy. Soc. London*, A 364, 1-28.

Longuet-Higgins, M.S., 1986, "Eulerian and Lagrangian aspects of surface waves", *J. Fluid Mech.*, 173, 683-707.

Longuet-Higgins, M.S., 1988, "Mechanisms of wave breaking in deep water", In: *Sea surface sound*, B.R. Kerman (ed), Kluwer, Boston MA, 1-30.

Lozano, C.J. and Liu, P.L., 1980, "Refraction-diffraction model for linear surface water waves", *J. Fluid Mech.*, 101, 705-720.

Lumley, J.L. and Panofsky, H.A., 1964, "The structure of atmospheric turbulence", *Interscience.*

Lygre, A. and Krogstad, H.E., 1986, "Maximum entropy estimation of the directional distribution in ocean wave spectra", *J. Phys. Oceanogr.*, 16, 2052-2060.

McLeish, W., and Ross, D.B., 1983, "Imaging radar observations of directional properties of ocean waves", *J. Geophys. Res.*, 88, 4407-4419.

Madsen, O.S. and Mei, C.C., 1969a, "Dispersive long waves of finite amplitude over an uneven bottom", *MIT, Hydrodyn. Lab*, 117.

Madsen, O.S. and Mei, C.C., 1969b, "The transformation of a solitary wave over an uneven bottom", *J. Fluid Mech.*, 39, 781-791.

Madsen, P.A., Murray, R. and Sørensen, 1991, "A new form of the Boussinesq equations with improved linear dispersion characteristics", *Coastal Engineering*, 15, 371-389.

Madsen, P.A., Sørensen, O.R., Schäffer, H.A., 1997, "Surf zone dynamics simulated by a Boussinesq type model, Part I. Model description and cross-shore motion of regular waves", *Coastal Engineering*, 32, 255-287.

Makin, V.K., Kudryavtsev, V.N. and Mastenbroek, C., 1995, "Drag of the sea surface", *Boundary-Layer Met.*, 73, 159-182.

Mansard, E.P.D. and Funke, E.R., 1990, "On fitting of JONSWAP spectra to measured sea states", *Proc. 22nd. Int. Coastal Eng. Conf.*, Delft, The Netherlands, 465-477.

Marchuk, G.I., 1974, "The numerical solution of problems of atmospheric and oceanic dynamics", *Gidrometeoizdat*, Leningrad, (Translated from Russian by Rainbow Systems Inc., Alexandria, Virginia), 387pp.

Masson, D., 1990, "Observations of the response of sea waves to veering winds", *J. Phys. Oceanogr.*, 20, 1876-1885.

Masson, D., 1993, "On the nonlinear coupling between swell and wind waves", *J. Phys. Oceanogr.*, 23, 1249-1258.

Masuda, A., 1980, "Nonlinear energy transfer between wind waves", *J. Phys. Oceanogr.*, 10, 2082-2093.

Masuda, A., and Kuo, Y., "A note on the imaginary part of bispectra", *Deep Sea Res.*, A, 28, 213-222.

Mei, C.C., 1989, "The applied dynamics of ocean surface waves", *World Scientific*, 740pp.

Melville, W.K., Loewen, M.R. and Lamarre, E., 1991, "Sound production and air entrainment by breaking waves: A review of recent laboratory experiments", In: 'Breaking waves', Banner, M.L. and Grimshaw, R.H.J. (eds), *Springer-Verlag*, 139-146.

Miche, M., 1944, "Mouvements ondulatoires de la mer en profondeur constante ou decroissante", *Annales des Ponts et Chaussees*.

Miles, J.W., 1957, "On the generation of surface waves by shear flows", *J. Fluid Mech.*, 3, 185-204.

Miles, J.W., 1959, "On the generation of surface waves by shear flows, Part 2", *J. Fluid Mech.*, 6, 568-582.

Miles, J.W., 1960, "On the generation of surface waves by turbulent shear flows", *J. Fluid Mech.*, 7, 469-478.

Miller, H.C. and Vincent, C.L., 1990, "FRF spectrum: TMA with Kitaigorodskii's f^{-4} scaling", *ASCE Jnl. Waterway, Port, Coastal and Ocean Div.*, 116, 57-78.

Mitsuyasu, H., 1966, "Interactions between water waves and wind (1)", *Rep. Inst. Appl. Mech.*, 14, 65-88.

Mitsuyasu, H., 1968, "On the growth of the spectrum of wind-generated waves. 1", *Rep. Res. Inst. Appl. Mech.*, Kyushu Univ., 16, 459pp.

Mitsuyasu, H., Tasai, F., Suhara, T., Mizuno, S., Onkusu, M., Honda, T. and Rukiiski, K. , 1975, "Observations of the directional spectrum of ocean waves using a cloverleaf buoy", *J. Phys. Oceanogr.*, 5, 751-761.

Mitsuyasu, H., Tasai, F., Suhara, T., Mizuno, S., Onkusu, M., Honda, T. and Rukiiski, K. , 1980, "Observations of the power spectrum of waves using a cloverleaf buoy", *J. Phys. Oceanogr.*, 10, 286-296.

Mognard, N.M., Campbell, W.J., Cheney, R.E. and Marsh, J.G., 1983, "Southern Ocean mean monthly wave and surface winds for winter 1978 by SEASAT radar altimeter", *J. Geophys. Res.*, 88, 1736-1744.

Monin, A.S. and Obukhov, A.M., 1954, "Basic relationships of turbulent mixing in the surface layer of the atmosphere", *Akad. Nauk. SSSR Trud. Geofiz. Inst.*, No. 24 (151), 163-187.

Monin, A.S. and Yaglom, A.M., 1971, "Statistical fluid mechanics", *MIT Press*, 769pp.

Motzfeld, H., 1937, "Die turbulente Strömung an welligen Wänden", *Z. Math. Mech.*, 17, 193-212.

Munk, W.H. and Arthur, R.S., 1952, "Wave intensity along a refracted ray in gravity waves", *Natl. Bur. Stand.*, Circ. 521, Washington, D.C.

Nelson, R.C., 1994, "Depth limited design wave heights in very flat regions", *Coastal Engineering*, 23, 43-59.

Nielsen, P., 1983, "Analytical determination of nearshore wave height variation due to refraction, shoaling and friction", *Coastal Eng.*, 7, 233-251.

NOAA Data Buoy Office, 1973, "Practical experience with buoys developed by the NOAA Data Buoy Office", *U.S. Dept. of Commerce.*

Nwogu, O., 1994, "Nonlinear evolution of directional wave spectra in shallow water", *24th Conf. on Coastal Eng.*, Kobe, 467-481.

Ochi, M.K. and E.N. Hubble, 1976, "On six-parameter wave spectra", *Proc. 15th Int. Conf. Coastal Eng.*, 1, 301-328.

Ochi, M.K. and M.H. Chiu, 1982, "Nearshore wave spectra measured during Hurricane David", *18th Int. Conf. Coastal Eng.*, Cape Town, 77-86.

Ogawa, K. and Yoshida, K., 1959, "A practical method for a determination of reflection of long gravitational waves", *Records Oceanogr. Works in Japan*, 5, 38-50.

Patterson, M.M., 1972, "Hindcasting hurricane waves in the Gulf of Mexico", *Soc. Pet. Eng.*, 321-328.

Pawka, S.S., 1983, "Island shadows in wave directional spectra", *J. Geophys. Res.*, 88, 2579-2591.

Penney, W.G. and Price, A.T., 1952, "The diffraction theory of sea waves and the shelter afforded by breakwaters", *Philos. Trans. Roy. Soc.*, A 244, 236-253.

Peregrine, D.H., 1967, "Long waves on a beach", *J. Fluid Mech.*, 27, 815-827.

Peregrine, D.H., 1972, "Equations for water waves and the approximation behind them", In: 'Waves on beaches and resulting sediment transport', Meyer, R.E. (ed.), *Academic Press*, New York, 95-121.

Peregrine, D.H., 1990, "Computations of breaking waves", In: 'Waterwave Kinematics', Torum, A. and Gudmestad, O.T. (eds.), *Kluwer Academic Publishers*, Amsterdam, 475-490.

Phillips, O.M., 1957, "On the generation of waves by turbulent wind", *J. Fluid Mech.*, 2, 417-445.

Phillips, O.M., 1958, "The equilibrium range in the spectrum of wind-generated waves", *J. Fluid Mech.*, 4, 426-434.

Phillips, O.M., 1960, "On the dynamics of unsteady gravity waves of finite amplitude, Part 1", *J. Fluid Mech.*, 9, 193-217.

Phillips, O.M., 1977, "The dynamics of the upper ocean", *Cambridge Univ. Press.*, 336pp.

Phillips, O.M., 1985, "Spectral and statistical properties of the equilibrium range in wind-generated gravity waves", *J. Fluid Mech.*, 156, 505-531.

Phillips, O.M. and Banner M.L., 1974, "Wave breaking in the presence of wind drift and swell", *J. Fluid Mech.*, 66, 625-640.

Pierson, W.J. (ed.) 1960, "The directional spectrum of a wind-generated sea as determined from data obtained by the Stereo Wave Observation Project", *New York Univ. Coll. Eng.*, Met. Pap. 2, 6.

Pierson, W.J. and Moskowitz, L., 1964, "A proposed spectral form for fully developed wind seas based on the similarity theory of S.A. Kitaigorodskii", *J. Geophys. Res.*, 69, 5181-5190.

Pierson, W.J., Tick, L.G. and Baer, L., 1966, "Computer-based procedures for predicting global wave forecasts and wind field analyses capable of using wave data obtained by spacecraft", *6th Naval Hydrodynamics Symposium*, Washington.

Plant, W.J., 1982, "A relationship between wind stress and wave slope", *J. Geophys. Res.*, 87, 1961-1967.

Plant, W.J., 1987, "The microwave measurement of ocean-wave directional spectra", *APL Tech. Digest*, 8, 1, Johns Hopkins Univ., 55-59.

Powell, M.D., 1980, "Evaluations of diagnostic marine boundary layer models applied to hurricanes, *Mon. Weather Rev.*, 108, 758-766.

Priestly, C.H.B., 1959, "Turbulent transfer in the lower atmosphere", *Univ. Chicago Press*.

Priestley, J.T., 1965, "Correlation studies of pressure fluctuations on the ground beneath a turbulent boundary layer", *U.S. Nat. Bureau Standards*, Rep. 8942, 92pp.

Putnam, J.A., and Johnson, J.W., 1949, "The dissipation of wave energy by bottom friction", *Trans. AGU*, 30, 67-74.

Radder, A.C., 1979, "On the parabolic equation method for water wave propagation", *J. Fluid Mech.*, 95, 159-176.

Raney, R.K., 1981, "Wave orbital velocity, fade, and SAR response to azimuth waves", *IEEE J. Oceanic Eng.*, OE-6, 140-146.

Rapp, R.J. and Melville, W.K., 1990, "Laboratory measurements of deep water breaking waves", *Philos. Trans. R. Soc. London*, A 331, 735-780.

Resio, D. and Perrie, W., 1991, "A numerical study of nonlinear energy fluxes due to wave-wave interactions. Part 1: Methodology and basic results", *J. Fluid Mech.*, 223, 609-629.

Rice, S.O., 1954, "Mathematical analysis of random noise", *Selected Paper on Noise and Stochastic Processes*, Dover Pub. Inc., 133-294.

Ross, D.B., 1976, "A simplified model for forecasting hurricane generated waves", *Bull. A.M.S.*, 113.

Rosseau, M., 1952, "Contribution à la théorie des ondes liquides de gravité en profondeur variable", *Publ. Sci. et Tech. du Ministère de l'Air*, 275, 90pp.

Sanders, J. W., 1976, "A growth-stage scaling model for the wind-driven sea", *Dtsch. Hydrogr. Z.*, 29, 136-161.

Schloemer, R.W., 1954, "Analysis and synthesis of hurricane wind patterns over Lake Okechobee, FL", *Hydromet Rep., Govt. Print Office*, 31, 49pp.

Sell, W. and Hasselmann, K., 1972, "Computations of nonlinear energy transfer for JONSWAP and empirical wind-wave spectra", *Rep. Inst. of Geophys., Univ. of Hamburg*, Germany.

Shapiro, L.J., 1983, "The asymmetric boundary layer flow under a translating hurricane", *J. Atmos. Sci.*, 40, 1984-1998.

Shea, D.J., and Gray, W.M., 1973, "The hurricane's inner core region, I: Symmetric and asymmetric structure", *J. Atmos. Sci.*, 30, 1544-1564.

Shemdin, O.H., 1969, "Instantaneous velocity and pressure measurements above propagating waves", *Dept. Coastal and Ocean. Eng.*, Tech. Rep. 4, Univ. of Florida, 105pp.

Shemdin, O.H., 1977, "Hurricane waves, storm surge and currents: An assessment of the state of the art", *U.S. - South East Asia Symp. on Eng. for Natural Hazards Protection*, Manila.

Shemdin, O.H., 1980, "Prediction of dominant wave properties ahead of hurricanes", *17th Int. Coastal Eng. Conf., ASCE*, Sydney, 600-609.

Shemdin, O.H. and Hsu, E.Y., 1967, "The dynamics of wind in the vicinity of progressive water waves", *J. Fluid Mech.*, 30, 403-416.

Simiu, E., Patel, V.C., and Nash, J.F., "Mean speed profiles of hurricane winds", *J. Eng. Mech. Div., Am. Soc. Civ. Eng.*, 102, 265-273.

Smith, P.C. and MacPherson, J.I., 1987, "Cross-shore variations of near-surface wind velocity and atmospheric turbulence at the land-sea boundary during CASP", *Atmosphere-Ocean*, 25, 279-303.

Smith, R.K., 1968, "The surface boundary layer of a hurricane", *Tellus*, 20, 473-484.

Smith, S.D., Katsaros, K.B., Oost, W.A. and Mestager, P.G., 1990, "Two major experiments in the humidity exchange over the sea (HEXOS) program", *Bull. Amer. Meteor. Soc.*, 71, 161-172.

Smith, S.D., Anderson, R.J., Oost, W.A., Kraan, C., Maat, N., DeCosmo, J., Katsaros, K.B., Davidson, K.L., Bumke, K., Hasse, L. and Chadwick, H.M., 1992, "Sea surface wind stress and drag coefficients: The HEXOS results", *Boundary-Layer Meteool.*, 60, 109-142.

Snodgrass, F.E. et al., 1966, "Propagation of ocean swell across the Pacific", *Philos. Trans. Roy. Soc.*, London, A, 259, 431-497.

Snyder, R.L., 1974, "A field study of wave-induced pressure fluctuations above surface gravity waves", *J. Marine Res.*, 32, 485-496.

Snyder, R.L. and Cox, C.S., 1966, "A field study of the wind generation of ocean waves", *J. Marine Res.*, 24, 141-178.

Snyder, R.L., Dobson, F.W., Elliot, J.A. and Long, R.B., 1981, "Array measurements of atmospheric pressure fluctuations above surface gravity waves", *J. Fluid Mech.*, 102, 1-59.

Sobey, R.J., 1986, "Wind-wave prediction", *Ann. rev. Fluid Mech.*, 16, 149-1172.

Sobey, R.J. and Young, I.R., 1986, "Hurricane wind waves – A discrete spectral model", *ASCE J. Waterw. Port Costal Ocean Eng.*, 112, 370-389.

Srokosz, M.A. and Longuet-Higgins, M.S., 1986, "On the skewness of sea-surface elevation", *J. Fluid Mech.*, 164, 487-497.

Stanton, T.E., Marshall, D. and Houghton, R., 1932, "The growth of waves on water due to the action of the wind", *Proc. Roy. Soc.*, A, 137, 283-293.

Stoker, J.J., 1957, "Water waves", *Interscience*, New York.

Stokes, G.G., 1847, "On the theory of oscillatory waves", *Trans. Cambridge Philos. Soc.*, 8, 441-473.

Stokes, G.G., 1880, "On the theory of oscillatory waves", *Cambridge University Press*, London, England.

Svendsen, I.A. and Jonsson, I.G., 1976, "Hydrodynamics of coastal regions", *Technical University of Denmark*, 282pp.

Sverdrup, H.V. and Munk, W.H., 1944a, "Wind waves and swell: Principles in forecasting", *Hydrographic Office, U.S. Navy*, Misc. 11, 275.

Sverdrup, H.V. and Munk, W.H., 1944b, "Breakers and surf: Principles in forecasting", *Hydrographic Office, U.S. Navy*, Publ. No.235.

Sverdrup, H.V. and Munk, W.H., 1946, "Empirical and theoretical relations between wind, sea, and swell", *Trans. Amer. Geophys. Union*, 27, 823-827.

Sverdrup, H.V. and Munk, W.H., 1947, "Wind, sea, and swell: Theory of relation for forecasting", *Hydrographic Office, U.S. Navy*, Publ. No. 601.

SWAMP Group, 1985, "Ocean wave modeling", *Plenum Press*, New York, 256pp.

Swart, D.H., 1974, "Offshore sediment transport and equilibrium beach profiles", *Delft Hydraul. Lab.*, Publ. 131, Delft, Netherlands.

Taylor, P.A. and Lee, R.J., 1984, "Simple guidelines for estimating wind speed variations due to small-scale topographic features", *Climatol. Bull.*, 18, 3-32.

Terray, E.A., Donelan, M.A., Agrawal, Y.C., Drennan, W.M., Kahma, K.K., Williams, A.J., Hwang, P.A. and Kitaigorodskii, S.A., 1996, "Estimates of kinetic energy dissipation under breaking waves", *J. Phys. Oceanogr.*, 26, 792-807.

Thijsse, J.Th., 1949, "Dimensions of wind-generated waves", *General Assembly of Association d'Océanographie Physique, Procés-Verbaux*, 4, Oslo. 80-81.

Thijsse, J.T., 1951, "Growth of wind-generated waves and energy transfer", *Gravity Waves*, Nat. Bur. Stand. Washington Circular, 521, 281-287.

Thomas, J.P., 1988, "Retrieval of energy spectra from measured data for assimilation into a wave model", *Q. J. Roy. Meteorol. Soc.*, 114, 781-800.

Thornton, E.B., 1977, "Rederivation of the saturation range in the frequency spectrum of wind-generated gravity waves", *J. Phys. Oceanogr.*, 7, 137-140.

Toba, Y., 1973, "Local balance in the air-sea boundary process", *J. Oceanogr. Soc. Japan*, 29, 209-220.

Toba, Y., Okada, K. and Jones, I.S.F., 1988, "The response of wind-wave spectra to changing winds, Part I: Increasing winds", *J. Phys. Oceanogr.*, 18, 1231-1240.

Toba, Y, Iida, N., Kawamura, H., Ebuchi, N. and Jones, I.S.F., 1990, "Wave dependence of sea-surface wind stress", *J. Phys. Oceanogr.*, 20, 705-721.

Tolman, H.L., 1991, "A third-generation model for wind on slowly varying unsteady and inhomogeneous depths and currents", *J. Phys. Oceanogr.*, 21, 782-797.

Tolman, H.L., 1992, "Effects of numerics on the physics in a third-generation wind-wave model", *J. Phys. Oceanogr.*, 22, 1095-1111.

Tolman, H.L., 1994, "Wind-waves and moveable-bed bottom friction", *J. Phys. Oceanogr.*, 24, 994-1009.

Townsend, A.A., 1980, "Sheared turbulence and additional distortion", *J. Fluid Mech.*, 98, 171-191.

Townsend, M. and Fenton, J.D., 1995, "Numerical comparison of wave analysis methods", *12th Aust. Coastal & Ocean Eng. Conf.*, Melbourne, 169-173.

Tracy, B.A. and Resio, D.T., 1982, "Theory and calculation of the nonlinear energy transfer between sea waves in deep water", *U.S. Army Engineer Waterways Experiment Station*, Rep. No. 11, Vicksburg, U.S.A.

Tsanis, I.K. and Brissette, F.P., 1992, "Methods for directional spectra measurements by small arrays", *3rd Int. Workshop on Wave Hindcasting and Forecasting*, Montreal, 12-23.

Tucker, M.J., 1991, "Waves in ocean engineering: Measurement, analysis, interpretation", *Ellis Horwood*, 431pp.

Uji, T., 1975, "Numerical estimation of the sea wave in a typhoon area", *Papers in Meteorology and Geophysics*, 26, 199-217.

U.S. Army Corps of Engineers, 1955, "Waves and wind tides in shallow lakes and reservoirs", *Summary Report*, Project CW-167, Jacksonville District, Fla.

Ursell, F. 1956, "Wave generation by wind", *Surveys in Mechanics* (ed. G.K. Batchelor), Cambridge Univ. Press, 216-249.

Van Vledder, G. Ph., 1990, "Directional response of wind waves to turning wind", *Faculty of Civil Engineering, Delft University of Tech.*, Rep. No. 90-2, The Netherlands.

Van Vledder, G. Ph. and Holthuijsen, L.H., 1993, "The directional response of ocean waves to turning winds", *J. Phys. Oceanogr.*, 23, 177-192.

Vogel, J.A., Radder, A.C. and de Reus, J.H., 1988, "Verification of numerical wave propagation models in tidal inlets", *21st Int. Conf. Coastal Eng.*, Costa del Sol-Malaga, Spain, 433-447.

Volkov, Yu.,A., 1968, "Analysis of the spectra of sea excitation", *Izv. AN SSSR, Fizika Atkosfery i Okeana*, IV, 9.

The WAMDI Group (Hasselmann, S., Hasselmann, K., Bauer, E., Janssen, P.A.E.M., Komen, G.J., Bertotti, L., Lionello, P., Guillaume, A., Cardone, V.C., Greenwood, J.A., Reistad, M., Zambresky, L. and Ewing, J.A.), 1988, "The WAM model – a third generation ocean wave prediction model", *J. Phys. Oceanogr.*, 18, 1775-1810.

Wang, G.C., 1978, "Sea level pressure profile and gusts within a typhoon circulation", *Mon. Weather Rev.*, 106, 954-960.

Ward, E.G., 1974, "Ocean Data Gathering Program - An overview", *6th Offshore Technology Conf.*, Houston, OTC 2108-B.

Webb, D.J., 1978, "Non-linear transfers between sea waves", *Deep-Sea Res.*, 25, 279-298.

Webb, E.K., 1970, "Profile relationships: the log-linear range, and extension to strong stability" *Q. J. R. Meteorol. Soc.*, 96, 67-90.

Weber, S.L., 1978, "The energy balance of finite depth gravity waves", *J. Geophys. Res.*, 93, 3601-3607.

Weber, N., 1991, "Bottom friction for wind sea and swell in extreme depth-limited situations", *J. Phys. Oceanogr.*, 21, 149-172.

Whalen, J.E. and M.K. Ochi, 1978, "Variability of wave spectral shapes associated with hurricanes", *10th Offshore Tech. Conf.*, 1515-1522.

Whitham, G.B., 1974, "Linear and nonlinear waves", *J. Wiley & Sons*, New York, 636pp.

Wiegel, R.L., 1960, "A presentation of cnoidal wave theory for practical applications", *J. Fluid. Mech.*, 7, 273-286.

Wiegel, R.L., 1961, "Wind waves and swell", *Proc. 7th Conf. Coastal Eng.*, The Engineering Foundation, Council on Wave Research, Berkeley, Calf., 1-40.

Wilson, B.W., 1965, "Numerical prediction of ocean waves in the North Atlantic for December, 1959", *Dtsch. Hydrogr. Z.*, 18, 114-130.

Withee, G.W., and Johnson, A., 1975, "Data report: Buoy observations during Hurricane Eloise (September 19 to October 11, 1975)", *Environmental Science Div., Data Buoy Office, NOAA*, U.S. Dept. of Commerce.

Wu, H.Y., Hsu, E.Y. and Street, R.L., 1977, "The energy transfer due to air-input, non-linear wave-wave interaction and white cap dissipation associated with wind-generated waves", *Stanford University*, Tech. Rep. 207, Stanford, California, 158pp.

Wu, H.Y., Hsu, E.Y. and Street, R.L., 1979, "Experimental study of nonlinear wave-wave interaction and white-cap dissipation of wind-generated waves", *Dyn. Atmos. Oceans*, 3, 55-78.

Wu, J., 1980, "Wind-stress coefficient over sea surface near neutral conditions – A revisit", *J. Phys. Oceanogr.*, 10, 727-740.

Young, I.R., 1986, "Probability Distribution of Spectral Integrals", *ASCE Journal of Waterway, Ports, Coastal and Ocean Engineering*, 112, 338-341.

Young, I.R., 1988a, "A Shallow Water Spectral Wave Model", *J. Geophys. Res.*, 93, 5113-5129.

Young, I.R., 1988b, "A parametric hurricane wave prediction model", *J. Waterw. Port Coastal Ocean Eng., Am. Soc. Civ. Eng.*, 114, 637-652.

Young, I.R., 1993, "An estimate of the Geosat altimeter wind speed algorithm at high wind speeds", *J. Geophys. Res.*, 98, 20275-20285.

Young, I.R., 1994a, "On the measurement of directional wave spectra", *App. Ocean Res.*, 16, 283-294.

Young, I.R., 1994b, "Global Ocean Wave Statistics Obtained from Satellite Observations", *Applied Ocean Research*, 16, 235-248.

Young, I.R., 1994c, "On the Measurement of Directional Wave Spectra", *Applied Ocean Research*, 16, 283-294.

Young, I.R., 1995, "The Determination of Confidence Limits Associated with Estimates of the Spectral Peak Frequency", *Ocean Engineering*, 22, 7, 669-686.

Young, I.R., 1997a, "Observations of the Spectra of Hurricane Generated Waves", *Ocean Engineering*, 25, 261-276.

Young, I.R., 1997b, "The Growth Rate of Finite Depth Wind-Generated Waves", *Coastal Engineering*, 32, 181-195.

Young, I.R., 1998a, "Experimental investigation of the role of atmospheric stability in wind wave growth", *Coastal Engineering*, 34, 23-33.

Young, I.R., 1998b, "An intercomparison of GEOSAT, TOPEX and ERS1 measurements of wind speed and wave height", *Ocean Engineering*, 26, 67-81.

Young, I.R. and Sobey, R.J., 1980, "Discussion of Hunt, J.N., 'Direct solution of the wave dispersion equation', *J. Waterway, Port, Coastal and Ocean Div., ASCE*, 106, WW4, 499-501.

Young, I.R., Hasselmann, S. and Hasselmann, K., 1985, "Calculation of the nonlinear wave-wave interactions in cross seas", *Hamburger Geophys. Einzel.*, 74, 50 pp.

Young, I.R. and Sobey, R.J., 1985, "Measurements of the wind-wave energy flux in an opposing wind", *J. Fluid Mech.*, 123, 427-442.

Young, I.R., Hasselmann, S. and Hasselmann, K., 1987, "Computations of the response of a wave spectrum to a sudden change in the wind direction", *J. Phys. Oceanogr.*, 17, 1317-1338.

Young, I.R. and van Vledder, G.Ph., 1993, "The Central Role of Nonlinear Interactions in Wind-Wave Evolution", *Philos. Trans. R. Soc. Lond.*, A 342, 505-524.

Young, I.R. and Gorman, R.M., 1995, "Measurements of the evolution of ocean wave spectra due to bottom friction", *J. Geophys. Res.*, 100, 10,987-11,004.

Young, I.R. and Burchell, G.P., 1996, "Hurricane Generated Waves as Observed by Satellite", *Ocean Engineering*, 23, 761-776.

Young, I.R. and Glowacki, T.J., 1996, "Assimilation of altimeter wave height data into a spectral wave model using statistical interpolation", *Ocean Eng.*, 23, 667-689.

Young, I.R. and Holland, G.J., 1996, "Atlas of the Oceans: Wind and Wave Climate", *Pergamon Press*, ISBN 0-08-042519-4, 241pp.

Young, I.R. and Holland, G.J., 1996, "A Multi-media Atlas of the Oceans Wind and Wave Climate, Version 1.0", CD-ROM, *Pergamon Press*, ISBN 0-08-042434-1.

Young, I.R. and Verhagen, L.A., 1996a, "The growth of fetch limited waves in water of finite depth. Part I: Total energy and peak frequency", *Coastal Engineering*, 28, 47-78.

Young, I.R. and Verhagen, L.A., 1996b, "The growth of fetch limited waves in water of finite depth. Part II: Spectral evolution", *Coastal Engineering*, 28, 79-100.

Young, I.R., Verhagen, L.A. and Khatri, S.K., 1996, "The Growth of Fetch Limited Waves in Water of Finite Depth. Part III: Directional Spectra", *Coastal Engineering*, 28, 101-122.

Young, I.R., Dalton, M.A., McMahon, P.J. and Verhagen, L.A., 1997, "Design of an integrated shallow water wave experiment", *IEEE J. Oceanic Eng.*, 22, 184-188.

Young, I.R. and Van Agthoven, A., 1998, "The Response of waves to a sudden change in wind speed", in Perrie, W. (ed.), 'Nonlinear Ocean Waves', *Advances in Fluid Mechanics Series*, Computational Mechanics Pubs.

Young, I.R. and Eldeberky, Y., 1998, "Observations of Triad Coupling of Finite Depth Wind-Waves", *Coastal Engineering*, 33, 137-154.

Young, I.R. and Holland, G.J., 1998, "A Multi-media Atlas of the Oceans Wind and Wave Climate, Version 2.0", CD-ROM, *Pergamon Press.*

Yuan, Y., Tung, C.C. and Huang, N.E., 1986, "Statistical characteristics of breaking waves", In *Wave dynamics and radio probing of the ocean surface*, O.M. Phillips and K. Hasselmann (eds.), Plenum, New York, 265-272.

Zakharov, V.E., 1968, "Stability of periodic waves of finite amplitude on the surface of a deep fluid", *Zhurnal Prikladnoi Mekhaniki Tekhnicheskoi Fiziki*, 3, 80-94.

Zwarts, C.M.G., 1974, "Transmission line wave height transducer", *Int. Symp. on Wave Measurement and Analysis*, 605-620.

Index

T